Frontiers in Quantum Information Research

Decoherence, Entanglement, Entropy, MPS and DMRG

Kinki University Series on Quantum Computing

Editor-in-Chief: Mikio Nakahara *(Kinki University, Japan)*

ISSN: 1793-7299

Published

Vol. 1 Mathematical Aspects of Quantum Computing 2007
edited by Mikio Nakahara, Robabeh Rahimi (Kinki Univ., Japan) & Akira SaiToh (Osaka Univ., Japan)

Vol. 2 Molecular Realizations of Quantum Computing 2007
edited by Mikio Nakahara, Yukihiro Ota, Robabeh Rahimi, Yasushi Kondo & Masahito Tada-Umezaki (Kinki Univ., Japan)

Vol. 3 Decoherence Suppression in Quantum Systems 2008
edited by Mikio Nakahara, Robabeh Rahimi & Akira SaiToh (Kinki Univ., Japan)

Vol. 4 Frontiers in Quantum Information Research: Decoherence, Entanglement, Entropy, MPS and DMRG
edited by Mikio Nakahara (Kinki Univ., Japan) & Shu Tanaka (Univ. of Tokyo, Japan)

Kinki University Series on Quantum Computing – Vol. 4

editors

Mikio Nakahara
Kinki University, Japan

Shu Tanaka
University of Tokyo, Japan

Frontiers in Quantum Information Research

Decoherence, Entanglement, Entropy, MPS and DMRG

World Scientific

NEW JERSEY • LONDON • SINGAPORE • BEIJING • SHANGHAI • HONG KONG • TAIPEI • CHENNAI

Published by

World Scientific Publishing Co. Pte. Ltd.
5 Toh Tuck Link, Singapore 596224
USA office: 27 Warren Street, Suite 401-402, Hackensack, NJ 07601
UK office: 57 Shelton Street, Covent Garden, London WC2H 9HE

British Library Cataloguing-in-Publication Data
A catalogue record for this book is available from the British Library.

Kinki University Series on Quantum Computing — Vol. 4
FRONTIERS IN QUANTUM INFORMATION RESEARCH
Proceedings of the Summer School on Decoherence, Entanglement, Entropy, MPS and DMRG

ISBN-13 978-981-4407-18-2
ISBN-10 981-4407-18-6

Printed in Singapore by B & Jo Enterprise Pte Ltd

PREFACE

This volume contains lecture notes and contributions presented at the Summer School on Decoherence, Entanglement and Entropy, held from 8 to 11 August, 2009 and Workshop on Matrix Product State Formulation and Density Matrix Renormalization Group Simulations, held from 12 to 13, August, 2009 both held at the Oxford Kobe Institute, Kobe, Japan. The aims of the summer school were to exchange and share ideas among researchers working in various fields of quantum information theory and to motivate students to tackle advanced topics in this research field. In contrast, the aim of the workshop is limited to exchange new developments in MPS and DMRG among the specialists working in these fields.

Invited speakers in both summer school and workshop were asked to prepare their lecture notes in a self-contained way and therefore we expect that they will be useful for students and researchers even with less background in the subject.

The two meetings were supported by "Open Research Center" Project for Private Universities: matching fund subsidy from MEXT (Ministry of Education, Culture, Sports, Science and Technology). These activities were impossible without the generous support form the project.

We would like to thank all the lecturers, contributors and participants, who made these two meeting exciting. We are also grateful to Koji Kobashi, Director of Oxford-Kobe Institute, for invaluable support. We would like to thank Ryota Mizuno for TeXnical assistance. Finally, we would like to thank Ms Zhang Ji of World Scientific for her excellent editorial work.

Mikio Nakahara
Shu Tanaka
Osaka and Tokyo, January 2012

Summer School on Decoherence, Entanglement and Entropy

Oxford Kobe Institute (Kobe, Japan)

---

8 – 11 August 2009

Invited Lecturers

Michael Duff (Imperial College London, England)
Black Holes, Qubits and Octonions

Akio Hosoya (Tokyo Institute of Technology, Japan)
Weak Value with Decoherence

Vahid Karimipour (Sharif University of Technology, Iran)
Matrix Product Representations of States

Miguel Angel Martin-Delgado (University of Complutense Madrid, Spain)
Topological Quantum Computation

Contributed Talks

Ayumu Sugita (Osaka City University, Japan)
Systematic generation of entanglement measures from group representation theory

Tsubasa Ichikawa (Kinki University, Japan)
On a Possible Definition of Entanglement in Antisymmetric States

Toshihiko Sasaki (KEK, Japan)
Entanglement measures for intermediate separability

William Rubens (Imperial College London, England and University of Tokyo, Japan)
Introduction the the black-hole/qubit correspondence for the quantum information-ist

Kin-ya Oda (Osaka University, Japan)
Plasma Entropy from Higher-Dimensional Black Hole

Yasuhiro Sekino (Okayama Institute for Quantum Physics, Japan)
Black holes as fast scramblers

Kazuki Hasebe (Takuma National College of Technology, Japan)
The Split-Algebras and Non-compact Hopf Maps

Kazutomu Shiokawa (Institute of Physics Academia Sinica and National Center for Theoretical Sciences, Taiwan)
Unruh Effect on Entanglement Dynamics and Teleportation

Toshiki Ide (Kinki University, Japan)
Telecloning of a photonic qubit via thermal noise

Poster Presentations

Shogo Tanimura (Kyoto University, Japan)
A new version of Bell's inequality

Keisuke Fujii (Kyoto University, Japan)
Highly efficient energy transfer in photosynthetic complexes and quantum anti-Zeno effect

Yoshitaka Sasaki (Kinki University, Japan)
On zeros of exponential polynomials and quantum algorithms

Takahiro Sagawa (University of Tokyo, Japan)
Minimal Energy Cost for Quantum Information Processing

i : 0 to d-1
C^d = Spanned by {|0>, |1>, ... |d-1>}

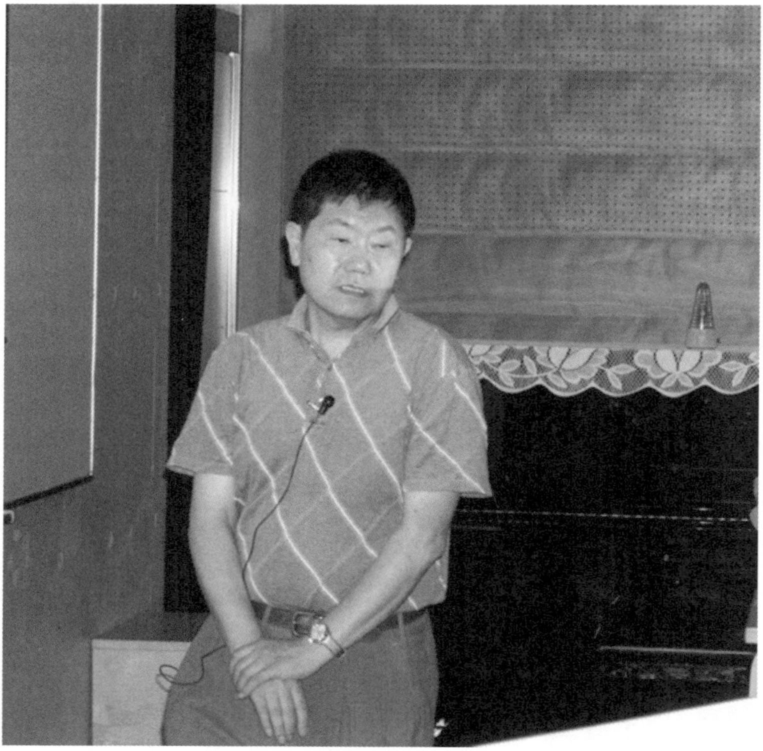

Workshop on
Matrix Product State Formulation and Density Matrix Renormalization Group Simulations (MPS&DMRG)

Oxford Kobe Institute (Kobe, Japan)

---

12 – 13 August 2009

Invited Speakers

Hiroaki Matsueda (Sendai National College of Technology, Japan)
Application of density matrix renormalization group method to photoinduced phenomena in strongly correlated electron systems

Kenji Harada (Kyoto University, Japan)
Multi-scale entanglement renormalization of the spin-1/2 Heisenberg model on a triangular lattice

Vahid Karimipour (Sharif University of Technology, Iran)
Matrix product representation of all valence bond states

Miguel Angel Martin-Delgado (University of Complutense Madrid, Spain)
New Directions for DMRG Exploration

Makoto Yamashita (NTT, Japan)
DMRG study of ultracold fermionic atoms in optical lattices

Naokazu Shibata (Tohoku University, Japan)
Fractional quantum Hall states in 2D electron systems

Contributed Talks

Shigetoshi Sota (Kyoto University, Japan)
Low-temperature density matrix renormalization method and its application to low-dimensional frustrated spin systems

Naokazu Shibata (Tohoku University, Japan)
Fractional quantum Hall states in 2D electron systems

Toshiya Hikihara (Hokkaido University, Japan)
DMRG for Tomonaga-Luttinger liquids

Keisuke Totsuka (Kyoto University, Japan)
Supersymmetric VBS models – hidden order and dynamics –

Hosho Katsura (RIKEN, Japan)
Zamolodchikov-Faddeev Algebra: From Matrix-Product Bethe Ansatz to Organic Spin-Peierls System

Yasuhiro Hieida (Saga University, Japan)
A systematic way to find and construct exact finite-dimensional matrix product stationary states

Akira SaiToh (Kinki University, Japan)
Matrix-product-state simulation of a bulk-ensemble quantum database search as a solver of NP problems

Shunsuke Furukawa (RIKEN, Japan)
Characterization of Tomonaga-Luttinger Liquid via Ground-State Entanglement

Koichi Okunishi (Niigata University, Japan)
On connection between Wilson's NRG and DMRG

Hiroshi Ueda (Osaka University, Japan)
Hyperbolic Deformation on Quantum Spin Chains – for gap estimation –

Tomotoshi Nishino (Kobe University, Japan)
Geometric Deformation

PRINCETON

LIST OF PARTICIPANTS

Asoudeh, Marzieh	Shahid Beheshti University, Iran
Bando, Masamitsu	Osaka University, Japan
Duff, Michael	Imperial College London, England
Fujii, Keisuke	Kyoto University, Japan
Furukawa, Shunsuke	RIKEN, Japan
Harada, Kenji	Kyoto University, Japan
Hasebe, Kazuki	Takuma National College of Technology, Japan
Hieida, Yasuhiro	Saga University, Japan
Hikihara, Toshiya	Hokkaido University, Japan
Hosoya, Akio	Tokyo Institute of Technology, Japan
Hosseini, Elham	Kinki University, Japan
Hotta, Chisa	Kyoto Sangyo University, Japan
Ichikawa, Tsubasa	Kinki University, Japan
Ide, Toshiki	Kinki University, Japan
Imai, Masayuki	Kinki University, Japan
Karimipour, Vahid	Sharif University of Technology, Iran
Katsura, Hosho	RIKEN, Japan
Kobayashi, Shingo	Tokyo Institute of Technology, Japan
Kondo, Yasushi	Kinki University, Japan
Kubo, Kazuhiro	Meiji University, Japan
Marmorini, Giacomo	Keio University, Japan
Martin-Delgado, Miguel	University of Complutense Madrid, Spain
Maruyama, Isao	Osaka University, Japan
Matsueda, Hiroaki	Sendai National College of Technology, Japan
Miyoshi, Hiroyuki	Kyoto Sangyo University, Japan
Morisawa, Yoshiyuki	Osaka University of Economics and Low
Nakahara, Mikio	Kinki University, Japan
Nishino, Tomotoshi	Kobe University, Japan
Oda, Kin-ya	Osaka University, Japan
Rahimi, Robabeh	Kinki University, Japan

Rubens, William	Imperial College London, England and University of Tokyo, Japan
Sagawa, Takahiro	University of Tokyo, Japan
SaiToh, Akira	Kinki University, Japan
Sasaki, Toshihiko	KEK, Japan
Sasaki, Yoshitaka	Kinki University, Japan
Sekino, Yasuhiro	Okayama Institute for Quantum Physics, Japan
Shibata, Naokazu	Tohoku University, Japan
Shimada, Tokuzo	Meiji University, Japan
Shiokawa, Kazutomu	National Center for Theoretical Sciences, Taiwan
Sota, Shigetoshi	Kyoto University, Japan
Sugita, Ayumu	Osaka City University, Japan
Tai, Ta-Sheng	RIKEN, Japan
Tanimura, Shogo	Kyoto University, Japan
Tohyama, Takami	Kyoto University, Japan
Tomita, Hiroyuki	Kinki University, Japan
Totsuka, Keisuke	Kyoto University, Japan
Ueda, Hiroshi	Osaka University, Japan
Yamashita, Makoto	NTT Basic Research Laboratories, Japan
Yamazaki, Chika	Kinki University, Japan
Masuda, Shumpei	Kwansei Gakuin University, Japan
Sugimoto, Takanori	Kyoto University, Japan
Noda, Shusaku	Osaka University, Japan
Nakao, Ken-ichi	Osaka City University, Japan

COMMITTEES

Summer School on Decoherence, Entanglement and Entropy

Mikio Nakahara	Kinki University (Chair)
Tomotoshi Nishino	Kobe University (Vice Chair)
Robabeh Rahimi	Kinki University
Akira SaiToh	Kinki University (technical support)

Workshop on Matrix Product State Formulation and Density Matrix Renormalization Group Simulations (MPS&DMRG)

Mikio Nakahara	Kinki University (Chair)
Tomotoshi Nishino	Kobe University (Vice Chair)
Toshiya Hikihara	Hokkaido University
Koichi Okunishi	Niigata University
Akira SaiToh	Kinki University

CONTENTS

PART A

Summer School on Decoherence, Entanglement and Entropy

BLACK HOLES AND QUBITS

L. BORSTEN, M. J. DUFF and W. RUBENS

Theoretical Physics, Blackett Laboratory, Imperial College London,
London SW7 2AZ, United Kingdom
E-mail: leron.borsten@imperial.ac.uk
E-mail: m.duff@imperial.ac.uk
E-mail: w.rubens@imperial.ac.uk

These notes have been compiled to accompany a series of four lectures given at the Kinki University Quantum Computing Series Summer School on Decoherence, Entanglement and Entropy, August 2009 at the Oxford Kobe Institute (Kobe, Japan).

Each of the four lectures focuses on a particular topic falling under the broad umbrella of the "black-hole/qubit correspondence". Lecture I introduces the first instance of the black-hole/qubit correspondence, the relationship between the entanglement of three qubits and the entropy of *STU* black holes. Lecture II develops this correspondence to the case of $\mathcal{N} = 8$ black holes and the tripartite entanglement of seven qubits. Lecture III examines the use of Jordan algebras and the Freudenthal triple system, which capture the U-duality symmetries of these black hole systems, in entanglement classification. Lecture IV introduces the superqubit, a natural candidate to represent supersymmetric quantum information.

These lectures draw on work done with D. Dahanayake, H. Ebrahim, S. Ferrara and A. Marrani whose efforts are most gratefully acknowledged.

Keywords: Black holes; entanglement.

Overview

This is a story with two leads: (i) black holes in M-theory and (ii) entanglement in quantum information theory (QIT). Both our characters can be capricious and uncooperative, but, in equal measure at least, they continue to surprise and enlighten. We are treated here to some glimpses of dialog between these normally independent protagonists.

Their first chance encounter occurred in 2006 when it was observed that the entropy formula for the *STU* black hole, S_{BH}, appearing in M-theory compactified to four dimensions, is given by an object called the hyperdeterminant,[1,2] a quantity first introduced by the mathematician Cayley in

1845. Remarkably, the 3-tangle, denoted τ_{ABC}, which measures the degree of tripartite entanglement shared by three qubits (Alice, Bob and Charlie), is also given by the hyperdeterminant.[3,4] This leads us to the rather curious mathematical identity:

$$S_{\rm BH} = \tfrac{\pi}{2}\sqrt{\tau_{ABC}}. \tag{1}$$

This initial observation has since inspired a growing body of work extending this "correspondence" in many varied directions.[5–22] Clearly it is not possible to properly treat all aspects in four lectures so we focus on just four points in this landscape. Some appear as rather isolated islands since, in part, it is not practical to build the bridges between them here and, in part, it is still not entirely clear where they lie in relation to each other. However, we hope that we have provided sufficient foundation for an intrepid graduate student to be able to build some bridges of their own. The lectures are organized as follows:

- **Lecture I: *The STU black-hole/qubit correspondence*** In this first lecture we introduce in some detail the first relationship between *STU* black holes and three qubits. In section 1 we review the basics of QIT with an emphasis on entanglement. Similarly, in section 2 we introduce the ideas behind M-theory with some special attention paid to black holes and the *STU* model. The final part section 3 brings these two subjects together, centered around the identity (1), and briefly explores some of their developments.
- **Lecture II: *Seven qubits, octonions and the Fano plane*** Here we extend these ideas to include the black holes of $\mathcal{N} = 8$ supergravity, which have an entropy invariant under E_7, as summarized in section 4. Consequently, the corresponding system of seven qubits enjoys a very special kind of entanglement discussed in section 5. In fact, the entanglement is described by the Fano plane, which in turn characterizes the imaginary octonions. The relationships between black holes, entanglement and the octonions is the subject of section 6.
- **Lecture III: *FTS classification of 3-qubit entanglement*** The Freudenthal triple system (FTS) has been known for some time now to underlie the symmetries of a large class of black holes, including the *STU* model. In this lecture we see how these structures can be exported to three qubits. In section 7 we introduce Jordan algebras and the FTS. Then in section 8 we apply these constructions to the entanglement classification of three qubits.

- **Lecture IV: *Superqubits*** In this last lecture we consider a possible supersymmetric extension of the qubit, the *superqubit.* In section 10 we give the necessary background on supergroups before introducing the superqubit in section 11. Finally, in section 12 we build supersymmetric measures of entanglement for two and three superqubits, the so-called superdeterminant and superhyperdeterminant, respectively.

Lecture I:
The STU black-hole/qubit correspondence

1. Qubits and entanglement

Quantum theory revolutionized physics perhaps like no other modern development in the sense that it challenges the very notion of what we mean by "reality". Of course, what is meant by reality is, and has always been, a philosophically perilous line of inquiry. Then there is the secondary question of how any particular notion of reality is to be put into practice, i.e. as a physical theory. In the context of classical physics, however, we may make a fairly uncontentious first approximation. The fundamental reality of objects in classical physics is constituted by their observable properties; "things" in themselves are defined as the bearers of said properties, which are in turn determinable from that object.[23,24] Essential to such a prescription of reality is the crucial assumption that all physical attributes do indeed possess definite, if not definitely known, values at all times.[23–26] These ideas form the crux of what may be called a realist philosophy of nature. This realist stance is naturally embodied by the mathematical (set-theoretic) formalism underpinning classical theory.

A second requirement that one might demand of a sensible description of reality is *locality* (or perhaps more precisely in this context *separability*). Indeed, this concept was important to Einstein's picture of the world, as he describes to Max Born,[27]

> "I just want to explain what I mean when I say that we should try to hold on to physical reality. We all of us have some idea of what the basic axioms of physics will turn out to be... whatever we regard as existing (real) should somehow be localised in time and space. That is, the real in part of space A should (in theory) somehow 'exist' independently of what is thought of as real in space B. When a system in physics extends over the parts of space A *and* B, then that which exists in B should somehow exist independently of that which exists in A. That which really exists in B should therefore not depend on what kind of measurement is carried out in part of space A; it should also be independent of whether or not any measurement at all is carried out in space A." - Einstein

The union of these ideas is typically referred to as *local realism*. The pinnacle of classical physics, Einstein's general theory of relativity, is a complete local realist theory. The orthodox Copenhagen interpretation of quantum physics rejects local realism in all its parts.

As for realism, it is not possible to speak of a physical system as actually "possessing" values for all their physical observables at a given time, and that these values are intrinsic and independent of any measurement setup used to reveal them. This is not merely a philosophical whim, it is a mathematical consequence of the Kochen-Specker theorem.[28] See[24,26] for introductory discussions. Rather, one is restricted to making probabilistic counter-factual statements, in which the act of measurement plays a fundamental role, "if measured, an observable A will have a value a with a probabilty p".

We also lose our grip on locality due to the quintessentially quantum phenomenon of *entanglement*, the main subject of this introduction to quantum information. Quantum entanglement is a phenomenon in which the quantum states of two or more objects must be described with reference to each other, even though the individual objects may be spatially separated.[29–34] This leads to classically unexplainable, but experimentally observable, quantum correlations between the spatially separated systems. A "spooky" action at a distance, as Einstein called it.

Given the startling nature of these two statements it comes as no surprise that over the years many a physicist and philosopher has taken exception, not least of all Einstein. This led to the now famous Bohr-Einstein dialog, with Einstein fighting the corner of local realism and Bohr that of the Copenhagen interpretation. Its culmination was the seminal 1935 work by Einstein, Podolsky, and Rosen (EPR).[29] They correctly concluded that, assuming local realism, the quantum mechanical wave function cannot be a complete description of reality. They speculated on the existence of a more fundamental underlying theory satisfying the principles of local realism. Subsequent attempts to develop such a hypothetical theory typically assumed that the mathematical representation of physical states, the wave function, is completed by a set of "hidden variables". The previously *fundamental* probabilistic nature of quantum mechanics would then admit a pedestrian classical ignorance interpretation. However, through an ingenious development of the EPR setup, Bell demonstrated that there are no local realist hidden variable models that can correctly reproduce the predictions of quantum mechanics.[31] The quantum mechanical predictions where experimentally verified in 1982 by Alain Aspect *et al.*[35] We will return to this remarkable subject in subsection 1.2 as it is essential to the proper understanding of entanglement. However, to this day many scientists feel uncomfortable, for important reasons such as the measurement problem and the trouble with quantum cosmology, with the operationalist interpretation of quantum mechanics and the debate rages on.

However, we can turn all this existential angst on its head, take quantum theory at face value, and ask not what we can do for quantum theory, but what quantum theory can do for us? What can we learn from and how can we use the elementary properties of quantum theory, such as the superposition principle and entanglement. This is the standpoint of many quantum information scientists: to embrace the basic mathematical formalism in an effort not only to better understand quantum theory, but also to go beyond what can be achieved in the field of classical information theory. In the following section we will introduce the basic ideas of quantum information theory with a particular emphasis on the central role of entanglement. There are a number of good introductions to the topics of QI and entanglement. We found[36–42] very useful and have used them throughout the reminder of this and the subsequent sections.

1.1. *A brief introduction to quantum information*

There is an expectation that quantum mechanics can be used to perform computational tasks beyond the capability of any, even idealistic, purely classical device[a].

While this is certainly a key motivation QIT, is not purely a technological enterprise. There is a view, which has already proved itself well-founded, that by adopting an information theoretic stance we are able to develop our understanding of quantum theory at a fundamental level, to clarify and sharpen our intuition about the features of quantum mechanics which may otherwise seem incomprehensible. This may take the form of developing the tools to better understand standard predictions or, to scrutinize, and potentially alter, the very foundations of the subject.

A key player in all these developments has been entanglement, which forms the focus of this introduction in anticipation of its surprising connection to stringy black holes.

Entanglement is both one of the most startling and exploitable characteristics of quantum mechanics and, as such, it is central to our fundamental understanding and as a quantum computing tool.

---

[a]Classical not in the sense that it does not utilize any quantum physics, such as semiconductor components, but that its algorithmic structures are classical, and, as such could be performed by any classical device whether it be made of the most advanced microchips or just sticks and twigs, geologically slow as it might be in that case.

Entanglement may be thought of as a quantum information *resource* in the same sense as entropy or energy are classical resources. However, its properties profoundly differ from the properties of those familiar concepts. We have, at best, an incomplete description. In order to fully understand entanglement we would like to be able to precisely describe its creation and transformation, to classify the distinct types of entanglement, to quantitatively measure it, to utilize it as a resource and to illustrate, precisely, how it differs from classical resources at a fundamental level.

1.1.1. *Qubits*

Digital information is conventionally represented by a bit string of "0"s and "1"s. The physical realisation of bits takes many forms. In real world technologies current, voltage or light pulses are predominately used. For example, a "0" is represented by no light pulse while "1" is represented by a short pulse.

One could imagine the length of the pulse being gradually reduced to a single photon - a genuine quantum mechanical object. In this case we could represent bits not by whether a photon was sent or not, but instead by its polarisation: horizontal for "0" and vertical for "1". However, unlike the (relatively) long lived light pulse, which was either there or not, a single photon can exist in a superposition of polarisation states. This is the idea of the quantum bit, or *qubit*; a quantum superposition of the binary digits "0" and "1". The particular physical realisation (there are many: photon polarisations, quantum dots, trapped ions, mode splitters, to name but a few) of the qubit is not important, any two state quantum system will do. Hence, qubits are simply denoted abstractly as elements of the 2-dimensional Hilbert space $\mathbb{C}^2$ equipped with the conventional norm, where the two basis states are labelled $|0\rangle$ and $|1\rangle$. An n-qubit bit string lives in the n-fold tensor product of $\mathbb{C}^2$.

One qubit: The 1-qubit system (Alice) is described by the state $|\Psi\rangle \in \mathbb{C}^2$,

$$|\Psi\rangle = a_A|A\rangle, \qquad \text{where} \qquad A = 0, 1 \tag{2}$$

so

$$|\Psi\rangle = a_0|0\rangle + a_1|1\rangle. \tag{3}$$

The density matrix ρ, defined by

$$\rho = |\Psi\rangle\langle\Psi|, \qquad \rho_{A_1A_2} = a_{A_1}a^*_{A_2}, \tag{4}$$

obeys

$$\operatorname{tr}\rho = \langle\Psi|\Psi\rangle. \tag{5}$$

Two qubits: The 2-qubit system (Alice and Bob) is described by the state $|\Psi\rangle \in \mathbb{C}^2 \otimes \mathbb{C}^2$,

$$|\Psi\rangle = a_{AB}|AB\rangle, \qquad \text{where} \qquad A, B = 0, 1 \tag{6}$$

so

$$|\Psi\rangle = a_{00}|00\rangle + a_{01}|01\rangle + a_{10}|10\rangle + a_{11}|11\rangle. \tag{7}$$

We defined the partially reduced density matrices

$$\begin{aligned} \rho_A &= \operatorname{Tr}_B |\Psi\rangle\langle\Psi|, \\ \rho_B &= \operatorname{Tr}_A |\Psi\rangle\langle\Psi|, \end{aligned} \tag{8}$$

or

$$\begin{aligned} (\rho_A)_{A_1A_2} &= \delta^{B_1B_2} a_{A_1B_1} a^*_{A_2B_2} = (\rho_A)_{A_2A_1}, \\ (\rho_B)_{B_1B_2} &= \delta^{A_1A_2} a_{A_1B_1} a^*_{A_2B_2} = (\rho_B)_{B_2B_1}. \end{aligned} \tag{9}$$

Explicitly,

$$\begin{aligned} \rho_A &= \begin{pmatrix} |a_0|^2 + |a_1|^2 & a_0a_2^* + a_1a_3^* \\ a_2a_0^* + a_3a_1^* & |a_2|^2 + |a_3|^2 \end{pmatrix}, \\ \rho_B &= \begin{pmatrix} |a_0|^2 + |a_2|^2 & a_0a_1^* + a_2a_3^* \\ a_1a_0^* + a_3a_2^* & |a_1|^2 + |a_3|^2 \end{pmatrix}. \end{aligned} \tag{10}$$

Note,

$$\operatorname{tr}\rho_A = \operatorname{tr}\rho_A = \langle\Psi|\Psi\rangle. \tag{11}$$

***n* qubits:** The n-qubit system ($\text{Alice}_1, \ldots, \text{Alice}_n$) is described by the state $|\Psi\rangle \in \mathbb{C}^2 \otimes \ldots \otimes \mathbb{C}^2$,

$$|\Psi\rangle = a_{A_1\ldots A_n}|A_1 \ldots A_n\rangle, \qquad \text{where} \qquad A_1, \ldots, A_n = 0, 1 \tag{12}$$

so

$$\begin{aligned} |\Psi\rangle = a_{00\ldots00}&|00\ldots00\rangle + a_{00\ldots01}|00\ldots01\rangle + \ldots a_{11\ldots10}|11\ldots10\rangle \\ &+ a_{11\ldots11}|11\ldots11\rangle. \end{aligned} \tag{13}$$

For any subset of qubits $X = \{A_i, A_j \ldots\} \subset \{A_1, \ldots, A_n\}$ we define the partially reduced density matrices

$$\begin{aligned} \rho_X &= \operatorname{Tr}_{\overline{X}} |\Psi\rangle\langle\Psi|, \\ \rho_{\overline{X}} &= \operatorname{Tr}_X |\Psi\rangle\langle\Psi|, \end{aligned} \tag{14}$$

where $\overline{X}$ denotes the set theoretic complement. Note,

$$\mathrm{Tr}_X\,\rho_{\overline{X}} = \mathrm{Tr}_{\overline{X}}\,\rho_X = \langle\Psi|\Psi\rangle. \tag{15}$$

1.2. *Entanglement and the Bell inequality*

Already in 1935 both EPR and Schrödinger had identified the phenomenon of quantum entanglement as central to quantum theory and were, at the very least, uncomfortable with its implications. While EPR had mathematically challenged the completeness of quantum theory on the basis of entanglement, it remained a conceptual matter, apparently not accessible to experiment. All this changed in 1965 when Bell introduced his famous inequality. In one fell swoop, entanglement had been elevated from a conceptual puzzle to an experimental observable confronting the very assumptions of local realism. Moreover, Bell had laid the foundations of entanglement as a quantum information theoretic resource. For example, in 1991 Ekert[43] used the Bell inequality as the basis for a new secure means of communication.

Essentially, if the state of a composite system cannot be written as a tensor product of its constituent subsystems then it is entangled. As a consequence, measurements on one component can affect the measurement results on another component, even if they are space-like separated. Consider, for example, the 2-qubit Bell state,

$$\frac{1}{\sqrt{2}}(|00\rangle + |11\rangle). \tag{16}$$

If Alice measures 0, Bob must also measure 0 and if Alice measures 1, Bob must also measure 1. Moreover, on tracing out either Alice or Bob we are left with a totally mixed state - while having maximal knowledge of the composite system we have minimal knowledge of it constituent pieces.

However, to understand properly the fundamental implications of entanglement we must first establish more precisely the philosophical stance of the quantum skeptic. Recall, he is a principled local realist. However, with good reason we have been rather vague about this concept until now. Like EPR we make no attempt to give a comprehensive definition, but rather adopt their set of minimal conditions (which are phrased in the context of a 2-particle experiment but generalise in an obvious way):[29]

(1) *Reality*: "If, without in any way disturbing a system, we can predict with certainty (i.e., with probability equal to unity) the value of a physical quantity, then there exists an element of physical reality corresponding to this physical quantity."

(2) *Locality*: "Since at the time of measurement the two systems no longer interact, no real change can take place in the second system in consequence of anything that may be done to the first system"
(3) *Completeness*: "Every element of the Physical Reality must have a counterpart in the physical theory"

However, while a staunch realist, our quantum skeptic does not doubt the experimental results and, moreover, is in full agreement that they are correctly predicted by the quantum formalism. Where he diverges from the quantum converts is that he maintains there is, in fact, a deeper theory which will give the correct experimental prognosis while adhering to the above principles of local realism. This theory would have well defined single measurement predictions; the probabilistic counter-factual statements, "*if measured* such and such observable will have this or that value *with probability* so and so", of quantum theory are simply an artifact of its incompleteness. The complete theory would depend on some additional physical elements currently out of reach. Perhaps our theoretical framework is deficient or our current experimental techniques inadequate. Whatever the case, these "hidden variables" would complete our theory providing a local and realistic description of nature.

However, as we shall see, quantum theory shows its doubters no mercy. The Bell inequalities[31] were the final nail in the coffin of local realism. The quantum mechanical violation of the Bell inequalities was verified in 1982 by the now famous *Orsay experiment*.[35] Here is the basic setup (the full experiment will be described below). An entangled photon pair is produced by some suitable atomic process. The emitted photons travel in opposite directions which define our z-axis. Eventually they each meet an analyser, denoted A and B, measuring their polarisations in perpendicular x- and y-directions. See Figure 1.

Experimentally one finds that only the pairs (h, h) and (v, v) are observed, where h and v denote horizontal and vertical polarisations respectively. Quantum mechanically this 2-photon system is represented by the Bell state,

$$|\Psi\rangle = \frac{1}{\sqrt{2}}(|hh\rangle + |vv\rangle). \tag{17}$$

So, the photon pair is totally correlated. EPR used these perfect correlations and the assumptions of *Reality*, *Locality* and *Completeness* to argue as follows: since we can predict with certainty the result at A by merely noting the result at B, by *Locality* and *Reality* there is an element of reality

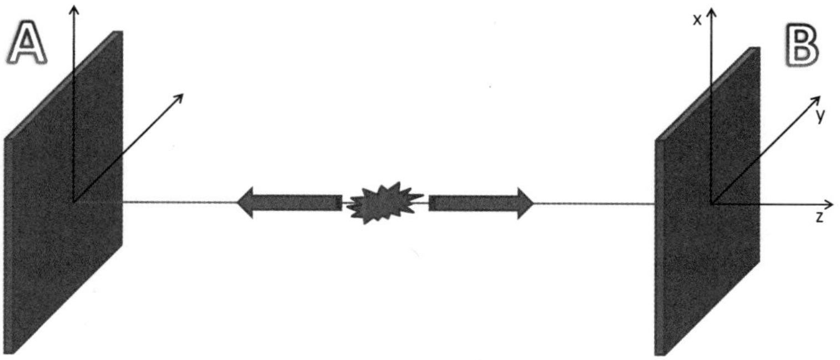

Fig. 1: Entangled photon pair are measured by the horizontal/vertical polarisation analysers, A and B.

corresponding to the polarisation of the photon at A. This is true regardless of which specific $x - y$ axises we choose to perform the experiment with. Hence, there is an element of reality corresponding to the polarisation of the photon at A in all possible $x - y$ orientations. However, there is no photon quantum state for which its polarisation in all directions has a definite value. Hence, by the *Completeness* criterion, quantum mechanics is incomplete.

Are such definite correlations enough to convince our quantum skeptic? Not at all, the EPR argument has demonstrated that quantum mechanics cannot be considered complete under the assumptions of local realism - it says nothing about the possible existence or not of a complete local realist theory. We must work a little harder before we declare experimental victory over classical reality. This was Bell's great insight - to derive from EPR's criteria something which could be used to experimentally check the phenomenological viability of local realism.

Bell had to consider two points. First, the experimental setup had to be tuned to best reveal its true quantum nature. Second, he had to formulate a sufficiently generic hidden variable model to be tested against the subsequent results.

As for the experiment we will use the Clauser-Horne-Shimony-Holt (CHSH)[44] variant of Bell's original arrangement, as this is the setup that was experimentally reproduced in.[35] The analysers are now allowed to be individually rotated about the z-axis into four possible orientations differing in steps of a rotation angle θ. See Figure 2. The experiment is performed with four out of the sixteen possible configurations, denoted $\{x_1, x_2\}, \{x_2, x_3\}, \{x_3, x_4\}, \{x_1, x_4\}$, where the first (second) slot specifies

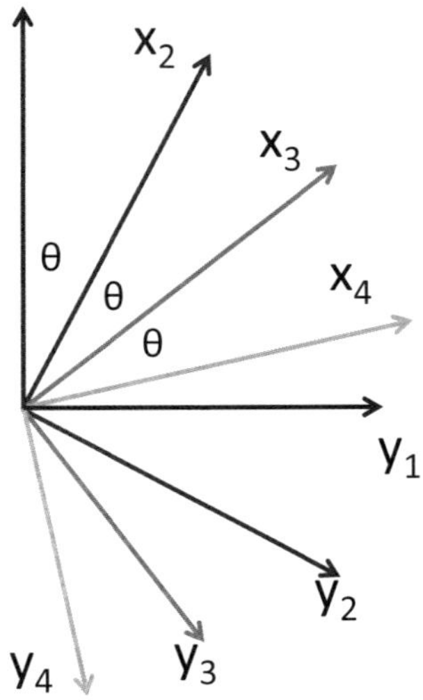

Fig. 2: Each polarisation analyser may be rotated about the z-axis to measure the polarisation along one of four sets of x-y axises differing by steps of θ.

the orientation of analyser A (B). Each of the four configurations has four possible measurement outcomes. For example, the possible measurement pairs for $\{x_2, x_3\}$ are $(h_2, h_3), (h_2, v_3), (v_2, h_3), (v_2, v_3)$. We define the functions σ_A^i such that

$$\sigma_A^i = \begin{cases} +1 & \text{for measurement outcome} \quad h_i \\ -1 & \text{for measurement outcome} \quad v_i \end{cases} \tag{18}$$

where h_i and v_i are the measurement outcomes at analyser A for orientation $i \in \{1, \ldots, 4\}$. The functions σ_B^i is defined similarly for analyser B. The experiment is run for many 2-photon pairs and the results are used to calculate the correlation coefficient,

$$S(\theta) := E(3,2) + E(3,4) + E(1,2) - E(1,4), \tag{19}$$

where

$$E(i,j) := \overline{\sigma_A^i \sigma_B^j}. \tag{20}$$

It is experimentally observed that the maximum value is $S_{\max}(\theta_0) = 2.697 \pm 0.015$, where $\theta_0 = 22.5°$.[35] Note, thus far we have made no reference to any particular physical theory, we have only been dealing with experimental data.

There are now two questions: 1) is quantum theory consistent with these results? 2) is there a local hidden variables model consistent with these results? We are assuming from the outset that quantum mechanics correctly predicts the experimental data and a standard quantum mechanical calculation shows this is indeed the case.[35] So, let us now address the question of hidden variables.

As tradition dictates our hidden variables will be denoted λ. The full state of our hypothetical complete theory might be written $|\Psi, \lambda\rangle$. The particular form of λ is not of importance. Our outcome functions $\sigma_A^i(\lambda)$ and $\sigma_B^j(\lambda)$ now depend on the hidden variables. The essential assumption of locality is that the result $\sigma_A^i(\lambda)$ at analyser A does not depend on the setting j at analyser B and vice-versa. The characteristics of the photon source are captured by a probability density $\rho(\lambda)$. A 2-photon state with parameter $\lambda \in [\lambda, \lambda + \mathrm{d}\lambda]$ is produced with probability $\rho(\lambda)\mathrm{d}\lambda$. As Bell noted, this setup subsumes the case in which σ_A and σ_B depend on two distinct sets of hidden variables.

The expectation value of a given configuration $\{x_i, x_j\}$ is given by,

$$E(i,j) = \int \sigma_A^i(\lambda)\sigma_B^j(\lambda)\rho(\lambda)\mathrm{d}\lambda. \tag{21}$$

Hence, the hidden variable correlation coefficient is given by,

$$S_{\mathrm{HV}}(\theta) = \int [\sigma_A^3(\lambda)\sigma_B^2(\lambda)+\sigma_A^3(\lambda)\sigma_B^4(\lambda)+\sigma_A^1(\lambda)\sigma_B^2(\lambda)-\sigma_A^1(\lambda)\sigma_B^4(\lambda)]\rho(\lambda)\mathrm{d}\lambda. \tag{22}$$

Since $\int \rho(\lambda)\mathrm{d}\lambda = 1$ and $\sigma = \pm 1$ it is clear that,

$$S_{\mathrm{HV}}(\theta) \leq 2. \tag{23}$$

This is the famous (generalised) Bell inequality.[44] The correlation coefficient for a local realistic hidden variables model cannot exceed 2. The Orsay experiment obtained a maximum of 2.697 ± 0.015. Reality is dead. Long live Reality!

1.2.1. *Entanglement dependent quantum information*

This experiment and its subsequent refinements have established entanglement as a now accepted feature of nature. We claimed that the quantum information theorists regards entanglement as one of his key resources. Let us spend a moment justifying this claim. The literature on this subject is vast and we can only give the briefest of glimpses.

Sheep yes, qubits no A single quantum cannot be cloned.[45] What would happen were quantum copying possible? Wootters and Zurek argue as follows.[45] If it were possible to copy a quantum state then one could ascertain the precise state of the original. This would facilitate super-luminal communication. Consider a Bell pair consisting of two photons. Once one photon has been measured the other will be in an eigenstate of the same polarisation. If the second photon could be copied and, hence, have its state precisely determined, the measurement basis used at the first photon would be known, allowing for a faster-than-light transmission of information encoded in the choice of measurement basis.

The logic negating this possibility is surprisingly simple and essentially relies only on the linearity of quantum mechanics. Suppose we do have a quantum copying device. We denote the copier's basic pre-copy state by $|A_\bullet\rangle$. We use use it to replicate a qubit state $|\Psi\rangle$. Mathematically,

$$|A_\bullet\rangle \otimes |\Psi\rangle \otimes |\Phi\rangle \rightarrow |A_\Psi\rangle \otimes |\Psi\rangle \otimes |\Psi\rangle, \tag{24}$$

where the arrow denotes some unitary evolution and $|A_\Psi\rangle$ is the post-copying state of the device, which may or may not actually depend on $|\Psi\rangle$. Now consider the case where $|\Psi\rangle = |0\rangle$,

$$|A_\bullet\rangle \otimes |0\rangle \otimes |\Phi\rangle \rightarrow |A_0\rangle \otimes |0\rangle \otimes |0\rangle. \tag{25}$$

However, for a state $|\Psi\rangle = a_0|0\rangle + a_1|1\rangle$ we find, by linearity,

$$|A_\bullet\rangle \otimes (a_0|0\rangle + a_1|1\rangle) \otimes |\Phi\rangle \rightarrow a_0|A_0\rangle \otimes |0\rangle \otimes |0\rangle + a_1|A_1\rangle \otimes |1\rangle \otimes |1\rangle. \tag{26}$$

If the device states $|A_0\rangle, |A_1\rangle$ differ then the qubits will emerge in a mixed state, if $|A_0\rangle, |A_1\rangle$ are the same then the qubits will emerge as a Bell-pair. In either case we do not obtain the desired copy state,

$$(a_0|0\rangle + a_1|1\rangle) \otimes (a_0|0\rangle + a_1|1\rangle). \tag{27}$$

Quantum cryptography The art of cryptography as a long history, from protecting Mesopotamian bakery recipes to modern internet security. The central ingredient in almost all contemporary cryptographic schemes is the *key*. To send a message, the plain text and a key are fed as inputs to the encryption algorithm. The receiver then feeds the encrypted message and his key into the decrypting algorithm. The cryptogram and encryption/decryption algorithms may be known by anyone. The security depends entirely on the secrecy of the keys. Historically, a *symmetric-key* algorithm has been the principal method employed. In this case there is essentially one key used to both encrypt and decrypt. Alice invents a key and secretly shares it with her intended correspondent, Bob. When the day comes Alice uses her key to encode her message and sends it to Bob who is then able to decrypt it using the same key. The tricky part is the sharing of the key. Back in the day, Alice and Bod could exchange keys face-to-face in some clandestine moonlit clearing, deep in the local forest. However, despite their best efforts, Amazon.co.uk have not been able to make this particular key distribution protocol workable in todays consumer driven society.

The problem of symmetric-key distribution was circumvented in the mid 70's with the invention of *asymmetric-key* cryptography. In this case Alice creates two keys. One *public* key which she shares with the world, one *private* key which she keeps entirely to herself. Now anyone can send Alice a message using the public key, which only she can decode using the private key. Obviously, the public and private keys must be mathematically related. The crucial point is that the private key cannot *feasibly* be deduced from the public key. A much used example is the Rivest, Shamir and Adleman (RSA) scheme which relies on the fact that, while multiplication is easy, factorisation is not - there is no known efficient classical factorisation algorithm. There is, however, an efficient quantum algorithm.[46] The practical realisation of a quantum computer would mean abandoning RSA.

There is a third way. The symmetric-key protocol would be completely secure if both Alice and Bob could check whether their shared secret key had been intercepted during transmission. Any intercepted keys would be discarded. Alice and Bob would only proceed when they are both sure their key is truly secret. This is the approach taken by quantum cryptography. Any interception method must involve some measurement procedure performed by the eavesdropper. Classically, this can, in principle, be done without disturbing the message. Quantum mechanically, this is not so, which suggests the possibility of a totally secure quantum key distribution method. The first entanglement based quantum cryptography scheme was introduced by Ekert.[43] There is a source of Bell pairs from which one qubit in each pair is sent to Alice and the other to Bob. So, Alice and Bob receive a string of qubits which they measure in some basis independently chosen at random from a set of previously agreed upon possibilities. Once complete, they publicly share their choice of basis for each measurement. For those instances in which they happen to choose the same basis their results will be perfectly correlated and can be used to form a bit string, which is then used as the private key. The remaining results can be used to check the Bell inequalities, which, only if no-one has tried to measure the qubits, will be violated. If the Bell inequalities are not violated they can be sure that someone was listening. This scheme was experimentally implemented in 1999.[47,48]

Quantum teleportation Can Alice send an unknown qubit state to Bob? No-cloning implies that this is not possible using only classical channels. However, Bennett *et al.* showed that a qubit can be *teleported* using only two bits of classical information together with the correlations of a Bell-pair.[49] The use of "teleportation" emphasizes the fact that, once Bob has received the qubit, Alice is left with no trace of the teleported state.

The teleportation protocol proceeds as follows. Alice has a qubit in an unknown state $|\Psi\rangle$ she wishes to send to Bob. Alice and Bob share a Bell-pair. The total composite system is represented by the state,

$$|\Psi\rangle = |\Psi\rangle \otimes \frac{1}{\sqrt{2}}(|00\rangle + |11\rangle). \tag{28}$$

Note, the first two slots in the tensor product (the unknown state and one half of the Bell-pair) are owned by Alice, the third slot is Bob's half of the Bell-pair. This state may be rewritten using the Bell basis,

$$|\Psi\rangle = \frac{1}{2}(|\Phi^+\rangle|\Psi\rangle + |\Phi^-\rangle\sigma_3|\Psi\rangle + |\Psi^+\rangle\sigma_1|\Psi\rangle + |\Psi^-\rangle i\sigma_2|\Psi\rangle), \tag{29}$$

where,

$$\begin{aligned} \Psi^{\pm} &= \frac{1}{\sqrt{2}}(|01\rangle \pm |10\rangle), \\ \Phi^{\pm} &= \frac{1}{\sqrt{2}}(|00\rangle \pm |11\rangle), \end{aligned} \tag{30}$$

and σ_i are the Pauli matrices. Alice then simply measures her pair of qubits in the Bell basis and communicates the result to Bob, which requires two classical bits of information. Finally, Bob performs a simple unitary rotation according to Alice's result and is left with precisely $|\Psi\rangle$ as required. This procedure, or variations of it, is now key to many quantum communication protocols.[40] The first actual quantum teleporation was perform in a pioneering experiment by Bouwmeester *et al.* in 1997.[50]

1.3. *Entanglement classification*

It ought to be clear by now that entanglement has striking foundational implications as well as important technological applications. Consequently, a proper understanding of entanglement is central to quantum information science as well as quantum theory in general. How can we manipulate entanglement? Are there different forms of entanglement? If so how are they distinguished and how do we quantify them?

1.3.1. *Bell inequalities without the inequality*

Let us now address the first question: are there qualitatively different forms of entanglement?

It is argued here that this is indeed the case. The crucial observation is that by generalising the EPR setup to a *tripartite* system the principles of local realism are unequivocally shattered by a *single* run of the (thought) experiment. This remarkable experiment was introduced by Greenberger, Horne and Zeilinger (GHZ).[51] The very special tripartite state upon which their argument relies has come to be known as the GHZ-state. The quantum nature of entanglement is so dramatically demonstrated that no appeal to inequalities is necessary. This seems to us sufficient justification for the claim that the GHZ-state possesses a qualitatively greater degree of entanglement compared to the Bell-state[b].

Actually, as emphasized by GHZ[51] this is the first instance that directly confronts the particular case when predictions can be made with certainty, as described in EPR's *Reality* condition. The requirement of certainty is not covered by Bell's analysis which is necessarily statistical. Only when both analyzers are aligned does the result on one photon fix with certainty the result on the other. For this particular sub-arrangement one is able to construct a local realist theory which correctly reproduces the quantum mechanical results.[31] Can one always find classical models for these subcases of experimental certainty? In a sense this is the most physically intriguing question. Cases in which one can with certainty predict the value of a physical property one might expect, more so than in any other case, that there really is an actual physical object possessing that specific property with that precise value. The one run certitude of the GHZ elaboration does away with any such possibility.

GHZ originally considered a four-particle system,[51] while the experimental test was performed with three entangled photons.[53] Here, we present a simplified version involving three spin-1/2 particles introduced by Mermin.[33] Imagine the three spin-1/2 particles leaving a common source in the entangled state,

$$|\Psi\rangle_{\mathrm{GHZ}} = \frac{1}{\sqrt{2}}(|\uparrow_z\uparrow_z\uparrow_z\rangle + |\downarrow_z\downarrow_z\downarrow_z\rangle), \tag{31}$$

where, $\uparrow_z$ and $\downarrow_z$ denote spin-up or -down along the z-axis as defined by the flight of the particles.

---

[b]For an interesting discussion on the subtleties of this matter see 52.

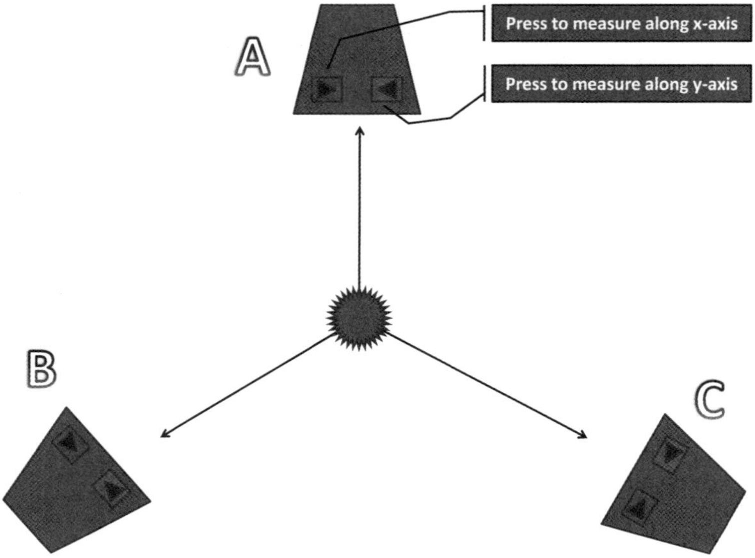

Fig. 3: Three Stern-Gerlach device, denoted A, B, C, which can be individually set to measure spin along their respective x- or y-axis.

This is the famous GHZ-state. Each particle eventually meets a Stern-Gerlach device, denoted A, B, C, which may be individually orientated to measure spin along the x- or y-axis. See Figure 3. The experiment is performed with the measuring devices randomly set in one of four configurations (out of the eight possible). These are the three configurations where two devices measure along the y-axis and one along the x-axis, denoted xyy, yxy, yyx, and the single configuration where all three devices measure along the x-axis, denoted xxx.

Let us consider the case with device A set to measure along the x-axis and devices B, C along the y-axis. Using

$$|\uparrow_z\rangle = \frac{1}{\sqrt{2}}(|\uparrow_x\rangle + |\downarrow_x\rangle), \quad |\downarrow_z\rangle = -\frac{1}{\sqrt{2}}(|\uparrow_x\rangle - |\downarrow_x\rangle) \tag{32}$$

and

$$|\uparrow_z\rangle = \frac{1}{\sqrt{2}}(|\uparrow_y\rangle + |\downarrow_y\rangle), \quad |\downarrow_z\rangle = -\frac{i}{\sqrt{2}}(|\uparrow_y\rangle - |\downarrow_y\rangle) \tag{33}$$

we can write (31) in the measurement eigenbasis,

$$|\Psi\rangle_{\text{GHZ}} = \frac{1}{2}(|\uparrow_x\uparrow_y\uparrow_y\rangle + |\downarrow_x\downarrow_y\uparrow_y\rangle + |\downarrow_x\uparrow_y\downarrow_y\rangle + |\uparrow_x\downarrow_y\downarrow_y\rangle), \tag{34}$$

from which it is clear that, with the detectors in the xyy orientation, only an odd number of particles will be measured as spin-up. By symmetry, this is also true for both the yxy and yyx orientations. On the other hand, for the final configuration with all three devices set to measure along the x-axis, (31) in the measurement eigenbasis is given by,

$$|\Psi\rangle_{\mathrm{GHZ}} = \frac{1}{2}(|\downarrow_x\downarrow_x\downarrow_x\rangle + |\downarrow_x\uparrow_x\uparrow_x\rangle + |\uparrow_x\downarrow_x\uparrow_x\rangle + |\uparrow_x\uparrow_x\downarrow_x\rangle), \tag{35}$$

so that only an even number of particles are found to be spin-up. Note, in all cases we can with certainty predict the measurement result of any one particle by simply noting the outcomes of the other two particles.

The question now, as before, is not whether quantum mechanics gets it right (we are assuming this from the outset), but whether there is a local realist theory that can reproduce these (hypothetical) experimental observations.

Mermin[33] gave a particularly clear reasoning as to why there is no such theory. Following EPR Mermin made the crucial inference that since there are no connections between the detectors, the highly coordinated experimental results must derive from the common source of the particles. The information determining the spin observed at A, which must be consistent with the spins observed at B and C, must be carried by the particle triggering A. This argument can be applied equally to any of the three particles. Moreover, since the particles have no knowledge of which direction the detectors are set to measure and the setting can change during flight, it would seem essential that each particle carries information specifying what spin will be measured for either of the detector settings.

In summary, we ought to be able to label the particles with the spin that will be observed given the detector is set to measure along the x- or y-axis. For example, the particle triggering detector A could be labeled $(\uparrow_x, \downarrow_y)_A$ where the first (second) slot denotes the spin that will be measured with the detector set to measure along the x-axis (y-axis). For the detector orientation xyy a possible labeling of the three particles is given by,

$$(\uparrow_x, \downarrow_y)_A(\uparrow_x, \uparrow_y)_B(\uparrow_x, \uparrow_y)_C \tag{36}$$

as all three detectors would register spin-up consistent with the experimental observation that only an odd number of spin-up particles are observed for xyy. However, this labeling is inconsistent with the yxy set-up as we would observe two spin-up results. A little careful thought shows that there are only eight labelings simultaneously consistent with xyy, yxy and yyx,

$$
\begin{aligned}
&(\uparrow_x, \uparrow_y)_A(\uparrow_x, \uparrow_y)_B(\uparrow_x, \uparrow_y)_C \\
&(\uparrow_x, \uparrow_y)_A(\downarrow_x, \downarrow_y)_B(\downarrow_x, \downarrow_y)_C \\
&(\downarrow_x, \downarrow_y)_A(\uparrow_x, \uparrow_y)_B(\downarrow_x, \downarrow_y)_C \\
&(\downarrow_x, \downarrow_y)_A(\downarrow_x, \downarrow_y)_B(\uparrow_x, \uparrow_y)_C \\
&(\uparrow_x, \downarrow_y)_A(\downarrow_x, \uparrow_y)_B(\downarrow_x, \uparrow_y)_C \\
&(\downarrow_x, \uparrow_y)_A(\uparrow_x, \downarrow_y)_B(\downarrow_x, \uparrow_y)_C \\
&(\downarrow_x, \uparrow_y)_A(\downarrow_x, \uparrow_y)_B(\uparrow_x, \downarrow_y)_C \\
&(\uparrow_x, \downarrow_y)_A(\uparrow_x, \downarrow_y)_B(\uparrow_x, \downarrow_y)_C.
\end{aligned}
\tag{37}
$$

However, a cursory glance reveals that all above labelings yield an odd number of spin-up measurements for xxx contradicting the observation, quantum mechanically derived from (35), that only an *even* number of spin-up states are ever recorded for this configuration.

Hence, a *single* run of the experiment in the xxx configuration suffices to put to bed the seemingly unavoidable idea, implied by local realism, that the particles carry labels. The fact that such a simple one run violation of local realism can be performed with the GHZ state, but not the Bell state, justifies our claim that there are qualitatively different forms of entanglement.

1.3.2. *The SLOCC paradigm*

Let us reconsider what we mean by entanglement. Conventionally, the state of a composite system is said to be entangled if it cannot be written as a tensor product of states of the constituent subsystems. However, this particular definition is perhaps insufficient to really capture the various subtleties of entanglement. For example, there are two totally non-separable 3-qubit states that have distinct entanglement properties, as we shall see later. Is there a more illuminating notion of entanglement? Let us take our cues from experiment. We do not actually observe the tensor product structure, even though it underpins our theoretical understanding. What we do observe are correlations between spatially separated systems that admit no classical explanation. This motivates the more general and quantum information theoretic notion of entanglement as correlations between constituent pieces of a composite system that are of a quantum origin.[39]

The question now is, how does one differentiate between classical correlations and those correlations which may be attributed to genuine quantum phenomena. A quantum information theoretic perspective provides a precise solution: classical correlations are *defined* as those which may be generated by *Local Operations and Classical Communication* (LOCC).[39,54,55] Any classical correlation may be experimentally established using LOCC. Conversely, all correlations unobtainable via LOCC are regarded as *bona fide* quantum entanglement.

The LOCC paradigm is quite intuitive. Heuristically, given a composite quantum system with its components spread among different laboratories around the world we allow each experimenter to perform any quantum operation or measurement on their component locally in their lab. These are the local operations (LO), which clearly cannot establish any correlations, classical or quantum. The local operations may be supplemented by classical communication (CC): the experimenters may communicate any information (experimental procedures, measurement results, family history) they see fit via a classical channel (carrier pigeon, smoke signals, even e-mail). Any number of LO and CC rounds may be performed. It seems eminently reasonable to expect any classical correlation may be set-up in this manner. However, no quantum correlations could have been generated - all information exchanged between the separated parties at any point was intrinsically classical.

If, after some LOCC protocol, we are left with a multipartite state which may be used to perform some classically forbidden task, such as a Bell inequality violation, then the state properties facilitating this operation are not a result of LOCC - they must correspond to quantum correlations that were *already present* in the initial state before commencing the LOCC protocol.[39]

LOCC cannot create entanglement.

Allowing for multiple rounds of classical communication implies that LOCC protocols are not strictly local and actually have a rather complicated mathematical structure. Since the experimenters may classically communicate either before or after any round of local operations, any subsequent rounds may be conditional on previous outcomes. Due to this additional complexity, there is no simple universal characterisation of LOCC operations.[39] However, by focusing instead on LOCC/SLOCC *equivalence* (described in the subsequent sections) we need not make use of general LOCC operations, for which a good summary is given in.[40]

1.3.3. *Entanglement measures*

In the context of LOCC we can begin in earnest to consider the first essential question. How do we quantify the amount of entanglement contained in a given system? A *measure* of entanglement is a map from the state space to $\mathbb{R}$ which quantifies some aspect of entanglement. There are quite a number of entanglement measures in the literature. For reviews of such measures see, for example,[39,40,56] On the whole these are physically motivated by their relevance to some quantum information theoretic task. For example, the *entanglement cost* of a state ρ is defined as the rate at which one can generate copies of ρ from a set of 2-qubit Bell states.[54]

However, one would also like to have a more axiomatic scheme which identifies the minimal properties that any decent entanglement measure must possess. The axiomatic approach was originally initiated in.[57] They argued that any sensible entanglement measure E must fulfill the following criteria:

(1) $E(\rho) = 0$ if and only if ρ is separable.
(2) $E(\rho)$ is invariant under local unitaries.
(3) $E(\rho)$ is monotonically decreasing under LOCC protocols. That is,

$$E(\Lambda(\rho)) \leq E(\rho), \tag{38}$$

where Λ represents an arbitrary LOCC operation.

Monotonicity under LOCC operations is now often considered as *the* basic requirement.[40,58] Indeed, invariance under local unitaries is actually implied by monotonicity. The motivation is quite clear; LOCC cannot create entanglement, hence any good measure of entanglement should not increase under LOCC operations. Any map satisfying the monotonicity requirement is referred to as an *entanglement monotone.*

There are several extra postulates that one might consider:

(1) The stronger monotonicity condition that $E(\rho)$ decreases on *average* under LOCC. That is,

$$\sum_i p_i E(\rho_i) \leq E(\rho), \tag{39}$$

where the LOCC protocol maps ρ to ρ_i with probability p_i.
(2) Convexity,

$$E(\sum_i p_i \rho_i) \leq \sum_i p_i E(\rho_i). \tag{40}$$

(3) Additivity,

$$E(\rho_1 \otimes \rho_2) = E(\rho_1) + E(\rho_2). \tag{41}$$

More on entanglement measures may be found in.[39,59–67]

1.3.4. *Stochastic LOCC equivalence*

As emphasised LOCC cannot create entanglement. Consequently, from a quantum information theoretic perspective, any two states which may be interrelated using LOCC ought to be physically equivalent with respect to their entanglement properties. This motivates the concept of *Stocastic* LOCC equivalence, introduced in:[55,68]

Definition 1.1 (SLOCC-equivalence). *Two states lie in the same SLOCC-equivalence class if and only if they may be transformed into one another with some* non-zero probability *using LOCC operations.*

The crucial observation is that since LOCC cannot create entanglement any two SLOCC-equivalent states must necessarily possess the same entanglement, irrespective of the particular measure used. It is this property which make the SLOCC paradigm so suited to the task of classifying entanglement. It is also operationally motivated by the fact that any set of SLOCC-equivalent entangled states may be used to probabilistically perform the very same non-classical, entanglement dependent, operations.

The stronger concept of LOCC-equivalence is defined analogously:

Definition 1.2 (LOCC-equivalence). *Two states lie in the same LOCC-equivalence class if and only if they may be transformed into one another with* certainty *using LOCC operations.*

Clearly, LOCC-equivalence implies SLOCC-equivalence. From here-on-in we will make the drastic simplification of only considering pure states (which is difficult enough).

For an n-qudit system, two pure states,

$$|\Psi\rangle = a_{A_1\dots A_n}|A_1\dots A_n\rangle, \qquad |\Phi\rangle = b_{B_1\dots B_n}|B_1\dots B_n\rangle, \tag{42}$$

are SLOCC-equivalent if and only if they are related by an element of

$$\mathrm{SL}_1(d,\mathbb{C}) \times \mathrm{SL}_2(d,\mathbb{C}) \times \cdots \mathrm{SL}_n(d,\mathbb{C}), \tag{43}$$

under which $a_{A_1\dots A_n}$ transforms in the fundamental representation.[68] That is,

$$|\Psi\rangle \sim_{\mathrm{SLOCC}} |\Phi\rangle \tag{44}$$

if and only if there exist $d \times d$ matrices $M^i \in \mathrm{SL}_i(d, \mathbb{C})$ such that,

$$a_{A_1 A_2 \ldots A_n} = M^1{}_{A_1}{}^{B_1} M^2{}_{A_2}{}^{B_2} \ldots M^n{}_{A_n}{}^{B_n} b_{B_1 \ldots B_n}. \tag{45}$$

The set of transformations relating equivalent states will be referred to as the *SLOCC-equivalence group*. It may be thought of as a gauge group with respect to entanglement in the sense that it mods out the physically redundant (local or classical) information.

These SLOCC transformations partition the Hilbert space into equivalence classes or *orbits*. For the n-qudit system the space of SLOCC-equivalence classes is given by,[68]

$$\frac{\mathbb{C}^d \otimes \mathbb{C}^d \cdots \otimes \mathbb{C}^d}{\mathrm{SL}_1(d, \mathbb{C}) \times \mathrm{SL}_2(d, \mathbb{C}) \times \cdots \mathrm{SL}_n(d, \mathbb{C})}. \tag{46}$$

It is the space of physically distinct entanglement classes and, hence, its structure determines the classification of entanglement under the SLOCC. Consequently, it will play an essential role in our understanding of entanglement.

For the n-qubit system we have,

$$\frac{\mathbb{C}^2 \otimes \mathbb{C}^2 \cdots \otimes \mathbb{C}^2}{\mathrm{SL}_1(2, \mathbb{C}) \times \mathrm{SL}_2(2, \mathbb{C}) \times \cdots \mathrm{SL}_n(2, \mathbb{C})}. \tag{47}$$

and, in this case, the lower bound on the number of real continuous variables needed to parameterise the space of orbits is $2(2^n - 1) - 6n$.

For the stricter case of LOCC-equivalence it was shown in[55] that two states of a composite system are LOCC equivalent if and only if they may be transformed into one another using the group of *local unitaries* (LU), unitary transformations which factorise into separate transformations on the component parts.

For a n-qudit system the LOCC-equivalence group (up to a global phase) is given by,[60,69]

$$\mathrm{SU}_1(d, \mathbb{C}) \times \mathrm{SU}_2(d, \mathbb{C}) \times \cdots \mathrm{SU}_n(d, \mathbb{C}). \tag{48}$$

Hence, the space of orbits is given by,[60,69]

$$\frac{\mathbb{C}^d \otimes \mathbb{C}^d \cdots \otimes \mathbb{C}^d}{\mathrm{U}(1) \times \mathrm{SU}_1(d, \mathbb{C}) \times \mathrm{SU}_2(d, \mathbb{C}) \times \cdots \mathrm{SU}_n(d, \mathbb{C})}. \tag{49}$$

For the n-qubit system we have,

$$\frac{\mathbb{C}^2 \otimes \mathbb{C}^2 \cdots \otimes \mathbb{C}^2}{\mathrm{U}(1) \times \mathrm{SU}_1(2, \mathbb{C}) \times \mathrm{SU}_2(2, \mathbb{C}) \times \cdots \mathrm{SU}_n(2, \mathbb{C})} \tag{50}$$

and, in this case, the number of real continuous variables needed to parameterise the space of orbits is precisely $2^{n+1} - (3n+1)$.[60,69]

Given that we wish classify the physically distinct forms of entanglement, but do not really care whether it is Alice and Bob or Alice and Charlie who shares the entanglement, one could also to consider the SLOCC-equivalence group plus permutations of the qubits. We denote the SLOCC-equivalence semi-direct product the permutations by SLOCC*. In this case the entanglement measure must be permutation invariant.

The challenge now is to characterise the space (46) of SLOCC-equivalence classes. One possibility is the "covariant" approach which distinguishes the orbits by the vanishing or not of SLOCC-equivalence group covariants/invariants built out of the state vector. This philosophy is adopted for the 3-qubit case in section 8, which makes use of the Freudenthal triple system. Typically, one expects that the space of equivalence classes will contain several discrete (0-dimensional) pieces which are distinguished by a set of algebraically independent covariants. For example, for n qubits the 0-dimensional equivalence class of totally separable states is distinguished by the vanishing of *all* covariants. On the other hand, there will also be some finite dimensional pieces parametrised by some set of algebraically independent invariants. For instance, the 1-dimensional space of Bell states is parametrised by the *concurrence*,[70] which is both SLOCC invariant and a good entanglement measure. In fact, any SLOCC invariant so constructed is a good entanglement measure.[63]

1.4. *Bipartite entanglement*

1.4.1. *Generic finite-dimensional bipartite systems*

For bipartite states,

$$|\Psi\rangle = a_{AB}|AB\rangle, \tag{51}$$

where $A = 0, \dots m$ and $B = 0, \dots n$ (without loss of generality we assume $n \leq m$), one can always answer the question of whether a state is entangled or not. It is separable if and only if a_{AB} is rank one.

The SLOCC classification is particularly simple in this case. The set of local unitaries is contained in SLOCC and consequently, using the Schmidt decomposition, any state $|\Psi\rangle$ can be written as

$$|\Psi\rangle = \sum_{i=1}^{n_\Psi} \sqrt{\alpha_i}|ii\rangle, \quad \alpha_i > 0, \tag{52}$$

$n_\Psi \leq n$, where n is the dimension of the smaller of the two sub-systems. The Schmidt number n_Ψ is given by the rank of either one of the reduced density matrices (10). Since their rank cannot be changed using $\mathrm{SL}(m,\mathbb{C}) \times \mathrm{SL}(n,\mathbb{C})$ there are n entanglement classes under SLOCC.[68] A state of a given rank may be transformed into any state of a lower rank with some non-zero probability using non-invertible SLOCC operations. No SLOCC operation can increase the rank of a reduced density matrix. Hence, the SLOCC classification is stratified: the higher the rank the stronger the entanglement.[68]

1.4.2. *Two qubits*

For a two-qubit system there are then only two SLOCC classes: entangled and separable corresponding to rank 2 or rank 1 reduced density matrices, respectively. The bipartite entanglement is measured by the concurrence,[70]

$$C_{AB} := 2\sqrt{\det \rho_A} = 2\sqrt{\det \rho_B}. \tag{53}$$

Note that

$$|\det a|^2 = \tfrac{1}{4}\varepsilon^{A_1A_2}\varepsilon^{B_1B_2}a_{A_1B_1}a_{A_2B_2}\varepsilon^{A_3A_4}\varepsilon^{B_3B_4}a^*_{A_3B_3}a^*_{A_4B_4}. \tag{54}$$

Using the identity

$$\varepsilon^{A_1A_2}\varepsilon^{A_3A_4} = \delta^{A_1A_3}\delta^{A_2A_4} - \delta^{A_1A_4}\delta^{A_2A_3}, \tag{55}$$

we have

$$\begin{aligned} |\det a|^2 &= \tfrac{1}{4}(\delta^{A_1A_3}\delta^{A_2A_4} - \delta^{A_1A_4}\delta^{A_2A_3})\varepsilon^{B_1B_2}a_{A_1B_1}a_{A_2B_2}\varepsilon^{B_3B_4}a^*_{A_3B_3}a^*_{A_4B_4} \\ &= \tfrac{1}{4}\varepsilon^{B_1B_2}\varepsilon^{B_3B_4}(\rho_{B_1B_3}\rho_{B_2B_4} - \rho_{B_1B_4}\rho_{B_2B_3}) \\ &= \tfrac{1}{2}\varepsilon^{B_1B_2}\varepsilon^{B_3B_4}\rho_{B_1B_3}\rho_{B_2B_4} \\ &= \det \rho_B = \det \rho_A. \end{aligned} \tag{56}$$

Hence,

$$C_{AB} = 2\ |\det a|, \tag{57}$$

which relates the concurrence to the so-called 2-tangle, introduced in,[70]

$$\tau_{AB} := 4\ |\det a| = 4(|a_0|^2|a_3|^2 + |a_1|^2|a_2|^2 - (a_0a_2^*a_3a_1^* + a_1a_3^*a_2a_0^*)). \tag{58}$$

Recall that the eigenvalues of a 2×2 matrix obey the characteristic equation

$$\det \rho - \mathrm{tr}\ \rho\lambda + \lambda^2 = 0. \tag{59}$$

Hence the eigenvalues of ρ_A are

$$\lambda_0 = \tfrac{1}{2}[\operatorname{tr}\rho + \sqrt{(\operatorname{tr}\rho)^2 - 4\det\rho}], \tag{60a}$$

and

$$\lambda_1 = \tfrac{1}{2}[\operatorname{tr}\rho - \sqrt{(\operatorname{tr}\rho)^2 - 4\det\rho}], \tag{60b}$$

obeying

$$\begin{aligned} \lambda_0 + \lambda_1 &= \operatorname{tr}\rho, \\ \lambda_0\lambda_1 &= \det\rho. \end{aligned} \tag{60c}$$

so for an entangled state, with non-zero concurrence, ρ_A has rank 2 but only rank 1 for a product state as required by the SLOCC classification.

An example of a totally separable state is

$$|\Psi\rangle = \tfrac{1}{\sqrt{2}}(|00\rangle + |01\rangle), \tag{61}$$

since when Alice measures 0, Bob can measure either 0 or 1 with equal probability. Here

$$\tau_{AB} = 0. \tag{62}$$

An example of a maximally entangled state is the Bell state:

$$|\Psi\rangle = \tfrac{1}{\sqrt{2}}(|00\rangle + |11\rangle), \tag{63}$$

for which

$$\tau_{AB} = 1. \tag{64}$$

1.5. *Three qubit entanglement*

The case of three qubits (Alice, Bob, Charlie) is particularly interesting[60–62,68,69,71–74] since it provides the simplest example of inequivalently entangled states. It is by now well understood that there are seven entanglement classes: (0) Null, (1) Separable A-B-C, (2a) Biseparable A-BC, (2b) Biseparable B-CA, (2c) Biseparable C-AB, (3) W and (4) GHZ. Note, we have been a little careless with our language here: the GHZ class is really a 1-dimensional space of orbits parametrized by Cayley's hyperdeterminant, introduced below.

In the case of three qubits, the group of SLOCC transformations is $\mathrm{SL}_A(2,\mathbb{C}) \times \mathrm{SL}_B(2,\mathbb{C}) \times \mathrm{SL}_C(2,\mathbb{C})$. Tensors transforming under the Alice, Bob or Charlie $\mathrm{SL}(2,\mathbb{C})$ carry indices $A_1, A_2...$, $B_1, B_2...$ or $C_1, C_2...$, respectively, so a_{ABC} transforms as a $(\mathbf{2},\mathbf{2},\mathbf{2})$.

The first SLOCC classification of three qubit entanglement, presented in subsubsection 1.5.3, was performed in[68] using a subset of the algebraically independent local unitary invariants, which are described in the following section, and the so-called 3-tangle.[3] The 3-tangle is the unique SLOCC invariant and measures the genuine tripartite entanglement between three qubits. It is given by Cayley's hyperdeterminant,[1,4] which is described in subsubsection 1.5.2.

1.5.1. *Local unitary invariants*

The number of parameters needed to describe unnormalised LOCC (local unitary) inequivalent states is given by the dimension of the space of orbits,[69]

$$\frac{\mathbb{C}^2 \times \mathbb{C}^2 \times \mathbb{C}^2}{\mathrm{U}(1) \times \mathrm{SU}(2) \times \mathrm{SU}(2) \times \mathrm{SU}(2)}, \tag{65}$$

namely $16 - 10 = 6$. This is also equivalent to the number of algebraically independent LU invariants,[72]

(1) The norm squared:

$$|\Psi|^2 = \langle \Psi | \Psi \rangle. \tag{66}$$

(2) The local entropies:

$$S_A = 4 \det \rho_A, \qquad S_B = 4 \det \rho_B, \qquad S_C = 4 \det \rho_C, \tag{67}$$

where ρ_A, ρ_B, ρ_C are the doubly reduced density matrices:

$$\rho_A = \mathrm{Tr}_{BC} |\Psi\rangle\langle\Psi|, \qquad \rho_B = \mathrm{Tr}_{CA} |\Psi\rangle\langle\Psi|, \qquad \rho_C = \mathrm{Tr}_{AB} |\Psi\rangle\langle\Psi|. \tag{68}$$

(3) The Kempe invariant:[62]

$$\begin{aligned} K &= \mathrm{tr}(\rho_A \otimes \rho_B \rho_{AB}) - \mathrm{tr}(\rho_A^3) - \mathrm{tr}(\rho_B^3) \\ &= \mathrm{tr}(\rho_B \otimes \rho_C \rho_{BC}) - \mathrm{tr}(\rho_B^3) - \mathrm{tr}(\rho_C^3) \\ &= \mathrm{tr}(\rho_C \otimes \rho_A \rho_{CA}) - \mathrm{tr}(\rho_C^3) - \mathrm{tr}(\rho_A^3), \end{aligned} \tag{69}$$

where $\rho_{AB}, \rho_{BC}, \rho_{CA}$ are the singly reduced density matrices:

$$\rho_{AB} = \mathrm{Tr}_C |\Psi\rangle\langle\Psi|, \qquad \rho_{BC} = \mathrm{Tr}_A |\Psi\rangle\langle\Psi|, \qquad \rho_{CA} = \mathrm{Tr}_B |\Psi\rangle\langle\Psi|. \tag{70}$$

(4) The 3-tangle:[3]

$$\tau_{ABC} = 4 |\operatorname{Det} a_{ABC}| \tag{71}$$

where $|\Psi\rangle = a_{ABC}|ABC\rangle$ and $\operatorname{Det} a_{ABC}$ is Cayley's hyperdeterminant:[1,4]

$$\begin{aligned}\operatorname{Det} a_{ABC} :=& \tfrac{1}{2}\epsilon^{A_1A_2}\epsilon^{B_1B_2}\epsilon^{A_3A_4}\epsilon^{B_3B_4}\epsilon^{C_1C_4}\epsilon^{C_3C_2} \\ & \times a_{A_1B_1C_1}a_{A_2B_2C_2}a_{A_3B_3C_3}a_{A_4B_4C_4}.\end{aligned} \tag{72}$$

Here ϵ is the $\mathrm{SL}(2,\mathbb{C})$–invariant alternating tensor

$$\epsilon := \begin{pmatrix} 0 & 1 \\ -1 & 0 \end{pmatrix}. \tag{73}$$

Hence, under local unitary operations the most general state may be written as a six real parameter generating solution.[75] For subsequent comparison with the STU black hole we restrict our attention to states with *real* coefficients a_{ABC}. In this case, one can show that there are five algebraically independent LU invariants:[75] $\operatorname{Det} a$, S_A, S_B, S_C and the norm $\langle\Psi|\Psi\rangle$, corresponding to the dimension of

$$\frac{\mathbb{R}^2 \times \mathbb{R}^2 \times \mathbb{R}^2}{SO(2)\times SO(2)\times SO(2)}, \tag{74}$$

namely $8-3=5$. Hence, the most general real three-qubit state can be described by just five parameters,[75] conveniently taken as four real numbers N_0, N_1, N_2, N_3 and an angle θ:

$$\begin{aligned}|\Psi\rangle =& -N_3\cos^2\theta|001\rangle - N_2|010\rangle + N_3\sin\theta\cos\theta|011\rangle \\ & - N_1|100\rangle - N_3\sin\theta\cos\theta|101\rangle + (N_0 + N_3\sin^2\theta)|111\rangle.\end{aligned} \tag{75}$$

1.5.2. *Cayley's hyperdeterminant*

Cayley described a multi-indexed array, such as a_{ABC}, as a *hypermatrix*. The present $2\times2\times2$ case may be graphically represented as a cube, as in Figure 4. In 1845 he generalised the determinant of a 2×2 matrix to the

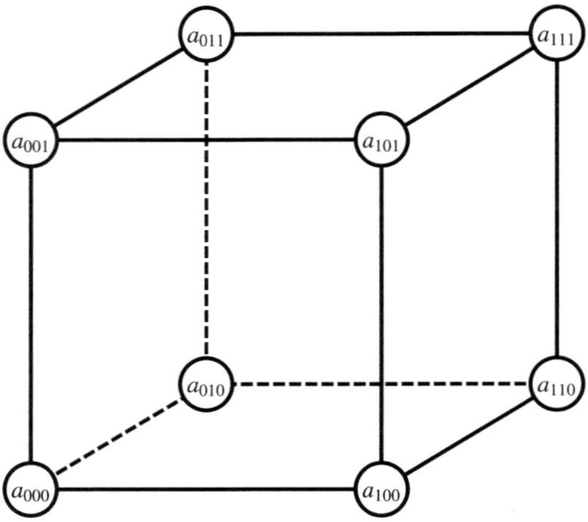

Fig. 4: The 3-index quantity a_{ABC} is an example of a hypermatrix, here depicted as a cube. In 1845 Cayley generalised the determinant of a 2×2 matrix to the hyperdeterminant of a $2 \times 2 \times 2$ hypermatrix.

hyperdeterminant of a $2 \times 2 \times 2$ hypermatrix a_{ABC}[1]

$$\begin{aligned}
\operatorname{Det} a := & -\tfrac{1}{2}\, \epsilon^{A_1A_2}\epsilon^{B_1B_2}\epsilon^{A_3A_4}\epsilon^{B_3B_4}\epsilon^{C_1C_4}\epsilon^{C_2C_3} \\
& \times a_{A_1B_1C_1}a_{A_2B_2C_2}a_{A_3B_3C_3}a_{A_4B_4C_4} \qquad (76a)
\end{aligned}$$

$$\begin{aligned}
= \; & a_{000}^2a_{111}^2 + a_{001}^2a_{110}^2 + a_{010}^2a_{101}^2 + a_{100}^2a_{011}^2 \\
& - 2\,(a_{000}a_{001}a_{110}a_{111} + a_{000}a_{010}a_{101}a_{111} \\
& \quad + a_{000}a_{100}a_{011}a_{111} + a_{001}a_{010}a_{101}a_{110} \\
& \quad + a_{001}a_{100}a_{011}a_{110} + a_{010}a_{100}a_{011}a_{101}) \\
& + 4\,(a_{000}a_{011}a_{101}a_{110} + a_{001}a_{010}a_{100}a_{111}) \qquad (76b)
\end{aligned}$$

$$\begin{aligned}
= \; & a_0^2a_7^2 + a_1^2a_6^2 + a_2^2a_5^2 + a_3^2a_4^2 \\
& - 2\,(a_0a_1a_6a_7 + a_0a_2a_5a_7 + a_0a_4a_3a_7 \\
& \quad + a_1a_2a_5a_6 + a_1a_3a_4a_6 + a_2a_3a_4a_5) \\
& + 4\,(a_0a_3a_5a_6 + a_1a_2a_4a_7), \qquad (76c)
\end{aligned}$$

where we have made the binary conversion 0, 1, 2, 3, 4, 5, 6, 7 for 000, 001, 010, 011, 100, 101, 110, 111. (The cyclic permutations are 0, 4, 1, 5, 2, 6, 3, 7 and 0, 2, 4, 6, 1, 3, 5, 7. See Table 1.) Crucially, the hyperdeterminant is invariant under $[\mathrm{SL}(2)]^3$ and under a triality that interchanges A, B and C, irrespective of whether the a_{ABC} are complex, real or integer.

Table 1: Three cyclic permutations of the binary notation. The hyperdeterminant, (76a), is invariant under this triality.

	Binary			Decimal	
ABC	CAB	BCA	ABC	CAB	BCA
000	000	000	0	0	0
001	100	010	1	4	2
010	001	100	2	1	4
011	101	110	3	5	6
100	010	001	4	2	1
101	110	011	5	6	3
110	011	101	6	3	5
111	111	111	7	7	7

One way to understand this triality is to think of having three different metrics (Alice, Bob and Charlie)[13]

$$\begin{aligned} \gamma^A(a)_{A_1A_2} &= \epsilon^{B_1B_2}\epsilon^{C_1C_2}a_{A_1B_1C_1}a_{A_2B_2C_2}, \\ \gamma^B(a)_{B_1B_2} &= \epsilon^{C_1C_2}\epsilon^{A_1A_2}a_{A_1B_1C_1}a_{A_2B_2C_2}, \\ \gamma^C(a)_{C_1C_2} &= \epsilon^{A_1A_2}\epsilon^{B_1B_2}a_{A_1B_1C_1}a_{A_2B_2C_2}. \end{aligned} \tag{77}$$

which rotate into each other under the A-B-C triality. Explicitly,

$$\begin{aligned} \gamma^A(a) &= \begin{pmatrix} 2(a_0a_3 - a_1a_2) & a_0a_7 - a_1a_6 + a_4a_3 - a_5a_2 \\ a_0a_7 - a_1a_6 + a_4a_3 - a_5a_2 & 2(a_4a_7 - a_5a_6) \end{pmatrix}, \\ \gamma^B(a) &= \begin{pmatrix} 2(a_0a_5 - a_4a_1) & a_0a_7 - a_4a_3 + a_2a_5 - a_6a_1 \\ a_0a_7 - a_4a_3 + a_2a_5 - a_6a_1 & 2(a_2a_7 - a_6a_3) \end{pmatrix}, \\ \gamma^C(a) &= \begin{pmatrix} 2(a_0a_6 - a_2a_4) & a_0a_7 - a_2a_5 + a_1a_6 - a_3a_4 \\ a_0a_7 - a_2a_5 + a_1a_6 - a_3a_4 & 2(a_1a_7 - a_3a_5) \end{pmatrix}. \end{aligned} \tag{78}$$

All are equivalent, however, since

$$\det \gamma^A(a) = \det \gamma^B(a) = \det \gamma^C(a) = -\operatorname{Det} a. \tag{79}$$

We may also isolate a single qubit to make the isomorphisms $\mathrm{SL}(2,\mathbb{R}) \times \mathrm{SL}(2,\mathbb{R}) \cong \mathrm{SO}(2,2)$ (or $\mathrm{SL}(2,\mathbb{C}) \times \mathrm{SL}(2,\mathbb{C}) \cong \mathrm{SO}(4,\mathbb{C})$ in the complex case) manifest. For example, selecting Charlie, the components a_{AB0} and a_{AB1} form two 4-vectors as follows,

$$\begin{aligned} a_0 &= \tfrac{1}{\sqrt{2}}(P^0 - P^2) & a_1 &= -\tfrac{1}{\sqrt{2}}(Q_0 + Q_2) \\ a_2 &= \tfrac{1}{\sqrt{2}}(P^1 - P^3) & a_3 &= -\tfrac{1}{\sqrt{2}}(Q_3 + Q_1) \\ a_4 &= \tfrac{1}{\sqrt{2}}(P^1 + P^3) & a_5 &= \tfrac{1}{\sqrt{2}}(Q_3 - Q_1) \\ a_6 &= -\tfrac{1}{\sqrt{2}}(P^0 + P^2) & a_7 &= \tfrac{1}{\sqrt{2}}(Q_0 - Q_2), \end{aligned} \tag{80}$$

or inversely,

$$
\begin{array}{llll}
P^0 = & \frac{1}{\sqrt{2}}(a_0 - a_6) & Q_0 = & \frac{1}{\sqrt{2}}(a_7 - a_1) \\
P^1 = & \frac{1}{\sqrt{2}}(a_4 + a_2) & Q_1 = & -\frac{1}{\sqrt{2}}(a_5 + a_3) \\
P^2 = & -\frac{1}{\sqrt{2}}(a_0 + a_6) & Q_2 = & -\frac{1}{\sqrt{2}}(a_7 + a_1) \\
P^3 = & \frac{1}{\sqrt{2}}(a_4 - a_2) & Q^3 = & \frac{1}{\sqrt{2}}(a_5 - a_3),
\end{array}
\tag{81}
$$

then

$$
\begin{array}{rcccl}
2(-a_0a_6 + a_2a_4) & = & P^2 & = & {P^0}^2 + {P^1}^2 - {P^2}^2 - {P^3}^2, \\
2(-a_1a_7 + a_3a_5) & = & Q^2 & = & {Q_0}^2 + {Q_1}^2 - {Q_2}^2 - {Q_3}^2, \\
a_0a_7 - a_2a_5 + a_1a_6 - a_3a_4 & = & P \cdot Q & = & P^0Q_0 + P^1Q_1 + P^2Q_2 + P^3Q_3,
\end{array}
\tag{82}
$$

and

$$
\gamma^3 = \begin{pmatrix} -P^2 & P \cdot Q \\ P \cdot Q & -Q^2 \end{pmatrix}. \tag{83}
$$

Hence

$$
\operatorname{Det} a = -P^2Q^2 + (P \cdot Q)^2. \tag{84}
$$

This basis makes the $SO(2,2)_{AB} \times SL(2,\mathbb{R})_C$ symmetry clear, P and Q together transform as an $SL(2,\mathbb{R})_C$ doublet while individually they transform as vectors of $SO(2,2)_{AB}$. In two component spinor notation

$$
\begin{aligned}
P^{AB} &= \frac{1}{\sqrt{2}} \begin{pmatrix} P^0 - P^2 & P^1 - P^3 \\ P^1 + P^3 & -P^0 - P^2 \end{pmatrix}, \\
Q^{AB} &= \frac{1}{\sqrt{2}} \begin{pmatrix} -Q_0 - Q_2 & -Q_1 - Q_3 \\ -Q_1 + Q_3 & Q_0 - Q_2 \end{pmatrix},
\end{aligned}
\tag{85}
$$

we have

$$
a_{ABC} = \begin{pmatrix} P^{AB} \\ Q^{AB} \end{pmatrix}, \tag{86}
$$

and

$$
\gamma^C = \epsilon^{A_1A_2}\epsilon^{B_1B_2} \begin{pmatrix} P_{A_1B_1}P_{A_2B_2} & P_{A_1B_1}Q_{A_2B_2} \\ Q_{A_1B_1}P_{A_2B_2} & Q_{A_1B_1}Q_{A_2B_2} \end{pmatrix}. \tag{87}
$$

This is manifestly invariant under $\mathrm{SO}(2,2)_{AB}$ and transforms as a $\mathbf{3}$ under $\mathrm{SL}(2)_C$. Group theoretically this corresponds to the $(\mathbf{1},\mathbf{1},\mathbf{3})$ in the tensor product,

$$\begin{aligned}(\mathbf{2},\mathbf{2},\mathbf{2}) \times (\mathbf{2},\mathbf{2},\mathbf{2}) = \quad & (\mathbf{1},\mathbf{1},\mathbf{1}) \\ & + (\mathbf{1},\mathbf{1},\mathbf{3}) + (\mathbf{1},\mathbf{3},\mathbf{1}) + (\mathbf{3},\mathbf{1},\mathbf{1}) \\ & + (\mathbf{3},\mathbf{3},\mathbf{1}) + (\mathbf{3},\mathbf{1},\mathbf{3}) + (\mathbf{1},\mathbf{3},\mathbf{3}) \\ & + (\mathbf{3},\mathbf{3},\mathbf{3}).\end{aligned} \tag{88}$$

Similarly, γ^A and γ^B correspond to the $(\mathbf{3},\mathbf{1},\mathbf{1})$ and $(\mathbf{1},\mathbf{3},\mathbf{1})$ respectively.

1.5.3. *Entanglement classification*

Dür et al.[68] used simple arguments concerning the conservation of ranks of reduced density matrices to show that there are only six types of 3-qubit equivalence classes (or seven if we count the null state); only two of which show *genuine* tripartite entanglement. They are as follows:

Null: The trivial zero entanglement orbit corresponding to vanishing states,

$$\text{Null}: \quad 0. \tag{89}$$

Separable: Another zero entanglement orbit for completely factorisable product states,

$$A\text{-}B\text{-}C: \quad |000\rangle. \tag{90}$$

Biseparable: Three classes of bipartite entanglement,

$$\begin{aligned} A\text{-}BC: &\quad |010\rangle + |001\rangle, \\ B\text{-}CA: &\quad |100\rangle + |001\rangle, \\ C\text{-}AB: &\quad |010\rangle + |100\rangle. \end{aligned} \tag{91}$$

Note, these three classes are identified under SLOCC*.

W: Three-way entangled states that do not maximally violate Bell-type inequalities in the same way as the GHZ class. However, they are robust in the sense that tracing out a subsystem generically results in a bipartite mixed state that is maximally entangled under a number of criteria,[68]

$$\text{W}: \quad |100\rangle + |010\rangle + |001\rangle. \tag{92}$$

GHZ: Genuinely tripartite entangled Greenberger-Horne-Zeilinger[51] states. These maximally violate Bell-type inequalities but, in contrast to class W, are fragile under the tracing out of a subsystem since the resultant state is completely unentangled,

$$\text{GHZ}: \quad |000\rangle + |111\rangle. \tag{93}$$

These classes and their representative states are summarised in Table 2. They are characterised[68] by the vanishing or not of the invariants listed in the table. Note that the Kempe invariant is redundant in this SLOCC classification. A visual representation of these SLOCC orbits is provided by the onion-like classification[4] of Figure 5a.

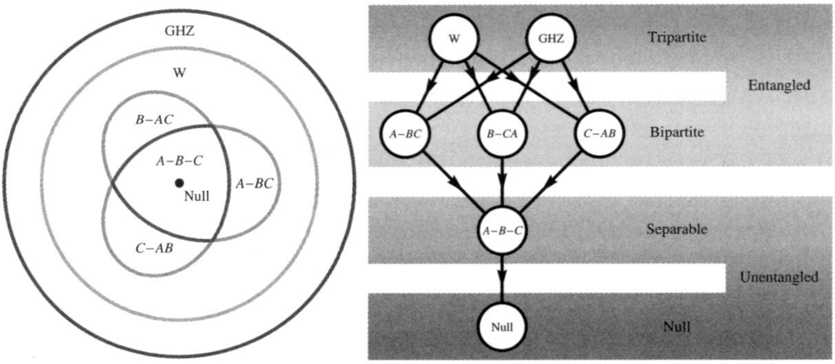

(a) Onion structure (b) Hierarchy

Fig. 5: (a) Onion-like classification of SLOCC orbits. (b) Stratification. The arrows are non-invertible SLOCC transformations between classes that generate the entanglement hierarchy. The partial order defined by the arrows is transitive, so we may omit e.g. GHZ $\to$ A-B-C and A-BC $\to$ Null arrows for clarity.

Table 2: The values of the local entropies S_A, S_B, and S_C and the hyperdeterminant $\operatorname{Det} a$ are used to partition three-qubit states into entanglement classes.

Class	Representative	Condition				
		Ψ	S_A	S_B	S_C	$\operatorname{Det} a$
Null	0	$=0$	$=0$	$=0$	$=0$	$=0$
A-B-C	$\|000\rangle$	$\neq 0$	$=0$	$=0$	$=0$	$=0$
A-BC	$\|010\rangle + \|001\rangle$	$\neq 0$	$=0$	$\neq 0$	$\neq 0$	$=0$
B-CA	$\|100\rangle + \|001\rangle$	$\neq 0$	$\neq 0$	$=0$	$\neq 0$	$=0$
C-AB	$\|010\rangle + \|100\rangle$	$\neq 0$	$\neq 0$	$\neq 0$	$=0$	$=0$
W	$\|100\rangle + \|010\rangle + \|001\rangle$	$\neq 0$	$\neq 0$	$\neq 0$	$\neq 0$	$=0$
GHZ	$\|000\rangle + \|111\rangle$	$\neq 0$	$\neq 0$	$\neq 0$	$\neq 0$	$\neq 0$

These SLOCC equivalence classes are then stratified by *non-invertible* SLOCC operations into an entanglement hierarchy[68] as depicted in Figure 5b. Note that no SLOCC operations (invertible or not) relate the GHZ and W classes; they are genuinely distinct classes of tripartite entanglement. However, from either the GHZ class or W class one may use non-invertible SLOCC transformations to descend to one of the biseparable or separable classes and hence we have a hierarchical entanglement structure. For more on three qubit entanglement see.[74,76–80]

2. Black holes in M-theory

The dawn of the 20th century was a magnificent time for physics. Perhaps in no other period have so many profoundly important concepts emerged concurrently. The atomic foundation of matter, the unification of space and time, the uncertainty principle, the fundamentally probabilistic nature of reality - each one a milestone in our understanding of the world around us.

As the dust settled, two conceptual pillars were left standing; the foundations of physics today, general relativity and quantum theory. Each of which has independently revolutionised our basic understanding and heralded an age of unprecedented predictive power. However, these developments have largely evolved along separate lines. While general relativity deals with the large scale structures of our universe, quantum theory reigns over the microscopic world. Yet, at high enough energy scales both quantum and gravitational effects become simultaneously significant. Our understanding of the fundamental laws of nature is surely lacking until we have a consistent framework which marries quantum theory with gravity.

Direct attempts at quantizing gravity using perturbative quantum field theory are plagued by uncontrollable infinities. Order by order in perturbation theory the ultraviolet divergences, corresponding to radiative corrections to gravitation, become increasingly severe. The standard renormalisation methods fail, as can be seen by simple power counting. It seems that we need a more radical perspcetive if we are to ever understand the quantum mechanical nature of spacetime. There are many possible avenues one could try ranging from the non-perturbative approaches of Loop quantum gravity and causal dynamical triangulations through to the fundamentally discrete picture offered by causal set theory. The particular path we choose here is that of string theory or rather its bigger brother M-theory.

2.1. *The road to M-theory*

This journey begins with a relatively simple conceptual leap: to relinquish the long-cherished and most profitable notion that the fundamental constituents of matter are (mathematically) point like. While this possibility had certainly been previously entertained,[81] it only really found its feet with the advent of *string theory*, the theory of propagating 1-dimensional extended objects. Originally formulated, in its bosonic guise, as a theory of the strong nuclear force in 1968,[82] it was quickly superseded by quantum chromodynamics (QCD). See the introductory sections of[83] for a brief summary of these developments. However, just as the stringy theory of the strong force was being eclipsed by QCD it emerged from the shadows, re-conceived as a candidate theory of quantum gravity, with, moreover, grand aspirations of unification. The vibrational modes of the string are to be thought of as representing different particles. In particular, the closed string spectrum included a massless spin two particle, which coupled to the other modes in the manner of general relativity. Was this the correct realisation of the graviton? Furthermore, quantum mechanical consistency required that strings propagate in a 26-dimensional spacetime, an undesirable feature in the context of strong interactions, but a welcome addition in the setting of Kaluza-Klein unification.

While all this was certainly an enticing prospect, it was not without its own serious faults. Principally, fermions were entirely absent, a real set back for any theory purporting to describe matter! Moreover, the bosonic closed string spectrum contains a tachyonic mode, which, to this day, has not been resolved.

These potentially fatal flaws were surmounted during what would come to be called *the first string revolution.* In 1975 the *a priori* unrelated concept of supersymmetry was discovered.[84] Supersymmetry requires that for each known boson there is a fermion of equal mass. However, no such superpartners have yet been found. The symmetry, if it exists at all, must be broken, so that the postulated particles do not have the same mass as known ones but instead are too heavy to be seen in current accelerators.

The introduction of supersymmetry allowed for the consistent inclusion of fermions living on the string world-sheet resulting in the Ramond-Neveu-Schwarz (RNS) string. Following on from the work of Gliozzi, Scherk and Olive, in which they showed that the RNS string could be freed of the tachyon, Green and Schwarz discovered a formulation with spacetime supersymmetry. The *superstring* was born and with it a great new hope. Superstrings propagate in ten spacetime dimensions for quantum mechanical consistency and it turns out there are five anomaly-free theories: the $E_8 \times E_8$ and SO(32) heterotic strings, the SO(32) type I, and the type IIA and type IIB strings. It was shown that the superstring is 1-loop finite and, while there is as yet no complete universally accepted proof, it is expected to be finite at all orders. Superstring theory would appear to be the first example of a perturbatively finite theory of quantum gravity. Moreover, via Kaluza-Klein compactification on some Calabi-Yau manifold,[85] the heterotic superstring seemed capable, in principle at least, of correctly accounting for the standard model.

Meanwhile, in (not entirely) separate developments there had been a section of the quantum gravity community looking to 11-dimensional supergravity for answers. Supergravity is the supersymmetric generalization of general relativity. In fact, the low energy limit of each string theory is a particular supergravity theory, but strangely unlike strings, which insist on ten spacetime dimensions, supergravity allows upto eleven.[86] Not only is eleven dimensions the upper limit, it is where it takes its most elegant form.[87] Moreover, it is *unique* (at the two derivative level, assuming diffeomorphism invariance, local Lorentz invariance, local supersymmetry and Abelian gauge invariance). Initially 11-dimensional supergravity was seen as more of a tool for obtaining, via dimensional reduction, the various extended supergravity theories existing in four dimensions.[88] In particular, the maximally extended $\mathcal{N} = 8$ theory was thus first derived.[89] However, it was not long before the extra seven dimensions started to be taken seriously. In 1981 Witten considered the possibility of obtaining a realistic Kaluza-Klein model of particle physics starting from 11-dimensional supergravity concluding that, remarkably, seven extra dimensions are in fact the minimum required.[90] See[91] for a review of the Kaluza-Klein program in the context of supergravity.

A further important 11-dimensional development followed from the 1986 discovery, by Hughes, Liu and Polchinski, that, in $D = 6$, not only strings but also *membranes* could be consistently made supersymmetric.[92] The following year Bergshoeff, Sezgin and Townsend described their supermembrane propagating in an 11-dimensional supergravity background.[93] The eleven dimensional supermembrane was subsequently derived as a solution to $D = 11$ supergravity preserving one half of the supersymmetries.[94] An important distinction was made in this work. The supermembrane of Hughes, Liu and Polchinski was a *solitonic* solution of a $D = 6$ guage theory. However, the supermembrane solution to $D = 11$ supergravity is singular and carries Noether charges so should not be considered solitonic, but rather *elementary* or *fundamental.* The proper solitonic solution of 11-dimensional supergravity was later found to be the superfivebrane.[95]

Despite these clear successes, the world of eleven dimensions went, to some degree at least, unnoticed by the 10-dimension universe of superstrings, which was making substantial progress in its own right. Interest in 11-dimensional supergravity was also eroded by the realisation that a 4-dimensional theory with the requisite chiral fermions could not be obtained from a non-chiral theory in $D = 11$ using the traditional Kaluza-Klein approach.[96]

In fact, the separate worlds of superstrings and 11-dimensional supergravity would appear to be *curiously* at odds with one-another.[97] First, strings demanded ten dimensions while supersymmetry had chosen eleven. Second, once one had contradicted the orthodoxy of point particles in favour of strings, why stop there, especially in light of the emerging work on supersymmetric p-branes. Finally, one might have had certain misgivings about the superstring itself. There were open questions which called for a non-perturbative solution, such as the paradoxes of quantum black holes. Could the non-perturbative solitonic p-branes play a role? What about the existence of five superstring theories, could they be consistently incorporated into a single framework? Does eleven dimensions have anything to do with this?

With hindsight, these apparent discrepancies make an eventual synthesis seem inevitable. Indeed, substantial evidence had been steadily accumulating in favour of some kind of integration.[88] In 1987 it was shown that the perturbative type IIA string could be obtained from $D = 11$ supergravity with $\mathbb{R}^{1,9} \times S^1$ topology by wrapping a membrane around the circle and taking the zero radius limit.[98] It was later conjectured that this correspondence goes beyond the perturbative regime. The previously unwanted Kaluza-Klein modes, associated with the non-zero radius, are to be identified as charged extreme black holes states of the $D = 10$ theory.[99–101] There was also good evidence that compactifying a further dimension of the $D = 11$ theory on a second circle yields the type IIB string on a single circle, its non-perturbative S-duality is then given by the modular group of the $D = 11$ torus.[102] The heterotic string was thrown into the mix by compactifying on not a circle but a line segment.[103] Moveover, a number of surprising dualities relating these various theories had been discovered,[104–109] creating a web of relations between the five string theories and 11-dimensional supergravity. For the many more intriguing aspects of these remarkable developments the reader is referred to.[88]

All this progress culminated in the landmark work by Witten,[110] which collected the five superstring theories together with the work on eleven dimensions under a unified framework, which he rather mysteriously dubbed "M-theory". The five superstrings and 11-dimensional supergravity are to be thought of as merely corners of the M-theory. However, the basic objects were no longer strings, but, rather, the eleven dimensional membrane and fivebrane, henceforth denoted as the M2-brane and M5-brane, respectively, so as to emphasize their new found significance. The second string revolution was well under way and the 11-dimensional M-branes were leading the charge.

Subsequently, it was realised that certain p-brane solutions, carrying R-R charge, of the type IIA/B theories could be interpreted as surfaces on which open strings, obeying Dirichlet boundary conditions, can end.[111] These *Dp-branes* have complemented the theory of black holes as intersecting black-branes wrapped around the six (seven) compact dimensions of string theory (M-theory),[99,101,112–119] providing a much sought after framework for a quantum mechanical understanding of black hole entropy. Indeed, in 1996 Strominger and Vafa reproduced the Bekenstein-Hawking black hole entropy formula by counting D-brane quantum states.[120]

Despite these substantial developments it is fair to say M-theory is a work in progress. Black holes and, more generally, black p-branes offer unique insights into its deep structure and will clearly play an important role in its development. In the words of 't Hooft,[121]

> "If we wish to know how to do non-perturbative gravity, it is the black holes that we must study." - 't Hooft.

2.2. *Black holes*

It is believed that black hole physics provides a testing ground for our attempts at a theory of quantum gravity, such as M-theory. This rather challenging expectation derives largely from their intriguing thermodynamic properties, which were described in the 1970s by Bekenstein and Bardeen, Carter and Hawking.[122,123] It was not long before Hawking demonstrated that black holes quantum mechanically radiate as if they are indeed black bodies.[124] The entropy of a black hole is given by its horizon area, as described in the famous Bekenstein-Hawking black hole entropy formula,

$$S_{\rm BH} = \frac{A}{4\hbar G_4}, \tag{94}$$

where G_4 is the four dimensional Newton constant. This fascinating equality, which relates the thermodynamic quantity $S_{\rm BH}$ to the geometric property A, has motivated much theoretical work over the decades. It raises an important question: what are the quantum microstates underlying the statistical interpretation of $S_{\rm BH}$? Any candidate theory of quantum gravity ought to provide an answer. The non-pertubative nature of black holes leaves the peturbative superstring lacking in this respect. However, non-perturbative developments using D-branes have provided a partial answer for the special subclass of *extremal* black holes.

2.2.1. *Extremal black holes*

The black holes of supergravity typically carry electromagnetic charges generated by Abelian gauge potentials. The simplest example, and our starting point, is given by the Reissner- Nordström solution of the Einstein–Maxwell action in 4 dimensions:

$$S = \frac{1}{16\pi G_4}\int d^4x\sqrt{-\det g}\left[R - \frac{1}{4}F^{\mu\nu}F_{\mu\nu}\right]. \tag{95}$$

Here $F^{\mu\nu}$ is the electromagnetic field strength, $g_{\mu\nu}$ is the metric and R is the Ricci scalar. To obtain a solution we make the simplifying metric ansatz,

$$ds^2 = -e^{2U(r)}dt^2 + e^{-2U(r)}dr^2 + r^2 d\Omega^2, \tag{96}$$

which follows from imposing that the solution be static and spherically symmetric. Here $d\Omega^2 = d\theta^2 + \sin^2\theta\, d\phi^2$ is the line element on a two-sphere and U is an undetermined function of only the radial coordinate, as required by spherical symmetry. Similarly, the gauge potential field strength is required to have the form

$$F = P\sin\theta\, d\theta\wedge d\phi + Q\, dt\wedge\frac{dr}{r^2}. \tag{97}$$

Integrating over a sphere at radial infinity we obtain the electromagnetic charges Q, P carried by the black hole,

$$\frac{1}{4\pi}\int_{S^2} F = P, \qquad \frac{1}{4\pi}\int_{S^2}\star F = Q. \tag{98}$$

The final undetermined function U may also be solved for and one finds,

$$e^{2U(r)} = 1 - \frac{2M}{r} + \frac{P^2+Q^2}{r^2}, \tag{99}$$

where M is the mass of the black hole. Note, this reduces to the Schwarzschild solution when the EM charges are absent, $P = Q = 0$.

Superficially it would seem that the above solution has a singularity at the zeros of $e^{U(r)}$ given by,

$$r_{\pm} = M \pm \sqrt{M^2 - (P^2 + Q^2)}. \tag{100}$$

In fact, this is merely a coordinate singularity, where r_+ and r_- are the radial locations of the event and Cauchy horizons, respectively. There is, however, a genuine curvature singularity as r tends to zero as revealed by the square of the Ricci tensor,

$$R_{\mu\nu}R^{\mu\nu} = 4\frac{(Q^2 + P^2)^2}{r^8}. \tag{101}$$

We conclude that in order for the this singularity to be hidden $r_{\pm}$ must be real valued implying,

$$M^2 \geq P^2 + Q^2. \tag{102}$$

For a given charge there is a minimum mass such that the singularity is surrounded by the horizons. Saturating this bound we see that the double horizon collapses to a single horizon at $r_* = r_+ = r_-$. Such black holes are called *extremal.* The extremality parameter c, measuring how close we are to extremality, is defined as,

$$c = r_+ - r_- = \sqrt{M^2 - (P^2 + Q^2)}. \tag{103}$$

The extremal Reissner–Nordström black hole interpolates between a conformally flat $AdS_2 \times S^2$ geometry near the horizon and Minkowski spacetime at radial infinity, as can be seen from the line element (setting $P = 0$ we do not loss anything but save on clutter),

$$ds^2 = dt^2 \left(1 + \frac{Q}{\rho}\right)^{-2} - \left(1 + \frac{Q}{\rho}\right)^2 (d\rho^2 + \rho^2\, d\Omega^2) \tag{104}$$

where $r = \rho + M$.

There are two other quantities, surface gravity and the horizon area, which are of particular importance to our discussion. Surface gravity κ_S is the acceleration as measured from infinity of a static particle held above the horizon in the limit approaching the horizon. For our ($P = 0$) Reissner-Nordström solution they are respectively given by

$$\kappa_S = \frac{\sqrt{M^2 - Q^2}}{2M(M + \sqrt{M^2 - Q^2}) - Q^2}, \qquad A = 4\pi(M + \sqrt{M^2 - Q^2})^2. \tag{105}$$

Note, in the extremal limit the surface gravity vanishes while the horizon is fixed by the charge alone,

$$A = 4\pi Q^2. \tag{106}$$

2.3. *Black hole thermodynamics*

Black hole physics provides a nice relationship between geometrical properties of space-time and thermodynamic properties of a statistical system. This relation is embodied by a set of laws, called the laws of black hole mechanics, in analogy with the laws of thermodynamics.[123] See Table 3. The zeroth law states that the surface gravity κ_S of a stationary black hole is constant over the event horizon. Comparing to the first law of thermodynamics one can conclude that the surface gravity acts like the temperature. In fact if one analyses black holes in quantum field theory in a curved background, where gravity is considered classically and matter fields are treated quantum mechanically, one can see that they emit radiation, called Hawking radiation. In comparison with the black body spectrum one finds that the temperature of the radiation is

$$T_{\mathrm{H}} = \frac{\hbar \kappa_S}{2\pi} . \tag{107}$$

For a stationary black hole, the first law of black hole mechanics relates the variation of the energy to the other conserved quantities of the black hole such as angular momentum J and charge Q. It states that

$$\delta M = \frac{\kappa_S}{8\pi} \delta A + \mu \delta Q + \Omega \delta J , \tag{108}$$

where μ is the electric potential, Ω is the angular velocity of rotation and A is the area of the event horizon. Having the temperature given by (107) and comparing the relation (108) with the first law of thermodynamics suggests

Table 3: The Laws of black hole mechanics and classical thermodynamics.

Law	Thermodynamics	Black Holes
0th	If two thermodynamic systems are each in thermal equilibrium with a third, then they are in thermal equilibrium with each other.	The horizon has constant surface gravity for a stationary black hole.
1st	$dU = TdS - pdV + \mu dN$	$dM = \frac{\kappa}{8\pi} dA + \Omega dJ = \Phi dQ$
2nd	$dS \geq 0$	$dA \geq 0$
3rd	It is impossible to reach absolute zero temperature in a physical process.	It is impossible to form a black hole with vanishing surface gravity.

taking the area as the entropy of the black hole:

$$S_{\mathrm{BH}} = \frac{A}{4\hbar G_4}\,, \tag{109}$$

where G_4 is the four dimensional Newton constant. This is called the Bekenstein-Hawking entropy. The analogy of the area and the entropy is confirmed by the second law of black hole mechanics which states the total area of all event horizons is non-decreasing. This is a statement about non-stationary processes in space-time such as black hole collisions.

Recall, in the extremal limit of the Reissner–Nordström black hole the surface gravity vanished and, hence, the Hawking temperature also goes to zero. Extremal Reissner–Nordström black holes do not radiate and are stable. Also recall that A was non-zero. Extremal Reissner–Nordström black holes have zero temperature but non-zero entropy.

2.4. *Black holes in supergravity*

Classical supergravity is, unsurprisingly, a theory of gravity in the sense of general relativity. It has black hole solutions much like the Reissner–Nordström example described above. The key additional ingredient is that supergravity has a local symmetry relating each bosonic degree of freedom to a fermionic degree of freedom. It is because this symmetry relates bosons to fermions, unlike the conventional symmetries of field theory, that is it called supersymmetry. In four dimensions there can be $\mathcal{N} = 1, 2 \ldots 8$ supersymmetry generators and we speak of "$\mathcal{N}$ extended supergravity". The greater the supersymmetry the more constrained is the structure of the theory. By structure we mean both the boson/fermion content of the theory and the specific form of their couplings in the supergravity Lagrangian. For example, the maximally extended $\mathcal{N} = 8$ theory is unique. In this case all fields transform in the same representation, or *multiplet*, of the supersymmetry algebra. In addition to the metric (graviton) there are 70 scalar fields and 28 Abelian gauge potentials plus the corresponding fermions. Since this multiplet contains the graviton it is referred to as the "gravity multiplet". It is the *only* multiplet in this case. The couplings are also unique and take a very special form. In particular, the scalars parametrize the coset manifold $E_{7(7)}/\,\mathrm{SU}(8)$. Here $E_{7(7)}$ is, in fact, the global symmetry of the equations of motion. It is referred to as the "U-duality" group because it is the continuous limit of the full string theory dualities, such as the S- and T-dualities, which are Unified into U-duality.[107] The maximal compact subgroup SU(8) is the automorphism group of the supersymmetry algebra. The 28 electric gauge potentials plus their 28 magnetic duals transform linearly as a **56** under the $E_{7(7)}$ U-duality. This structure is quite generic for the extended supergavities: the scalars parametrize a homogeneous space G/H, where G is the U-duality group, and H its maximal compact subgroup[c]. The gauge potentials and their duals transform linearly in a symplectic representation of G. We will only deal with this class of theories here. This includes all $\mathcal{N} \geq 4$ theories, and an important class of $\mathcal{N} = 2$ theories, called $\mathcal{N} = 2$ Einstein-Maxwell supergravities, of which the STU model is an example. For $\mathcal{N} \leq 4$ the theory can, in addition to the gravity multiplet, couple to other independent multiplets. For example, the $\mathcal{N} = 2$ Einstein-Maxwell supergravities have a gravity multiplet, which contains the graviton, two spin-3/2 fermions or "gravitinos" and one spin-1 gauge potential A_μ, the so-called "graviphoton". However, one may add an arbitrary number of vector[d] multiplets, each consisting of one gauge potential, two spin-1/2 fermions or "gauginos" and a single complex scalar.

There is a vast body of literature cataloging the allowed forms of supergravity and examining their various properties. See, for example, Salam and Sezgin[125] for the tip of the iceberg. For our purposes, though, the take home message is that the black hole solutions will be close to the familiar Reissner–Nordström case, but with (typically) more electromagnetic charges in the game. One must also pay due diligence to the scalar equations of motion. For the static, spherically symmetric extremal solutions the similarity to Reissner–Nordström black hole is even more striking, thanks to a remarkable property of the scalars in this case called the "attractor mechanism".[126–130] One finds that, assuming regularity of the solution at the horizon, the scalars at the horizon must be fixed in terms of the charges independent of their asymptotic values. Consequently, the horizon area loses all memory of the scalars and is a (non-polynomial) quadratic function of only the charges, just as in the Reissner–Nordström case (106).

This quadratic function of the charges has a very important property. Since the electromagnetic field strengths transform linearly under U-duality and the charges are given by integrating them over a 2-sphere, they must also transform linearly under U-duality in the same representation. However, the 4-dimensional metric (in the Einstein frame) is inert under U-duality and so the horizon must also be U-duality invariant. We therefore

[c]Note, for $\mathcal{N} \leq 4$ supergravity can couple to matter multiplets, in which case $H = H_{\text{Aut}} \times H_{\text{matter}}$, where H_{Aut} is the automorphism group of the supersymmetry algebra and H_{matter} is the isotropy group of the matter multiplets. For $\mathcal{N} > 4$ $H = H_{\text{Aut}}$.
[d]Named "vector multiplet" because the highest spin appearing is 1.

conclude that the quadratic function of charges must be a U-duality invariant built out of the appropriate representation. In fact, for the theories we consider this function is the square root of a quartic polynomial invariant of the U-duality group, denoted $I_4(p,q)$, where p,q are the generalized electromagnetic charges.

To make these ideas a little more concrete lets look at a schematic supergravity Lagrangian which covers the cases of interest here,

$$\sqrt{-g}\left(\frac{1}{4}R - \frac{1}{2}\gamma_{ab}\partial_\mu\phi^a\partial^\mu\phi^b - \frac{1}{4}m_{\Lambda\Sigma}F^{\Lambda\mu\nu}F^{\Sigma}_{\mu\nu} - \frac{1}{8}a_{\Lambda\Sigma}\varepsilon^{\mu\nu\rho\sigma}F^{\Lambda}_{\mu\nu}F^{\Sigma}_{\rho\sigma}\right) \tag{110}$$

where we have set $\hbar = G_4 = 1$. Here $\gamma_{ab}(\phi)$ is the metric of the scalar manifold G/H. There are $\Lambda = 1, \ldots k$ gauge potentials appearing through their field strengths $F^{\Lambda}_{\mu\nu} = \partial_\mu A^{\Lambda}_{\nu} - \partial_\nu A^{\Lambda}_{\mu}$. In the $\mathcal{N} = 8$ case these are the $k = 28$ potentials in the gravity multiplet. In the $\mathcal{N} = 2$ theory coupled to $k-1$ vector multiplets, one would belong in the gravity multiplet and $k-1$ in the vector multiplets. The scalar dependent vector couplings $m_{\Lambda\Sigma}(\phi), a_{\Lambda\Sigma}(\phi)$ may be put into a positive definite Hermitian matrix $m_{\Lambda\Sigma}(\phi) + ia_{\Lambda\Sigma}(\phi)$.

The U-duality group is necessarily a subgroup of $\mathrm{Sp}(2k,\mathbb{R})$.[131] It acts on ϕ through the isometries of the scalar manifold G/H.

Making the static, spherically symmetric metric ansatz as in the pure Einstein-Maxwell case we find the black hole solutions carry charges defined by,

$$p^\Lambda = \frac{1}{4\pi}\int_{S^2} F^\Lambda, \qquad q_\Lambda = \frac{1}{4\pi}\int_{S^2} m_{\Lambda\Sigma}\star F^\Sigma + a_{\Lambda\Sigma}F^\Sigma. \tag{111}$$

These charges may be combined into a $2k$-vector,

$$Q = \begin{pmatrix} p^\Lambda \\ q_\Lambda \end{pmatrix}, \tag{112}$$

which transformations linearly in a symplectic representation under the U-duality group, $Q \mapsto MQ$ for $M \in G \subset \mathrm{Sp}(2k,\mathbb{R})$.

For extremal black holes, assuming the regularity of the scalars and geometry at the black hole horizon, the area is completely fixed by the so-called *black hole potential*,[130]

$$\frac{A}{4\pi} = \mathcal{V}_{BH}(p,q,\phi^i_h), \tag{113}$$

where ϕ^i_h are the scalars evaluated at the horizon and

$$\mathcal{V}_{\mathrm{BH}}(p,q,\phi^i) = \frac{1}{2}Q^T\begin{pmatrix} m + am^{-1}a & am \\ m^{-1}a & m^{-1}\end{pmatrix}Q. \tag{114}$$

One also finds that regularity implies,[130]

$$\left.\frac{\partial \mathcal{V}_{\rm BH}}{\partial \phi^i}\right|_h = 0. \tag{115}$$

Consequently, the scalars are fixed at the horizon in terms of the charges Q and they drop out of the entropy formula,

$$S_{\rm BH} = \pi \mathcal{V}_{BH}(p,q). \tag{116}$$

The entropy is only a function of the charges as claimed.

Recall, the charges transform as a symplectic representation of the U-duality group. The entropy must be given by a U-duality invariant expression built out of p and q.[132] This is indeed the case.[114,132] One finds, for the class of theories considered here, that the entropy is actually given by a quartic U-duality invariant $I_4(p,q)$,

$$S_{\rm BH} = \pi\sqrt{|I_4(p,q)|}. \tag{117}$$

For the STU model $-I_4$ is nothing but Cayley's hyperdeterminant!

A particular solution to supergravity may break some or all of the supersymmetry. A black hole solution which preserves some degree of supersymmetry is said to be BPS (Bogomolnyi-Prasad-Sommefield) and non-BPS otherwise. A black hole may preserve at most 1/2 of the supersymmetries. Note, all BPS black holes are extremal but extremal black holes may be either BPS or non-BPS. Black holes with $I_4 < 0$ are always non-BPS. For more details on black holes in supergravity see, for example,[133] and the reference therein.

2.5. *The STU model*

2.5.1. *The Lagrangian*

An interesting subsector of string compactification to four dimensions is provided by the STU model whose low energy limit is described by $\mathcal{N} = 2$ supergravity coupled to three vector multiplets.[108,134–136] One may regard it as a truncation of an $\mathcal{N} = 2$ theory obtained by compactifying the Type IIA string on Calabi-Yau. Alternatively, one may regard it as a truncation of an $\mathcal{N} = 4$ theory obtained by compactifying the heterotic string on T^6, where S, T, U correspond to the dilaton/axion, complex Kähler form and complex structure fields respectively. It exhibits an $SL(2,\mathbb{Z})_S$ strong/weak coupling duality and an $SL(2,\mathbb{Z})_T \times SL(2,\mathbb{Z})_U$ target space duality[e]. By

[e]This $SL(2,\mathbb{Z})_U$ should not be confused with U-duality.

string/string duality, this is equivalent to a Type IIA string on $K3 \times T^2$ with S and T exchanging roles.[105,107,109] Moreover, by mirror symmetry this is in turn equivalent to a Type IIB string on the mirror manifold with T and U exchanging roles. Hence the truncated theory has a combined $[SL(2,\mathbb{Z})]^3$ duality and complete S-T-U triality symmetry.[108] Alternatively, one may simply start with this $\mathcal{N}=2$ theory directly as an interesting four-dimensional supergravity in its own right, as described below.

The model admits extremal black holes solutions carrying four electric and four magnetic charges.[108,114] Below we organise these 8 charges into the $2\times2\times2$ *hypermatrix* and display the S-T-U symmetric Bogomol'nyi mass spectrum.[108] Associated with this hypermatrix is a *hyperdeterminant*, discussed in subsubsection 1.5.2. The black hole entropy, first calculated in,[137] is quartic in the charges and must be invariant under $[SL(2,\mathbb{Z})]^3$ and under triality. As we shall see, the important observation for its link to entanglement is that it is given by the square root of Cayley's hyperdeterminant.

Consider the three complex scalars axion/dilaton field S, the complex Kähler form field T and the complex structure field U

$$\begin{aligned} S &= S_1 + iS_2, \\ T &= T_1 + iT_2, \\ U &= U_1 + iU_2. \end{aligned} \tag{118}$$

This complex parametrisation allows for a natural transformation under the various $SL(2,\mathbb{Z})$ symmetries. The action of $SL(2,\mathbb{Z})_S$ is given by

$$S \to \frac{aS+b}{cS+d}, \tag{119}$$

where a,b,c,d are integers satisfying $ad-bc=1$, with similar expressions for $SL(2,\mathbb{Z})_T$ and $SL(2,\mathbb{Z})_U$. Defining the matrices $\mathcal{M}_S, \mathcal{M}_T$ and $\mathcal{M}_U$ via

$$\mathcal{M}_S = \frac{1}{S_2}\begin{pmatrix} 1 & S_1 \\ S_1 & |S|^2 \end{pmatrix}, \tag{120}$$

the action of $SL(2,\mathbb{Z})_S$ now takes the form

$$\mathcal{M}_S \to {\omega_S}^{\mathsf{T}} \mathcal{M}_S \omega_S, \tag{121}$$

where

$$\omega_S = \begin{pmatrix} d & b \\ c & a \end{pmatrix}, \tag{122}$$

with similar expressions for $\mathcal{M}_T$ and $\mathcal{M}_U$. We also define the $SL(2,\mathbb{Z})$ invariant tensors

$$\varepsilon_S = \varepsilon_T = \varepsilon_U = \begin{pmatrix} 0 & 1 \\ -1 & 0 \end{pmatrix}. \tag{123}$$

Starting from the heterotic string, the bosonic action for the graviton $g_{\mu\nu}$, dilaton η, two-form $B_{\mu\nu}$ four $U(1)$ gauge fields A_S^a and two complex scalars T and U is[108]

$$\begin{aligned} I_{STU} = \frac{1}{16\pi G}\int d^4x\sqrt{-g}e^{-\eta}\Big[R_g + g^{\mu\nu}\partial_\mu\eta\partial_\nu\eta - \tfrac{1}{12}g^{\mu\lambda}g^{\nu\tau}g^{\rho\sigma}H_{\mu\nu\rho}H_{\lambda\tau\sigma} \\ + \tfrac{1}{4}\,\mathrm{tr}(\partial{\mathcal{M}_T}^{-1}\partial\mathcal{M}_T) + \tfrac{1}{4}\,\mathrm{tr}(\partial{\mathcal{M}_U}^{-1}\partial\mathcal{M}_U) \\ - \tfrac{1}{4}{F_S}_{\mu\nu}^{\mathsf{T}}(\mathcal{M}_T \times \mathcal{M}_U){F_S}^{\mu\nu}\Big]. \end{aligned} \tag{124}$$

where the metric $g_{\mu\nu}$ is related to the four-dimensional canonical Einstein metric $g^c_{\mu\nu}$ by $g_{\mu\nu} = e^\eta {g^c}_{\mu\nu}$ and where

$$H_{\mu\nu\rho} = 3\left(\partial_{[\mu}B_{\nu\rho]} - \tfrac{1}{2}A^{\mathsf{T}}_{S[\mu}(\varepsilon_T \times \varepsilon_U){F_S}_{\nu\rho]}\right). \tag{125}$$

This action is manifestly invariant under T-duality and U-duality, with

$${F_S}_{\mu\nu} \to ({\omega_T}^{-1} \times {\omega_U}^{-1}){F_S}_{\mu\nu}, \qquad \mathcal{M}_{T/U} \to {\omega_{T/U}}^{\mathsf{T}}\mathcal{M}_{T/U}\,\omega_{T/U}, \tag{126}$$

and with η, $g_{\mu\nu}$ and $B_{\mu\nu}$ inert. Its equations of motion and Bianchi identities (but not the action itself) are also invariant under S-duality (119), with T and ${g^c}_{\mu\nu}$ inert and with

$$\begin{pmatrix} {F_S}_{\mu\nu}{}^a \\ \widetilde{F}_{S\mu\nu}{}^a \end{pmatrix} \to {\omega_S}^{-1}\begin{pmatrix} {F_S}_{\mu\nu}{}^a \\ \widetilde{F}_{S\mu\nu}{}^a \end{pmatrix}, \tag{127}$$

where

$$\widetilde{F}_{S\mu\nu}{}^a = -S_2[({\mathcal{M}_T}^{-1} \times {\mathcal{M}_U}^{-1})(\varepsilon_T \times \varepsilon_U)]^a{}_b \star {F_S}_{\mu\nu}{}^b - S_1{F_S}_{\mu\nu}{}^a, \tag{128}$$

where the axion field a is defined by

$$\varepsilon^{\mu\nu\rho\sigma}\partial_\sigma a = \sqrt{-g}e^{-\eta}g^{\mu\sigma}g^{\nu\lambda}g^{\rho\tau}H_{\sigma\lambda\tau}, \tag{129}$$

and where $S = S_1 + iS_2 = a + ie^{-\eta}$.

Thus T-duality transforms Kaluza-Klein electric charges $({F_S}^3, {F_S}^4)$ into winding electric charges $({F_S}^1, {F_S}^2)$ (and Kaluza-Klein magnetic charges into winding magnetic charges), U-duality transforms the Kaluza-Klein and winding electric charge of one circle $({F_S}^3, {F_S}^2)$ into those of the other

$({F_S}^4, {F_S}^1)$ (and similarly for the magnetic charges) but S-duality transforms Kaluza-Klein electric charge $({F_S}^3, {F_S}^4)$ into winding magnetic charge $(\widetilde{F}_S^3, \widetilde{F}_S^4)$ (and winding electric charge into Kaluza-Klein magnetic charge). In summary we have $SL(2,\mathbb{Z})_T \times SL(2,\mathbb{Z})_U$ and $T \leftrightarrow U$ off-shell but $SL(2,\mathbb{Z})_S \times SL(2,\mathbb{Z})_T \times SL(2,\mathbb{Z})_U$ and an S-T-U interchange on-shell.

One may also consider the Type IIA action I_{TUS} and the Type IIB action I_{UST} obtained by cyclic permutation of the fields S, T, U. Finally, one may consider an action[137] where the S, T and U fields enter democratically with a prepotential

$$F = STU, \tag{130}$$

which off-shell has the full STU interchange but none of the $SL(2,\mathbb{Z})$. All four versions are on-shell equivalent.

2.5.2. *The Bogomol'nyi spectrum*

Following,[108] it is now straightforward to write down an S-T-U symmetric Bogomol'nyi mass formula. Let us define electric and magnetic charge vectors α_S^a and β_S^a associated with the field strengths ${F_S}^a$ and $\widetilde{F_S}^a$ in the standard way. The electric and magnetic charges Q_S^a and P_S^a are given by

$${F_S}_{0r}{}^a \sim \frac{Q_S^a}{r^2}, \qquad \star {F_S}_{0r}{}^a \sim \frac{P_S^a}{r^2}, \tag{131}$$

giving rise to the charge vectors

$$\begin{pmatrix} \alpha_S^a \\ \beta_S^a \end{pmatrix} = \begin{pmatrix} S_2^{(0)} {\mathcal{M}_T}^{-1} \times {\mathcal{M}_U}^{-1} & S_1^{(0)} \varepsilon_T \times \varepsilon_U \\ 0 & -\varepsilon_T \times \varepsilon_U \end{pmatrix}^{ab} \begin{pmatrix} Q_S^b \\ P_S^b \end{pmatrix}. \tag{132}$$

Interestingly enough, as was shown in,[108,137] the 8 charges may usefully be represented as a $2 \times 2 \times 2$ hypermatrix a_{ABC}

$$\begin{aligned} &\Big(a_{000},\, a_{001},\, a_{010},\, a_{011},\, a_{100},\, a_{101},\, a_{110},\, a_{111}\Big) \\ =\ &\Big(-\beta_S^1,\, -\beta_S^2,\, -\beta_S^3,\, -\beta_S^4,\, \alpha_S^1,\, \alpha_S^2,\, \alpha_S^3,\, \alpha_S^4\Big), \end{aligned} \tag{133}$$

transforming as

$$a^{ABC} \to {\omega_S}^A{}_E {\omega_T}^B{}_F {\omega_U}^C{}_G a^{EFG}. \tag{134}$$

Then the mass formula is

$$\begin{aligned} M_{\text{BPS}}^2 = \tfrac{1}{8} a^{\mathsf{T}} \big(&\mathcal{M}_U^{-1} \otimes \mathcal{M}_T^{-1} \otimes \mathcal{M}_S^{-1} - \varepsilon_U \otimes \varepsilon_T \otimes \mathcal{M}_S^{-1} \\ &- \varepsilon_U \otimes \mathcal{M}_T^{-1} \otimes \varepsilon_S - \mathcal{M}_U^{-1} \otimes \varepsilon_T \otimes \varepsilon_S \big) a, \end{aligned} \tag{135}$$

This is consistent with the general $\mathcal{N} = 2$ Bogomol'nyi formula.[138] Although all theories have the same mass spectrum, there is clearly a difference of interpretation with electrically charged elementary states in one picture being solitonic monopole or dyon states in the other.

2.5.3. *Black hole entropy*

The *STU* model admits extremal black hole solutions satisfying the Bogomol'nyi mass formula. As usual, their entropy is given by one quarter the area of the event horizon. However, to calculate this area requires evaluating the mass not with the asymptotic values of the moduli, but with their frozen values on the horizon which are fixed in terms of the charges.[126] This ensures that the entropy is moduli-independent, as it should be. The relevant calculation was carried out in[137] for the model with the *STU* prepotential. The electric and magnetic charges of that paper are denoted (p^0, q_0), (p^1, q_1), (p^2, q_2), (p^3, q_3) with $O(2,2)$ scalar products

$$\begin{aligned} p^2 &= (p^0)^2 + (p^1)^2 - (p^2)^2 - (p^3)^2, \\ q^2 &= (q_0)^2 + (q_1)^2 - (q_2)^2 - (q_3)^2, \\ p \cdot q &= (p^0 q_0) + (p^1 q_1) + (p^2 q_2) + (p^3 q_3). \end{aligned} \tag{136}$$

This charge basis is related to the one above by[f]

$$\begin{aligned} &\big(p^0,\ p^1,\ p^2,\ p^3,\ q_0,\ q_1,\ q_2,\ q_3\big) \\ &= \big(a_0,\ -a_1,\ -a_2,\ -a_4,\ a_7,\ a_6,\ a_5,\ a_3\big), \end{aligned} \tag{137}$$

In these variables, the entropy is given by

$$S = \pi\sqrt{|D(p^\Lambda, q_\Lambda)|}, \tag{138}$$

where

$$\begin{aligned} D(p^\Lambda, q_\Lambda) = &-(p\cdot q)^2 + 4\big((p^1q_1)(p^2q_2) + (p^1q_1)(p^3q_3) + (p^3q_3)(p^2q_2)\big) \\ &- 4p^0q_1q_2q_3 + 4q_0p^1p^2p^3. \end{aligned} \tag{139}$$

The function $D(p^\Lambda, q_\Lambda)$ is symmetric under transformations: $p^1 \leftrightarrow p^2 \leftrightarrow p^3$ and $q_1 \leftrightarrow q_2 \leftrightarrow q_3$.

---

[f]Note that this is the convention of;[139] in[2] the sign of a_0, a_3, a_4, a_7 is flipped, which gives the same answer.

3. STU black holes and three qubits

3.1. *Entropy/entanglement correspondence*

The black hole/qubit correspondence now comes about by identifying[2] the black hole charge hypermatrix (137) with the three qubit state vector coefficients. Then we recognise from (76a) that

$$D(p^\Lambda, q_\Lambda) = -\operatorname{Det} a, \tag{140}$$

and hence the *STU* black hole entropy and Alice-Bob-Charlie 3-tangle are related by

$$S = \tfrac{\pi}{2}\sqrt{\tau_{ABC}}. \tag{141}$$

Thus Cayley's hyperdeterminant provides an interesting connection, at least at the level of mathematics, between string theory and quantum entanglement. Other mathematical similarities are provided by the division algebras[140] and by twistors.[73] What about physics? The near horizon geometry of the black holes is $AdS_2 \times S^2$ and one might expect a relation between the black hole entropy and the entanglement entropy of the conformal quantum mechanics that lives on the boundary,[141] although the nature of this particular AdS/CFT duality is not well-understood.[142] In any event, the 3-tangle is not the same as the entropy of entanglement.[72] So the appearance of the Cayley hyperdeterminant in these two different contexts of stringy black hole entropy and three-qubit quantum entanglement remains, for the moment, a purely mathematical coincidence.

To illustrate more these coincidences we compare some examples of $\mathcal{N} = 2$ black holes with examples of three-qubit states, following.[139]

3.2. *Rebits*

In the *STU* stringy black hole context[2,108,137,139] the a_{ABC} are integers (corresponding to quantised charges) and hence the symmetry group is $[SL(2,\mathbb{Z})]^3$ rather than $[SL(2,\mathbb{C})]^3$. However, as discussed by Levay,[5] it is possible within quantum information theory to focus on real qubits, called *rebits*, for which the a_{ABC} are real.[143,144] (One difference remains, however: one may normalise the wavefunction, whereas for black holes there is no such restriction on the charges a_{ABC}). It turns out that there are three reality classes which can be characterised by the hyperdeterminant

$$\operatorname{Det} a < 0, \tag{142a}$$

$$\operatorname{Det} a = 0, \tag{142b}$$

$$\operatorname{Det} a > 0. \tag{142c}$$

Case (142a) corresponds to the non-separable or GHZ class,[51] for example,

$$|\Psi\rangle = \tfrac{1}{2}(-|000\rangle + |011\rangle + |101\rangle + |110\rangle). \tag{143}$$

Case (142b) corresponds to the separable (A-B-C, A-BC, B-CA, C-AB) and W classes, for example

$$|\Psi\rangle = \tfrac{1}{\sqrt{3}}(|100\rangle + |010\rangle + |001\rangle). \tag{144}$$

In the string/supergravity interpretation,[2] cases (142a) and (142b) were shown to correspond to BPS black holes, for which half of the supersymmetry is preserved. Case (142a) has non-zero horizon area and entropy ("large" black holes), and case (142b) has vanishing horizon area and entropy ("small" black holes), at least at the semi-classical level. However, small black holes may acquire a non-zero entropy through higher order quantum effects. As discussed by Kallosh and Linde,[139] this entropy also has a quantum information interpretation involving bipartite entanglement of the three qubits.

Case (142c) is also GHZ, for example the above GHZ state (143) with a sign flip

$$|\Psi\rangle = \tfrac{1}{2}(|000\rangle + |011\rangle + |101\rangle + |110\rangle). \tag{145}$$

The canonical GHZ state

$$|\Psi\rangle = \tfrac{1}{\sqrt{2}}(|111\rangle + |000\rangle) \tag{146}$$

also belongs to case (142c).

3.3. *Classification of $\mathcal{N} = 2$ black holes and three-qubit states*

A-B-C states and singly charged black holes: A black hole with just one charge, say q_0 as in Figure 6, has vanishing entropy and corresponds to a completely separable A-B-C state

$$|\Psi\rangle = q_0|111\rangle, \tag{147}$$

with

$$\begin{gathered} S_A = S_B = S_C = 0, \\ \tau_{ABC} = 0. \end{gathered} \tag{148}$$

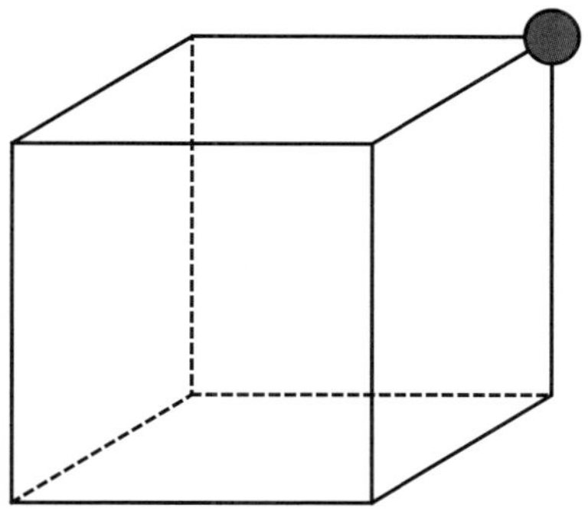

Fig. 6: The hypermatrix cube is restricted to correspond to the state (147) by retaining only a single nonzero entry at the q_0 vertex, denoted by the red disc. The state is completely separable and accordingly, the entropy vanishes for this cube.

Examples of singly charged supersymmetric black hole solutions[99] are provided by the electric Kaluza-Klein black hole with $\alpha = (1,0,0,0)$ and $\beta = (0,0,0,0)$; the electric winding black hole with $\alpha = (0,0,0,-1)$ and $\beta = (0,0,0,0)$; the magnetic Kaluza-Klein black hole with $\alpha = (0,0,0,0)$ and $\beta = (0,-1,0,0)$; the magnetic winding black hole with $\alpha = (0,0,0,0)$ and $\beta = (0,0,-1,0)$. These are characterised by a scalar-Maxwell coupling parameter $a = \sqrt{3}$.
By combining these 1-particle states, we may build up 2-, 3- and 4-particle bound states at threshold,[99,108,145] characterised by scalar-Maxwell coupling parameter $a = 1,1/\sqrt{3}$ and 0 respectively. The 1-, 2- and 3-particle states all yield vanishing contributions to $\operatorname{Det} a$. A non-zero value is obtained for the 4-particle example, however, which is just the Reissner-Nordström black hole, as explained below.
One could also consider a black hole with two charges, say q_0 and q_1 as in Figure 7, having vanishing entropy and corresponding to another completely separable state

$$|\Psi\rangle = q_0|111\rangle + q_1|110\rangle \tag{149}$$

also satisfying (148).

A-BC **states and doubly charged black holes:** A black hole with just two charges, say q_0 and p^1 as in Figure 8, has vanishing entropy and corresponds to a bipartite entangled state

$$|\Psi\rangle = q_0|111\rangle - p^1|001\rangle, \tag{150}$$

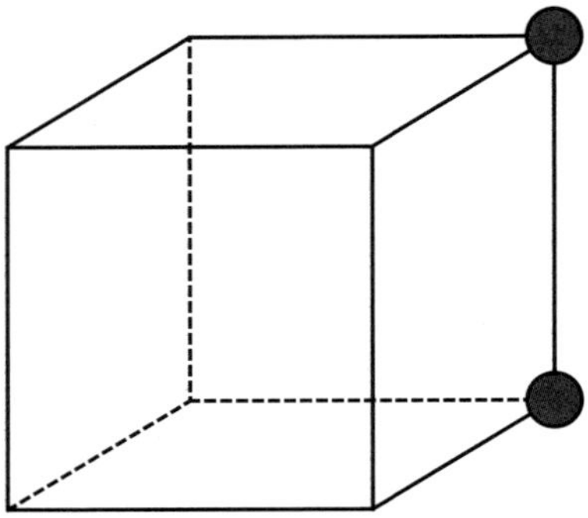

Fig. 7: The hypermatrix cube is restricted to correspond to the state (149) by retaining nonzero entries at the q_0 and q_1 vertices, denoted by the red discs. Despite having two nonzero vertices the cube's entropy vanishes since the state is again completely separable.

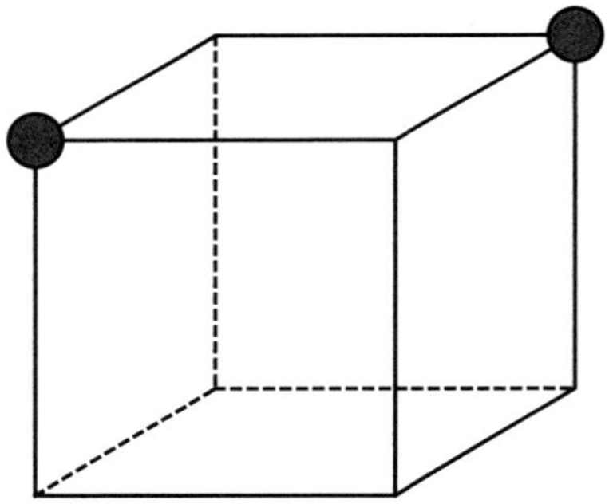

Fig. 8: The hypermatrix cube is restricted to correspond to the state (150) by retaining nonzero entries at the q_0 and p^1 vertices, denoted by the red discs. In this case the state is bi-separable, so the entropy vanishes once more.

with

$$\begin{aligned} S_A = S_B &= 4(q_0 p^1)^2, \\ S_C &= 0, \\ \operatorname{Det} a &= 0. \end{aligned} \tag{151}$$

W states and triply charged black holes: A black hole with just three charges, say q_0, p^1 and p^2 as in Figure 9, has vanishing entropy and corresponds to a W state

$$|\Psi\rangle = q_0|111\rangle - p^1|001\rangle - p^2|010\rangle, \tag{152}$$

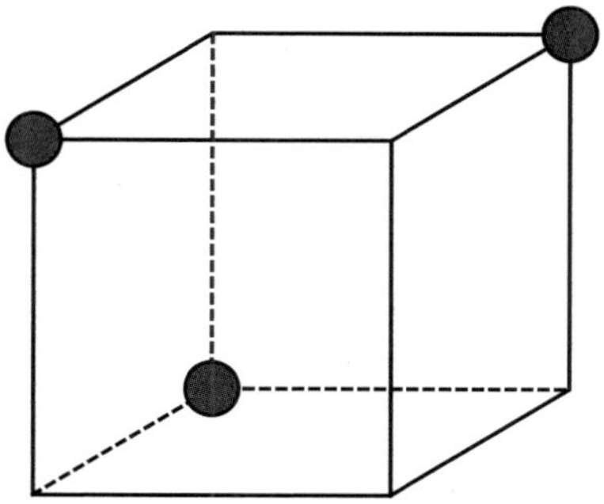

Fig. 9: The hypermatrix cube is restricted to correspond to the state (152) by retaining nonzero entries at the q_0, p^1 and p^2 vertices, denoted by the red discs. The entropy vanishes this time since the cube corresponds to a W state.

with

$$\begin{aligned} S_A &= 4(q_0)^2((p^1)^2 + (p^2)^2), \\ S_B &= 4(p^1)^2((q_0)^2 + (p^2)^2), \\ S_C &= 4(p^2)^2((q_0)^2 + (p^1)^2), \\ \operatorname{Det} a &= 0. \end{aligned} \tag{153}$$

GHZ and 4-charge BPS and non-BPS black holes: A black hole with just four charges, say q_0, p^1,p^2 and p^3 as in Figure 10, has non-vanishing entropy and corresponds to a GHZ state

$$|\Psi\rangle = q_0|111\rangle - p^1|001\rangle - p^2|010\rangle - p^3|100\rangle, \tag{154}$$

with

$$\begin{aligned} S_A &= 4((p^3)^2 + (q_0)^2)((p^1)^2 + (p^2)^2), \\ S_B &= 4((p^1)^2 + (p^3)^2)((q_0)^2 + (p^2)^2), \\ S_C &= 4((p^2)^2 + (p^3)^2)((q_0)^2 + (p^1)^2), \\ \operatorname{Det} a &= -4q_0p^1p^2p^3. \end{aligned} \tag{155}$$

This is a large BPS black hole if $\operatorname{Det} a < 0$ and a large non-BPS black hole if $\operatorname{Det} a > 0$.

GHZ and 2-charge non-BPS black holes: A black hole with just two charges Figure 11, say q_0 and p^0 as in Figure 11, has non-vanishing entropy and corresponds to a GHZ state

$$|\Psi\rangle = q_0|111\rangle + p^0|000\rangle, \tag{156}$$

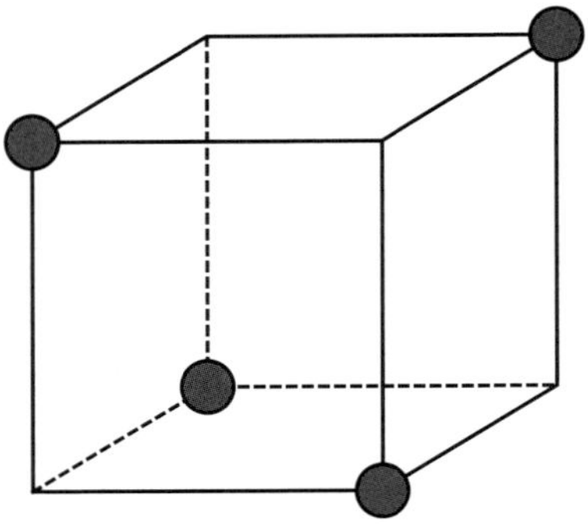

Fig. 10: The hypermatrix cube is restricted to correspond to the state (154) by retaining nonzero entries at the q_0, p^1, p^2 and p^3 vertices, denoted by the red discs. This is a GHZ state exhibiting genuine tripartite entanglement and accordingly the cube has nonzero entropy. While the previous cubes corresponded to small BPS black holes, this black hole is large and BPS/non-BPS for $\operatorname{Det} a < 0/\operatorname{Det} a > 0$.

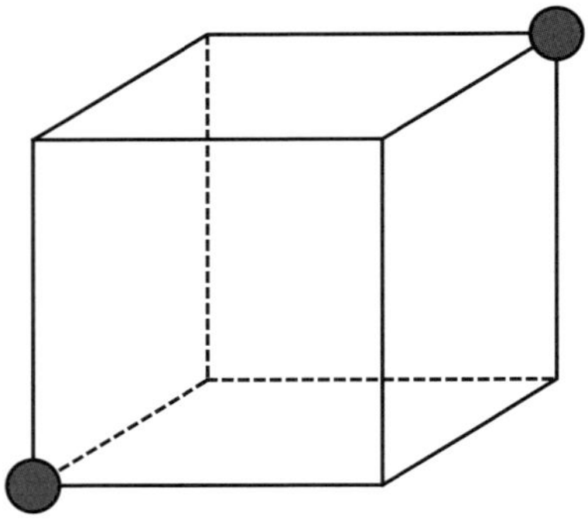

Fig. 11: The hypermatrix cube is restricted to correspond to the state (156) by retaining nonzero entries at the q_0 and p^0 vertices, denoted by the red discs. Despite having only two nonzero charges the cube has nonvanishing entropy since the state is GHZ. The corresponding black hole is large and always non-BPS.

with

$$S_A = S_B = S_C = 4(p^0)^2(q_0)^2, \tag{157}$$

and

$$\operatorname{Det} a = (p^0)^2(q_0)^2. \tag{158}$$

Since $\operatorname{Det} a > 0$, this describes a non-BPS large black hole.[146,147]

3.4. *Further developments*

In this section we briefly mention some further developments of the *STU* black-hole/qubit correspondence. In particular, we describe: (1) the microscopic interpretation of black holes as qubits[10] and (2) the classification of 4-qubit entanglement from *STU* black holes in three dimensions.[18]

3.4.1. *Microscopic interpretation*

The microscopic string-theoretic interpretation of the charges is given by configurations of intersecting D3-branes, wrapping around the six compact dimensions T^6. The 3-qubit basis vectors $|ABC\rangle$ are associated with the corresponding eight wrapping cycles. In particular, one can relate a well-known fact of quantum information theory, that the most general real three-qubit state up to local unitaries can be parameterised by four real numbers and an angle, to a well-known fact of string theory, that the most general STU black hole can be described by four D3-branes intersecting at an angle.

The microscopic analysis is not unique since there are many ways of embedding the *STU* model in string/M-theory, but a useful example from our point of view is that of four D3-branes of Type IIB wrapping the (579), (568), (478), (469) cycles of T^6 with wrapping numbers N_0, N_1, N_2, N_3 and intersecting over a string.[117] The wrapped circles are denoted by crosses and the unwrapped circles by noughts as shown in Table 4. This picture is consistent with the interpretation of the 4-charge black hole as bound state at threshold of four 1-charge black holes.[99,108,145] The fifth parameter θ is obtained[148,149] by allowing the N_3 brane to intersect at an angle which induces additional effective charges on the (579), (569), (479) cycles.

To make the black hole/qubit correspondence we associate the three T^2 with the $\mathrm{SL}(2)_A \times \mathrm{SL}(2)_B \times \mathrm{SL}(2)_C$ of the three qubits Alice, Bob, and Charlie. The eight different cycles then yield eight different basis vectors $|ABC\rangle$ as in the last column of Table 4, where $|0\rangle$ corresponds to xo and $|1\rangle$ to ox. To wrap or not to wrap; that is the qubit. We see immediately that we reproduce the five parameter three-qubit state $|\Psi\rangle$ of (75):

$$\begin{aligned}|\Psi\rangle = &-N_3\cos^2\theta|001\rangle - N_2|010\rangle + N_3\sin\theta\cos\theta|011\rangle \\ &- N_1|100\rangle - N_3\sin\theta\cos\theta|101\rangle + (N_0 + N_3\sin^2\theta)|111\rangle.\end{aligned} \tag{159}$$

Note the GHZ state describes four D3-branes intersecting over a string. By embedding this picture in $\mathcal{N} = 8$ supergravity we find an exact correspondence between the possible intersections preserving different degree of supersymmetry and the entanglement classes of 3-qubits as illustrated in Figure 12.

Table 4: Three qubit interpretation of the 8-charge $D = 4$ black hole from four D3-branes wrapping around the lower four cycles of T^6 with wrapping numbers N_0, N_1, N_2, N_3 and then allowing N_3 to intersect at an angle θ.

4	5	6	7	8	9	macro charges	micro charges	$\lvert ABC\rangle$
x	o	x	o	x	o	p^0	0	$\lvert 000\rangle$
o	x	o	x	x	o	q_1	0	$\lvert 110\rangle$
o	x	x	o	o	x	q_2	$-N_3 \sin\theta\cos\theta$	$\lvert 101\rangle$
x	o	o	x	o	x	q_3	$N_3 \sin\theta\cos\theta$	$\lvert 011\rangle$
o	x	o	x	o	x	q_0	$N_0 + N_3 \sin^2\theta$	$\lvert 111\rangle$
x	o	x	o	o	x	$-p^1$	$-N_3 \cos^2\theta$	$\lvert 001\rangle$
x	o	o	x	x	o	$-p^2$	$-N_2$	$\lvert 010\rangle$
o	x	x	o	x	o	$-p^3$	$-N_1$	$\lvert 100\rangle$

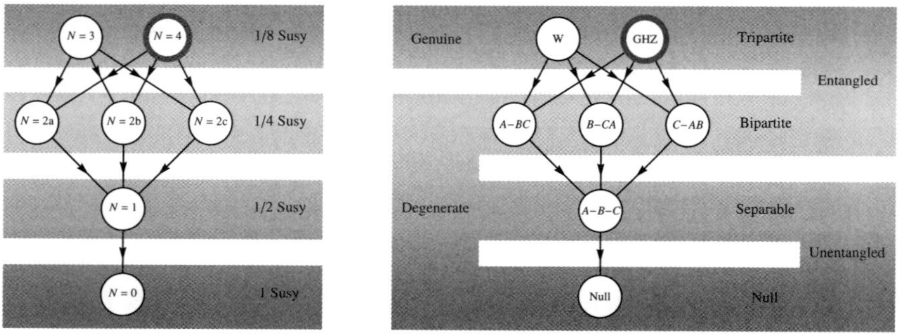

Fig. 12: Left: Classification of $\mathcal{N} = 8$ black holes according to susy. N denotes the number of intersecting D-branes in the microscopic picture. Right: The entanglement classification of 3-qubits. The arrows represent the removal of a D-brane or a non-invertible SLOCC operation.

3.4.2. *4-qubit entanglement and the STU model in $D = 3$*

For the STU model we found that the classification of the black hole solutions corresponded to the well known entanglement classification of three qubits. However, we can go one better and derive the much harder 4-qubit entanglement classification.[18,150] We consider $D = 4$ supergravity theories in which the moduli parameterize a symmetric space of the form $M_4 = G_4/H_4$, where G_4 is the global U-duality group and H_4 is its maximal compact subgroup. After a further time-like reduction to $D = 3$ the moduli space becomes a pseudo-Riemannian symmetric space $M_3^* = G_3/H_3^*$, where G_3 is the $D = 3$ duality group and H_3^* is a non-compact form of the maximal compact subgroup H_3. One finds that geodesic motion on M_3^* corresponds to stationary solutions of the $D = 4$ theory.[17,151–155] These geodesics are parameterized by the Lie algebra valued matrix of Noether charges Q and the problem of classifying the spherically symmetric extremal (non-extremal) black hole solutions consists of classifying the nilpotent (semisimple) orbits of Q (Nilpotent means $Q^n = 0$ for some sufficiently large n.)

In the case of the *STU* model the $D = 3$ moduli space G_3/H_3^* is $\mathrm{SO}(4,4)/[\mathrm{SL}(2,\mathbb{R})]^4$ (a para-quaternionic manifold), which yields the Lie algebra decomposition

$$\mathfrak{so}(4,4) \cong [\mathfrak{sl}(2,\mathbb{R})]^4 \oplus (\mathbf{2},\mathbf{2},\mathbf{2},\mathbf{2}). \tag{160}$$

The relevance of (160) to four qubits was pointed out in[13] and recently spelled out more clearly by Lévay[17] who relates four qubits to $D = 4$ *STU* black holes. The Kostant-Sekiguchi correspondence[156] then implies that the nilpotent orbits of $\mathrm{SO}(4,4)$ acting on the adjoint representation **28** are in one-to-one correspondence with the nilpotent orbits of $[\mathrm{SL}(2,\mathbb{C})]^4$ acting on the fundamental representation $(\mathbf{2},\mathbf{2},\mathbf{2},\mathbf{2})$ and hence with the classification of four-qubit entanglement. Note furthermore that it is the complex qubits that appear automatically, thereby relaxing the restriction to real qubits.

We found that there are 31 entanglement families which reduce to nine up to permutations of the four qubits. The nine agrees with[157,158] while the 31 is new.

Lecture II: *Seven qubits, octonions and the Fano plane*

4. Beyond the *STU* model

Despite its apparent successes, this investigation thus far has been limited to a very specific example, namely *STU* blacks holes and 3-qubit states. What about other supergravity theories and qubit systems? Interestingly, there are three related theories for which there exists a quartic invariant akin to Cayley's hyperdeterminant whose square root yields the corresponding black hole entropy:

(1) $\mathcal{N}=2$ supergravity coupled to $n+1$ vector multiplets where the symmetry is $\mathrm{SL}(2,\mathbb{Z})\times\mathrm{SO}(2,n;\mathbb{Z})$ and the black holes carry charges belonging to the $(2,n+2)$ representation ($n+2$ electric plus $n+2$ magnetic).
(2) $\mathcal{N}=4$ supergravity coupled to n vector multiplets where the symmetry is $\mathrm{SL}(2,\mathbb{Z})\times\mathrm{SO}(6,n;\mathbb{Z})$ where the black holes carry charges belonging to the $(2,6+n)$ representation ($n+6$ electric plus $n+6$ magnetic).
(3) $\mathcal{N}=8$ supergravity where the symmetry is the non-compact exceptional group $E_{7(7)}(\mathbb{Z})$ and the black holes carry charges belonging to the fundamental 56-dimensional representation (28 electric plus 28 magnetic).

Perhaps the most intriguing option is the case of $\mathcal{N}=8$ supergravity. This example was considered in[6] by Duff and Ferrara.

4.1. $\mathcal{N}=8$ *supergravity and black holes*

The black hole solutions of $\mathcal{N}=8$ supergravity depend on the $28+28$ electric/magnetic charges which form the fundamental **56** of $E_{7(7)}$.[89] The entropy is given by[132]

$$S=\pi\sqrt{|I_4|}, \tag{161}$$

where I_4 is the unique Cremmer-Julia quartic $E_{7(7)}$ invariant[89]

$$I_4=\mathrm{tr}(\bar{Z}Z)^2-\tfrac{1}{4}(\mathrm{tr}\,\bar{Z}Z)^2+4\left(\mathrm{Pf}\,Z+\mathrm{Pf}\,\bar{Z}\right), \tag{162}$$

and Z_{AB} is the central charge matrix. The relation between the entropy of stringy black holes and the Cartan-Cremmer-Julia $E_{7(7)}$ invariant was established in.[132] Alternatively, it may be written in terms of the quantized charges using the Cartan form[139,159,160]

$$I_4=-\,\mathrm{tr}(xy)^2+\tfrac{1}{4}(\mathrm{tr}\,xy)^2-4\left(\mathrm{Pf}\,x+\mathrm{Pf}\,y\right). \tag{163}$$

where the charges, x^{IJ} and y_{IJ}, are 8×8 antisymmetric matrices and Pf is the Pfaffian. The charges may be related to branes wrapping cycles of the T^7.[161] The exact relation between the Cartan invariant in (163) and Cremmer-Julia invariant[89] in (162) was established in,[161,162]

$$Z_{AB} = -\tfrac{1}{4\sqrt{2}}(x^{IJ} + iy_{IJ})(\gamma^{IJ})_{AB}, \tag{164}$$

and

$$x^{IJ} + iy_{IJ} = -\tfrac{\sqrt{2}}{4} Z_{AB}(\gamma^{AB})_{IJ}. \tag{165}$$

The matrices $(\gamma^{IJ})_{AB}$ form the SO(8) algebra. Here (I, J) are the 8 vector indices and (A, B) are the 8 spinor indices. The $(\gamma^{IJ})_{AB}$ matrices can be considered also as $(\gamma^{AB})_{IJ}$ matrices due to equivalence of the vector and spinor representations of the SO(8) Lie algebra.

Given that the exceptional Lie groups are anything but common place in the world of QI, we might expect that establishing an entanglement correspondence in this case would require some exotic arrangement of qudits. However, the work of Ferrara, Kallosh and Linde[139,160] suggested that we may need only to appeal to some generalisation of the, by now, familiar 3-tangle. This is due to the fact that we may bring Z_{AB} into a canonical form using an SU(8) transformation,

$$Z_{AB} = \begin{pmatrix} z_1 & 0 & 0 & 0 \\ 0 & z_2 & 0 & 0 \\ 0 & 0 & z_3 & 0 \\ 0 & 0 & 0 & z_4 \end{pmatrix} \otimes \begin{pmatrix} 0 & 1 \\ -1 & 0 \end{pmatrix}, \tag{166}$$

where $z_i = \rho_i e^{i\varphi_i}$ are complex. In this way the number of entries is reduced from 56 to 8. In a systematic treatment in,[163] the meaning of these parameters was clarified. From the four complex values of $z_i = \rho_i e^{i\varphi_i}$ one can remove three phases by an SU(8) rotation, but the overall phase cannot be removed; it is related to an extra parameter in the class of black hole solutions.[116,164] In this basis, the quartic invariant takes the form[132]

$$\begin{aligned} I_4 = \sum_i |z_i|^4 - 2\sum_{i<j} |z_i|^2|z_j|^2 + 4\left(z_1z_2z_3z_4 + \bar{z}_1\bar{z}_2\bar{z}_3\bar{z}_4\right) \\ = \quad & (\rho_1 + \rho_2 + \rho_3 + \rho_4) \\ & \times (\rho_1 + \rho_2 - \rho_3 - \rho_4) \\ & \times (\rho_1 - \rho_2 + \rho_3 - \rho_4) \\ & \times (\rho_1 - \rho_2 - \rho_3 + \rho_4) \\ & + \ 8\rho_1\rho_2\rho_3\rho_4\left(\cos\varphi - 1\right). \end{aligned} \tag{167}$$

This 5-parameter solution is called a generating solution for other black holes in $\mathcal{N}=8$ supergravity/M-theory.[165,166]

In terms of the charges the canonical form (166) is given by,[139,160]

$$(x^{IJ}+iy_{IJ})_{\text{can}} = \begin{pmatrix} 0 & \lambda_1 & 0 & 0 & 0 & 0 & 0 & 0 \\ -\lambda_1 & 0 & 0 & 0 & 0 & 0 & 0 & 0 \\ 0 & 0 & 0 & \lambda_2 & 0 & 0 & 0 & 0 \\ 0 & 0 & -\lambda_2 & 0 & 0 & 0 & 0 & 0 \\ 0 & 0 & 0 & 0 & 0 & \lambda_3 & 0 & 0 \\ 0 & 0 & 0 & 0 & -\lambda_3 & 0 & 0 & 0 \\ 0 & 0 & 0 & 0 & 0 & 0 & 0 & \lambda_4 \\ 0 & 0 & 0 & 0 & 0 & 0 & -\lambda_4 & 0 \end{pmatrix}. \tag{168}$$

Writing λ_α in terms of the parameters x^{IJ} and y_{IJ}, the matrix a_{ABD} and the black hole charges p^i and q_k,[2]

$$\begin{aligned} \lambda_1 &= x^{01} + iy_{01} = a_{111} + ia_{000} = q_0 + ip^0, \\ \lambda_2 &= x^{23} + iy_{23} = a_{100} + ia_{011} = -p^3 + q_3, \\ \lambda_3 &= x^{45} + iy_{45} = -a_{010} - ia_{101} = p^2 - iq_2, \\ \lambda_4 &= x^{56} + iy_{56} = -a_{001} - ia_{110} = p^1 - iq_1. \end{aligned} \tag{169}$$

we once again recover

$$I_4 = -\operatorname{Det} a. \tag{170}$$

We have identified the STU model invariant within the $\mathcal{N}=8$ invariant and, hence, a connection to 3-qubit entanglement. Consequently, we expect some relationship to 3-qubit entanglement when considering the full $\mathcal{N}=8$ generalization.

5. E_7 and the tripartite entanglement of seven qubits

How do we determine the precise qubit system corresponding to 56-charge black holes of $\mathcal{N}=8$ supergravity? The tripartite entanglement of three qubits is given by Cayley's hyperdeterminant, which follows from the SLOCC-equivalence group $[\mathrm{SL}(2,\mathbb{C})]^3$. However, a quantum information theoretic interpretation in the $\mathcal{N}=8$ theory cannot be naively obtained by adding more qubits. Recall, the n-qubit SLOCC-equivalence group is $[\mathrm{SL}(2,\mathbb{C})]^n$, which dramatically differs from E_7 and leaves little hope of reproducing the subtleties of $\mathcal{N}=8$ black holes. Lets tackle this conundrum by considering the decomposition of the fundamental 56-dimensional representation of $E_{7(7)}$ under seven factors of SL(2), the maximum number contained as a subgroup,

$$E_{7(7)}(\mathbb{Z}) \supset [SL(2,\mathbb{Z})]^7, \tag{171}$$

and

$$E_7(\mathbb{C}) \supset [SL(2,\mathbb{C})]^7. \tag{172}$$

We shall now show that the corresponding system in quantum information theory is that of seven qubits (Alice, Bob, Charlie, Daisy, Emma, Fred and George). However, the larger symmetry requires that they at most undergo a tripartite entanglement of a very specific kind. The entanglement measure will be given by the quartic Cartan $E_7(\mathbb{C})$ invariant.[89,132,159,167]

The crucial ingredient is the observation that E_7 contains seven copies of the single qubit SLOCC group SL(2) and that the **56** decomposes in a very particular way. We begin by considering the maximal subgroup $\mathrm{SL}(2)_A \times \mathrm{SO}(6,6)$,

$$\begin{array}{rl} E_{7(7)} & \supset \mathrm{SL}(2)_A \times \mathrm{SO}(6,6), \\ \mathbf{56} & \rightarrow (\mathbf{2},\mathbf{12}) + \ (\mathbf{1},\mathbf{32}). \end{array} \tag{173}$$

Further decomposing SO(6,6) in (173)

$$\begin{array}{rcccccccl} \mathrm{SL}(2)_A \times \mathrm{SO}(6,6) \supset & \mathrm{SL}(2)_A & \times & \mathrm{SL}(2)_B & \times & \mathrm{SL}(2)_D & \times & \mathrm{SO}(4,4), \\ (\mathbf{2},\mathbf{12}) + (\mathbf{1},\mathbf{32}) \rightarrow & (\mathbf{2},\mathbf{1},\mathbf{1},\mathbf{8_v}) & + & (\mathbf{1},\mathbf{2},\mathbf{1},\mathbf{8_s}) & + & (\mathbf{1},\mathbf{1},\mathbf{2},\mathbf{8_c}) & + & (\mathbf{2},\mathbf{2},\mathbf{2},\mathbf{1}). \end{array} \tag{174}$$

Further decomposing SO(4,4),

$$\begin{gathered} \mathrm{SL}(2)_A \times \mathrm{SL}(2)_B \times \mathrm{SL}(2)_D \times \mathrm{SO}(4,4) \\ \supset \mathrm{SL}(2)_A \times \mathrm{SL}(2)_B \times \mathrm{SL}(2)_D \times \mathrm{SO}(2,2) \times \mathrm{SO}(2,2) \\ (\mathbf{2},\mathbf{2},\mathbf{2},\mathbf{1}) + (\mathbf{2},\mathbf{1},\mathbf{1},\mathbf{8_v}) + (\mathbf{1},\mathbf{2},\mathbf{1},\mathbf{8_s}) + (\mathbf{1},\mathbf{1},\mathbf{2},\mathbf{8_c}) \\ \rightarrow (\mathbf{2},\mathbf{2},\mathbf{2},\mathbf{1},\mathbf{1}) + (\mathbf{2},\mathbf{1},\mathbf{1},\mathbf{4},\mathbf{1}) + (\mathbf{2},\mathbf{1},\mathbf{1},\mathbf{1},\mathbf{4}) \\ + (\mathbf{1},\mathbf{2},\mathbf{1},\mathbf{2},\mathbf{2}) + (\mathbf{1},\mathbf{2},\mathbf{1},\mathbf{2},\mathbf{2}) + (\mathbf{1},\mathbf{1},\mathbf{2},\mathbf{2},\mathbf{2}) + (\mathbf{1},\mathbf{1},\mathbf{2},\mathbf{2},\mathbf{2}). \end{gathered} \tag{175}$$

Finally, further decomposing each SO(2,2)

$$\begin{gathered} \mathrm{SL}(2)_A \times \mathrm{SL}(2)_B \times \mathrm{SL}(2)_D \times \mathrm{SO}(2,2) \times \mathrm{SO}(2,2) \\ \supset \mathrm{SL}(2)_A \times \mathrm{SL}(2)_B \times \mathrm{SL}(2)_D \times \mathrm{SL}(2)_C \times \mathrm{SL}(2)_G \times \mathrm{SL}(2)_F \times \mathrm{SL}(2)_E \\ (\mathbf{2},\mathbf{2},\mathbf{2},\mathbf{1},\mathbf{1}) + (\mathbf{2},\mathbf{1},\mathbf{1},\mathbf{4},\mathbf{1}) + (\mathbf{2},\mathbf{1},\mathbf{1},\mathbf{1},\mathbf{4}) \\ + (\mathbf{1},\mathbf{2},\mathbf{1},\mathbf{2},\mathbf{2}) + (\mathbf{1},\mathbf{2},\mathbf{1},\mathbf{2},\mathbf{2}) + (\mathbf{1},\mathbf{1},\mathbf{2},\mathbf{2},\mathbf{2}) + (\mathbf{1},\mathbf{1},\mathbf{2},\mathbf{2},\mathbf{2}) \\ \rightarrow (\mathbf{2},\mathbf{2},\mathbf{2},\mathbf{1},\mathbf{1},\mathbf{1},\mathbf{1}) + (\mathbf{2},\mathbf{1},\mathbf{1},\mathbf{2},\mathbf{2},\mathbf{1},\mathbf{1}) + (\mathbf{2},\mathbf{1},\mathbf{1},\mathbf{1},\mathbf{1},\mathbf{2},\mathbf{2}) \\ + (\mathbf{1},\mathbf{2},\mathbf{1},\mathbf{2},\mathbf{1},\mathbf{1},\mathbf{2}) + (\mathbf{1},\mathbf{2},\mathbf{1},\mathbf{1},\mathbf{2},\mathbf{2},\mathbf{1}) + (\mathbf{1},\mathbf{1},\mathbf{2},\mathbf{2},\mathbf{1},\mathbf{2},\mathbf{1}) \\ + (\mathbf{1},\mathbf{1},\mathbf{2},\mathbf{1},\mathbf{2},\mathbf{1},\mathbf{2}). \end{gathered} \tag{176}$$

In summary,

$$E_{7(7)} \supset \mathrm{SL}(2)_A \times \mathrm{SL}(2)_B \times \mathrm{SL}(2)_C \times \mathrm{SL}(2)_D \times \mathrm{SL}(2)_E \times \mathrm{SL}(2)_F \times \mathrm{SL}(2)_G, \tag{177}$$

and the **56** decomposes as

$$\begin{aligned} \mathbf{56} \rightarrow \quad & (\mathbf{2}, \mathbf{2}, \mathbf{1}, \mathbf{2}, \mathbf{1}, \mathbf{1}, \mathbf{1}) \\ & + (\mathbf{1}, \mathbf{2}, \mathbf{2}, \mathbf{1}, \mathbf{2}, \mathbf{1}, \mathbf{1}) \\ & + (\mathbf{1}, \mathbf{1}, \mathbf{2}, \mathbf{2}, \mathbf{1}, \mathbf{2}, \mathbf{1}) \\ & + (\mathbf{1}, \mathbf{1}, \mathbf{1}, \mathbf{2}, \mathbf{2}, \mathbf{1}, \mathbf{2}) \\ & + (\mathbf{2}, \mathbf{1}, \mathbf{1}, \mathbf{1}, \mathbf{2}, \mathbf{2}, \mathbf{1}) \\ & + (\mathbf{1}, \mathbf{2}, \mathbf{1}, \mathbf{1}, \mathbf{1}, \mathbf{2}, \mathbf{2}) \\ & + (\mathbf{2}, \mathbf{1}, \mathbf{2}, \mathbf{1}, \mathbf{1}, \mathbf{1}, \mathbf{2}). \end{aligned} \tag{178}$$

An analogous decomposition holds for

$$E_7(\mathbb{C}) \supset [\mathrm{SL}(2, \mathbb{C})]^7. \tag{179}$$

Notice that we find seven copies of the $(\mathbf{2}, \mathbf{2}, \mathbf{2})$ appearing in the *STU* model. This translates into seven copies of the 3-qubit Hilbert space:

$$\begin{aligned} |\Psi\rangle_{56} = \quad & a_{ABD}|ABD\rangle \\ & + b_{BCE}|BCE\rangle \\ & + c_{CDF}|CDF\rangle \\ & + d_{DEG}|DEG\rangle \\ & + e_{EFA}|EFA\rangle \\ & + f_{FGB}|FGB\rangle \\ & + g_{GAC}|GAC\rangle. \end{aligned} \tag{180}$$

Note that:

(1) Any pair of states has an individual in common
(2) Each individual is excluded from four out of the seven states
(3) Two given individuals are excluded from two out of the seven states
(4) Three given individuals are never excluded

So we have seven qubits (Alice, Bob, Charlie, Daisy, Emma, Fred and George) but where Alice has tripartite entanglement not only with Bob/Dave but also with Emma/Fred and also Charlie/George, and similarly for the other six individuals. So, in fact, each person has tripartite entanglement with each of the remaining three couples.

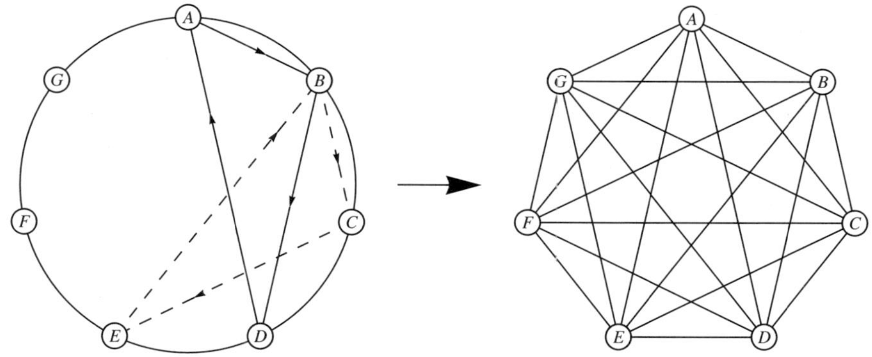

Fig. 13: The E_7 entanglement diagram corresponding to the decomposition (178) and the state (180). Each of the seven vertices A, B, C, D, E, F, G represents a qubit and each of the seven triangles $ABD, BCE, CDF, DEG, EFA, FGB, GAC$ describes a tripartite entanglement. As discussed in section 6 the oriented triangles correspond to quaternionic cycles in the multiplication table of imaginary octonions.

The entanglement may be represented by a heptagon as in Figure 13 with seven vertices A, B, C, D, E, F, G, and seven triangles

$$ABD, BCE, CDF, DEG, EFA, FGB, GAC.$$

Each of the seven states transforms as a $(\mathbf{2}, \mathbf{2}, \mathbf{2})$ under three of the SL(2)'s and are singlets under the remaining four. On restricting to just three of the seven qubits the entanglement ought to be given by Cayley's hyperdeterminant. Hence, we expect seven copies of Cayley's hyperdeterminant and consequently the entanglement measure must be quartic polynomial of the seven a, b, c, d, e, f, g. Taken together however, we see from (178) that they transform as a complex $\mathbf{56}$ of $E_7(\mathbb{C})$ and, hence, the entanglement measure must be invariant under $E_7(\mathbb{C})$. The unique possibility is the Cartan invariant I_4, and so the entanglement is given by

$$\tau_{ABCDEFG} = 4|I_4|. \tag{181}$$

It may be written as the sum of seven terms each of which is invariant under $[\mathrm{SL}(2)]^3$ plus cross terms. To see this, denote a $\mathbf{2}$ in one of the seven entries in (178) by A, B, C, D, E, F, G. So we may rewrite (178) symbolically as

$$\mathbf{56} = a + b + c + d + e + f + g. \tag{182}$$

Then I_4 is the singlet in $\mathbf{56} \times \mathbf{56} \times \mathbf{56} \times \mathbf{56}$:

$$
\begin{aligned}
I_4 = {} & a^4 + b^4 + c^4 + d^4 + e^4 + f^4 + g^4 \\
& + 2\Big[a^2b^2 + a^2c^2 + a^2d^2 + a^2e^2 + a^2f^2 + a^2g^2 \\
& \qquad + b^2c^2 + b^2d^2 + b^2e^2 + b^2f^2 + b^2g^2 \\
& \qquad + c^2d^2 + c^2e^2 + c^2f^2 + c^2g^2 \\
& \qquad + d^2e^2 + d^2f^2 + d^2g^2 \\
& \qquad + e^2f^2 + e^2g^2 \\
& \qquad + f^2g^2\Big] \\
& + 8\left[abce + bcdf + cdeg + defa + efgb + fgac + gabd\right],
\end{aligned}
\tag{183}
$$

where products like

$$
\begin{aligned}
a^4 &= (ABD)(ABD)(ABD)(ABD) \\
&= \tfrac{1}{2}\epsilon^{A_1A_2}\epsilon^{B_1B_2}\epsilon^{D_1D_4}\epsilon^{A_3A_4}\epsilon^{B_3B_4}\epsilon^{D_2D_3} \\
&\quad \times a_{A_1B_1D_1}a_{A_2B_2D_2}a_{A_3B_3D_3}a_{A_4B_4D_4},
\end{aligned}
\tag{184}
$$

exclude four individuals (here Charlie, Emma, Fred, and George), products like

$$
\begin{aligned}
a^2b^2 &= (ABD)(ABD)(BCE)(BCE) \\
&= \tfrac{1}{2}\epsilon^{A_1A_2}\epsilon^{B_1B_3}\epsilon^{D_1D_2}\epsilon^{B_2B_4}\epsilon^{C_3C_4}\epsilon^{E_3E_4} \\
&\quad \times a_{A_1B_1D_1}a_{A_2B_2D_2}b_{B_3C_3E_3}b_{B_4C_4E_4},
\end{aligned}
\tag{185}
$$

exclude two individuals (here Fred and George), and products like

$$
\begin{aligned}
abce &= (ABD)(BCE)(CDF)(EFA) \\
&= \tfrac{1}{2}\epsilon^{A_1A_4}\epsilon^{B_1B_2}\epsilon^{C_2C_3}\epsilon^{D_1D_3}\epsilon^{E_2E_4}\epsilon^{F_3F_4} \\
&\quad \times a_{A_1B_1D_1}b_{B_2C_2E_2}c_{C_3D_3F_3}e_{E_4F_4A_4},
\end{aligned}
\tag{186}
$$

exclude one individual (here George). These results may be verified using the dictionary between a, b, c, d, e, f, g and the x and y.[13]

6. Fano plane entanglement and the octonions

6.1. *Composition algebras*

The quaternions $\mathbb{H}$ were the unexpected prize of Hamilton's efforts to generalise the complex doublet, and its applications in 2-dimensional geometry, to a triplet of numbers that would play the same role in three dimensions. Of course, the original triplet was doomed from the outset, as we now know that the division algebras only exist in even dimensions. Upon hearing of Hamilton's magical discovery Graves, a close friend, asked,[168]

> "If with your alchemy you can make three pounds of gold, why should you stop there?" - Graves

Graves decided on some prospecting of his own and came up good, happening upon a rare nugget indeed, the octonions $\mathbb{O}$. The 8-dimensional octonions are the largest division algebra, completing the now famous sequence, $\mathbb{R}, \mathbb{C}, \mathbb{H}, \mathbb{O}$, but they are not the end of the story. In particular, the exceptional groups make equal use of their split signature cousins, which contain zero divisors. Jacobson in 1958 brought all these concepts together under the umbrella of *compsition algebras*.[169]

In the following we introduce the basic theory of composition algebras and some of their structural properties.

Much of the following will rely on a quadratic norm and its associated bilinear form:

Definition 6.1 (Quadratic norm). *A* quadratic norm *on a vector space* V *over a field* $\mathbb{F}$ *is a map* $\mathbf{n} : V \to \mathbb{F}$ *such that:*

(1) $\mathbf{n}(\lambda a) = \lambda^2 \mathbf{n}(a), \qquad \lambda \in \mathbb{F}, a \in V,$
(2) $\langle a, b\rangle := \mathbf{n}(a+b) - \mathbf{n}(a) - \mathbf{n}(b)$ *is bilinear.*

The quadratic norm is said to be non-degenerate if,

$$\langle a, b\rangle = 0 \quad \forall b \quad \Rightarrow a = 0. \tag{187}$$

Definition 6.2 (Composition algebra). *An algebra* $\mathbb{A}$ *defined over a field* $\mathbb{F}$ *with identity element* e*, is said to be* composition *if it has a non-degenerate quadratic form* $\mathbf{n} : \mathbb{A} \to \mathbb{F}$ *such that,*

$$\mathbf{n}(ab) = \mathbf{n}(a)\mathbf{n}(b), \quad \forall a, b \in \mathbb{A}, \tag{188}$$

where we denote multiplicative product of the algebra by juxtaposition.

Every composition algebra satisfies the quadratic equation,

$$a^2 - \langle a, e\rangle a + \mathbf{n}(a)e = 0. \tag{189}$$

Conjugation may be defined using the bilinear form, $\overline{a} := \langle a, e\rangle e - a$, and satisfies,

(1) $a\overline{a} = \overline{a}a = \mathbf{n}(a)e,$
(2) $\overline{ab} = \overline{b}\,\overline{a},$
(3) $\overline{\overline{a}} = a,$
(4) $\overline{a+b} = \overline{a} + \overline{b},$
(5) $\mathbf{n}(\overline{a}) = \mathbf{n}(a),$

(6) $\langle\bar{a}, \bar{b}\rangle = \langle a, b\rangle$.

We denote respectively the commutator and associator by $[a, b]$ and $[a, b, c]$,

$$\begin{aligned}[a, b] &= ab - ba, \\ [a, b, c] &= (ab)c - a(bc).\end{aligned} \tag{190}$$

An algebra is said to be associative if the associator vanishes and commutative if the commutator vanishes. If the associator is an alternating function of its arguments then the algebra is said to be *alternative.* All composition algebras are alternative.

By considering $\mathbb{F} \subset \mathbb{A}$ as the scalar multiples of the identity we may decompose $\mathbb{A}$ into its "real" and "imaginary" parts $\mathbb{A} = \mathbb{F} \oplus \mathbb{A}'$, where $\mathbb{A}' \subset \mathbb{A}$ is the subspace orthogonal to $\mathbb{F}$. An arbitrary element $a \in \mathbb{A}$ may be written $a = \Re(a) + \Im(a)$, where $\Re(a) \in \mathbb{F}$ and $\Im(a) \in \mathbb{A}'$.

Definition 6.3 (Division algebra). *A composition algebra* $\mathbb{A}$ *is said to be* division *if it contains no zero divisors,*

$$ab = 0 \Rightarrow \quad a = 0 \quad or \quad b = 0.$$

Hurwitz's celebrated theorem states that there are only four division algebras:[170] the reals, complexes, quaternions and octonions denoted respectively by $\mathbb{R}, \mathbb{C}, \mathbb{H}$ and $\mathbb{O}$. These algebras are obtained by the Cayley-Dickson process. With a slight modification one can also generate their split signature cousins $\mathbb{C}_s, \mathbb{H}_s$ and $\mathbb{O}_s$.[171]

We now recall some useful facts about the octonions:

- Octonions, $a, b, c \in \mathbb{R}^8$, are an 8-dimensional real vector space, with basis elements $e_0, \ldots, e_7$, where e_0 is the identity.
- Conjugation sends $e_0 \mapsto e_0$ and $e_i \mapsto -e_i$ for $i = 1, \ldots 7$.
- Non-commutative.
- Non-associative.

If we denote the imaginary octonions by e_i where $i = 1, \ldots 7$, the structure constants C_{ijk} are defined by

$$e_i e_j = -\delta_{ij} + C_{ijk} e_k, \tag{191}$$

and satisfy

$$C_{pmk}C_{qkn} + C_{qmk}C_{pkn} = \delta_{pm}\delta_{qn} + \delta_{pn}\delta_{qm} - 2\delta_{pq}\delta_{mn}. \tag{192}$$

Following[172] we define the "Jacobian" C_{ijkl}

$$\begin{aligned}&[e_i,[e_j,e_k]]+[e_k,[e_i,e_j]]+[e_j,[e_k,e_i]]\\&=4(C_{jkm}C_{imn}+C_{ijm}C_{kmn}+C_{kim}C_{jmn})e_n\\&\equiv 3C_{ijkl}e_l,\end{aligned} \tag{193}$$

where C_{ijkl} satisfies

$$C_{ijkl}=\tfrac{1}{6}\epsilon_{ijklmnp}C_{mnp}, \tag{194}$$

and

$$C_{ijkl}=-C_{mij}C_{mkl}-\delta_{il}\delta_{jk}+\delta_{ik}\delta_{jl}. \tag{195}$$

The norm of an arbitrary octonion $a=a_I e_I$, where $I=(0,i)$, is

$$\mathbf{n}(a)=(a_0)^2+(a_1)^2+(a_2)^2+(a_3)^2+(a_4)^2+(a_5)^2+(a_6)^2+(a_7)^2, \tag{196}$$

or, in the split case,

$$\mathbf{n}(a)=(a_0)^2+(a_1)^2+(a_2)^2+(a_3)^2-(a_4)^2-(a_5)^2-(a_6)^2-(a_7)^2. \tag{197}$$

6.2. *The octonionic tripartite entanglement of seven qubits*

Our special state,

$$\begin{aligned}|\Psi\rangle_{56}=\;&a_{ABD}|ABD\rangle\\&+b_{BCE}|BCE\rangle\\&+c_{CDF}|CDF\rangle\\&+d_{DEG}|DEG\rangle\\&+e_{EFA}|EFA\rangle\\&+f_{FGB}|FGB\rangle\\&+g_{GAC}|GAC\rangle,\end{aligned} \tag{198}$$

has some very distinctive structure. Observe,

(1) Two distinct qubits appear together in one and only one tripartite entanglement.
(2) Any two tripartite entanglements have at least one qubit in common.
(3) Every qubit belongs to three distinct tripartite entanglements.

On replacing the words qubit and tripartite entanglement with the words point and line, respectively, it becomes apparent that the state describes the projective plane of order 2. This is know as the Fano plane, which is depicted in Figure 14.

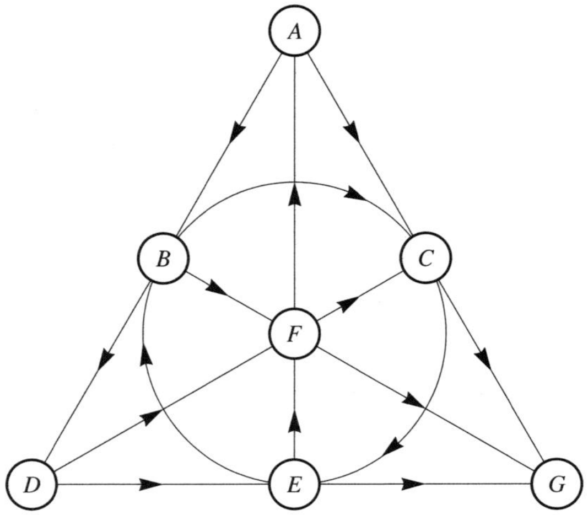

Fig. 14: The Fano plane is a projective plane with seven points and seven lines (the circle counts as a line). We may associate it to the state (180) by interpreting the points as the seven qubits A-G and the lines as the seven tripartite entanglements.

Table 5: Following the oriented lines of Figure 14 allows one to construct the Fano multiplication table, by identifying the qubits with the imaginary basis octonions. The minus signs arise when the Fano lines are followed counter to their orientation.

	A	B	C	D	E	F	G
A		D	G	$-B$	F	$-E$	$-C$
B	$-D$		E	A	$-C$	G	$-F$
C	$-G$	$-E$		F	B	$-D$	A
D	B	$-A$	$-F$		G	C	$-E$
E	$-F$	C	$-B$	$-G$		A	D
F	E	$-G$	D	$-C$	$-A$		B
G	C	F	$-A$	E	$-D$	$-B$	

That the Fano plane corresponds to the multiplication table of the imaginary octonions may be seen in Table 5. The non-vanishing independent components of the octonionic structure constants C_{ijk} and their duals C_{lmno} are then given in Table 6.

We may define a dual Fano plane by associating the lines to points and points to lines as in Figure 15.

Another way to understand the appearance of the dual Fano plane is to recognise the seven rows in (178) as the lines of the Fano plane and the

Table 6: From Table 5 one can read off which components of the Fano structure constants C_{ijk} are nonzero. Using (194) one then obtains the associator coefficients C_{lmno}.

i	j	k	l	m	n	o
1	2	4	3	5	6	7
2	3	5	4	6	7	1
3	4	6	5	7	1	2
4	5	7	6	1	2	3
5	6	1	7	2	3	4
6	7	2	1	3	4	5
7	1	3	2	4	5	6

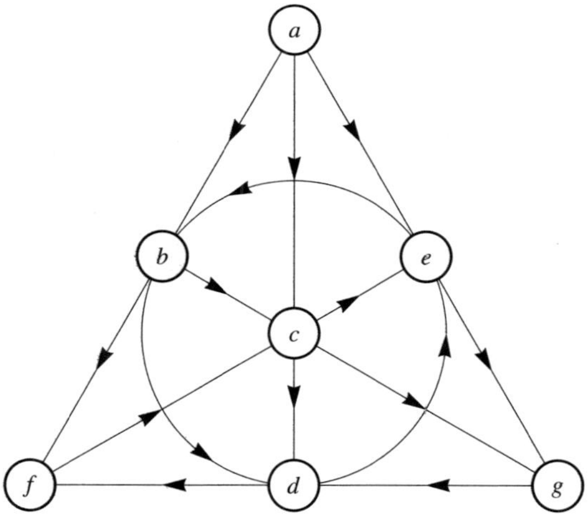

Fig. 15: Like the Fano plane of Figure 14 the dual Fano plane has seven points and seven lines, but this time the plane is associated with the dual state (199) interpreting the points as the seven tripartite entanglements and the lines as the seven qubits.

seven columns as vertices as in Table 7. The dual Fano plane corresponds to the transpose of this matrix and leads to a dual state

$$\begin{aligned} |\widetilde{\Psi}\rangle_{56} = \quad & A_{aeg}|aeg\rangle \\ &+ B_{bfa}|bfa\rangle \\ &+ C_{cgb}|cgb\rangle \\ &+ D_{dac}|dac\rangle \\ &+ E_{ebd}|ebd\rangle \\ &+ F_{fce}|fce\rangle \\ &+ G_{gdf}|gdf\rangle. \end{aligned} \tag{199}$$

Table 7: The seven terms in decomposition (178) may be written in a grid such that Fano lines and vertices are rows and columns.

	A	B	C	D	E	F	G
a	2	2	1	2	1	1	1
b	1	2	2	1	2	1	1
c	1	1	2	2	1	2	1
d	1	1	1	2	2	1	2
e	2	1	1	1	2	2	1
f	1	2	1	1	1	2	2
g	2	1	2	1	1	1	2

Table 8: In direct analogy with Figure 14 and Table 5, the oriented dual Fano plane of Figure 15 is used to write an octonionic multiplication table.

	a	b	c	d	e	f	g
a		f	d	$-c$	g	$-b$	$-e$
b	$-f$		g	e	$-d$	a	$-c$
c	$-d$	$-g$		a	f	$-e$	b
d	c	$-e$	$-a$		b	g	$-f$
e	$-g$	d	$-f$	$-b$		c	a
f	b	$-a$	e	$-g$	$-c$		d
g	e	c	$-b$	f	$-a$	$-d$	

Table 9: Dual Fano structure constants c_{ijk} and c_{lmno} are obtained from Table 8 in the same manner that Table 6 is obtained from Table 5.

i	j	k	l	m	n	o
1	2	6	3	4	5	7
2	3	7	4	5	6	1
3	4	1	5	6	7	2
4	5	2	6	7	1	3
5	6	3	7	1	2	4
6	7	4	1	2	3	5
7	1	5	2	3	4	6

The dual Fano plane corresponds to the multiplication table of the imaginary octonions given in Table 8. The non-vanishing independent components of the octonionic structure constants c_{ijk} and their duals c_{lmno} are then given by Table 9. Another way to understand the appearance of the dual Fano plane is to recognise the seven rows in (178) as the lines of the Fano plane and the seven columns as vertices as in Table 7.

6.3. *Subsectors*

Having understood the role of the Fano plane in the $\mathcal{N}=8$ tripartite entanglement of seven qubits, we now turn our attention to understanding the QI interpretation of its $\mathcal{N}=4$ truncation,[6,13] for which the black holes carry 12 + 12 electric/magnetic charges transforming as a $(\mathbf{2},\mathbf{12})$ under the $\mathrm{SL}(2)\times\mathrm{SO}(6,6)$ U-duality. As we shall see this corresponds to keeping a single line of the dual Fano plane Figure 15, which represents one of the seven possible reductions to the imaginary quaternions. We could also make the complementary truncation of the dual Fano plane, throwing away one line and keeping the remaining quadrangle. This is not actually a truncation of the $\mathcal{N}=8$ theory, but rather a particular black hole configuration with only R-R charges turned on. A quadrangle corresponds to the four imaginary octonions orthogonal to the quaternion subalgebra picked out by the complementary line and, hence, is not a closed subalgebra. Genuine consistent truncations correspond to closed subalgebras of the octonions. Finally, picking a single point of the dual Fano plane, i.e. an imaginary complex number, we recover the familiar $\mathcal{N}=2$ case with $\mathrm{SL}(2)\times\mathrm{SO}(2,2)\cong[\mathrm{SL}(2)]^3$.

First we recall the decomposition (173) of the fundamental 56-dimensional representation of $E_{7(7)}$ under its maximal subgroup,

$$\begin{aligned} E_{7(7)} &\supset \mathrm{SL}(2)_A\times\mathrm{SO}(6,6), \\ \mathbf{56} &\to (\mathbf{2},\mathbf{12}) + (\mathbf{1},\mathbf{32}). \end{aligned} \tag{200}$$

The $\mathcal{N}=4$ subsector consists of just the 24 states belonging to the $(\mathbf{2},\mathbf{12})$. Throwing away the $(\mathbf{1},\mathbf{32})$ and continuing the decomposition under the remaining six factors of SL(2) we find,

$$(\mathbf{2},\mathbf{12})\to(\mathbf{2},\mathbf{2},\mathbf{1},\mathbf{2},\mathbf{1},\mathbf{1},\mathbf{1})+(\mathbf{2},\mathbf{1},\mathbf{1},\mathbf{1},\mathbf{2},\mathbf{2},\mathbf{1})+(\mathbf{2},\mathbf{1},\mathbf{2},\mathbf{1},\mathbf{1},\mathbf{1},\mathbf{2}) \tag{201}$$

which corresponds to the state

$$|\Psi\rangle = a_{ABD}|ABD\rangle + e_{EFA}|EFA\rangle + g_{GAC}|GAC\rangle. \tag{202}$$

So only Alice talks to all the others. This is described by just those three lines passing through A in the Fano plane or the aeg line in the dual Fano plane. Then the equation analogous to (182) is

$$(\mathbf{2}, \mathbf{12}) = (ABD) + (EFA) + (GAC) = a + e + g, \tag{203}$$

and the corresponding quartic invariant, I_4, reduces to the singlet in $(\mathbf{2}, \mathbf{12}) \times (\mathbf{2}, \mathbf{12}) \times (\mathbf{2}, \mathbf{12}) \times (\mathbf{2}, \mathbf{12})$

$$I_4 \sim a^4 + e^4 + g^4 + 2[e^2g^2 + g^2a^2 + a^2e^2]. \tag{204}$$

If we identify the 24 numbers $(a_{ABD}, e_{EFA}, g_{GAC})$ with (P^μ, Q_ν) with $\mu, \nu = 0, \ldots, 11$ in a way analogous to (80) this becomes the SL(2)×SO(6, 6) invariant[108,116,164]

$$I_4 = P^2Q^2 - (P \cdot Q)^2. \tag{205}$$

So

$$I_4 = I_{aeg} \equiv \det(\gamma^1(a) + \gamma^2(g) + \gamma^3(e)). \tag{206}$$

This reduction from $\mathcal{N} = 8$ to $\mathcal{N} = 4$ corresponds to a reduction from the imaginary octonions of Table 8 to the imaginary quaternions of Table 10. This suggest that I_4 of (205) may be written as Cayley's hyperdeterminant over the imaginary quaternions, and this is indeed the case, as shown in.[13] From a stringy point of view, this subsector describes just the NS-NS charges.

For $\mathcal{N} = 2$, as may be seen from (174), we only have the $[\mathrm{SL}(2)]^3$ subgroup of the STU model where there are only eight states

$$|\Psi\rangle = a_{ABD}|ABD\rangle. \tag{207}$$

This is described by just the ABD line in the Fano plane or the a vertex in the dual Fano plane. This is simply the usual tripartite entanglement, for

Table 10: The aeg quaternionic multiplication table is obtained by selecting the aeg quaternionic cycle from Table 8.

	a	e	g
a		g	$-e$
e	$-g$		a
g	e	$-a$	

which

$$(\mathbf{2},\mathbf{2},\mathbf{2}) = (ABD) = a, \tag{208}$$

and the corresponding quartic invariant

$$I_4 \sim a^4, \tag{209}$$

is just Cayley's hyperdeterminant

$$I_4 = -\operatorname{Det} a. \tag{210}$$

Lecture III:
FTS classification of 3-qubit entanglement

7. Cubic Jordan algebras and the Freudenthal triple system

There is a remarkable connection between 4-dimensional supergravity theories and what have come to be known as "groups of type E_7".[173–181] These groups may be characterized by the FTS. In particular, the $\mathcal{N} = 8$ theory is related to the FTS defined over the Jordan algebra of 3×3 Hermitian matrices, which has an $E_{7(7)}$ automorphism symmetry. Since the STU model may be embedded in the $\mathcal{N} = 8$ theory there should be a corresponding STU Jordan algebra and FTS. There is and, as we shall see, it is particularly simple.

Consequently, we would expect the elegant mathematics of Jordan algebras and Freudenthal triple systems to naturally capture the entanglement classification of three qubits. Indeed, the FTS ranks, in a succinct algebraic manner, do indeed yield the correct classification. The entanglement classes correspond to FTS ranks 0, 1, 2a, 2b, 2c, 3 and 4, or, for SLOCC*, simply 0, 1, 2, 3, 4. In fact, we would argue that this is perhaps the most natural classification scheme. This is not only a matter of aesthetics. The classification of[68] based on the local entropies $S_{A,B,C}$ (see subsection 1.5 of subsection 1.3) did not make the $[\mathrm{SL}(2, \mathbb{C})]^3$ symmetry manifest. While the hyperdeteminant is SLOCC-covariant, the local entropies are not; they are natural objects for a classification based on local unitaries not SLOCC. This observation is important from the perspective of generalising to n qubits. Three qubits is the only non-trivial data point we have for a full SLOCC classification. If we seek to generalise this result, we should first formulate it in terms of the objects that will generalise, i.e. SLOCC-covariants. The FTS formulation is manifestly SLOCC-covariant since the automorphism group coincides with the SLOCC-equivalence group.

More speculatively, by studying the FTS classification, one might hope to identify those algebraic features which would usefully carry over to an n-qubit generalisation. There is in fact an n-qubit generalization of the FTS, but it is not yet clear how it captures the entanglement classification[182]

7.1. *Cubic Jordan algebras*

A Jordan algebra $\mathfrak{J}$ is vector space defined over a ground field $\mathbb{F}$ equipped with a bilinear product satisfying

$$X \circ Y = Y \circ X, \quad X^2 \circ (X \circ Y) = X \circ (X^2 \circ Y), \quad \forall\ X, Y \in \mathfrak{J}. \qquad (211)$$

The class of *cubic* Jordan algebras are constructed as follows.[183] Let V be a vector space equipped with a cubic norm, i.e. a homogeneous map of degree three, $N : V \to \mathbb{F}$ s.t. $N(\lambda X) = \lambda^3 N(X)$, $\forall \lambda \in \mathbb{F}, X \in V$, with trilinear linearisation

$$\begin{aligned} N(X,Y,Z) :=& \frac{1}{6}[N(X+Y+Z) - N(X+Y) - N(X+Z) - N(Y+Z) \\ &+ N(X) + N(Y) + N(Z)]. \end{aligned} \tag{212}$$

If V further contains a base point $N(c) = 1, c \in V$ one may define the following three maps,

$$\begin{aligned} \mathrm{Tr} : V \to \mathbb{F}; &\quad X \mapsto 3N(c,c,X), \\ S : V \times V \to \mathbb{F}; &\quad (X,Y) \mapsto 6N(X,Y,c), \\ \mathrm{Tr} : V \times V \to \mathbb{F}; &\quad (X,Y) \mapsto \mathrm{Tr}(X)\,\mathrm{Tr}(Y) - S(X,Y). \end{aligned} \tag{213}$$

A cubic Jordan algebra $\mathfrak{J}$, with multiplicative identity $\mathbb{1} = c$, may be derived from any such vector space if N is *Jordan cubic*. That is: (i) the trace bilinear form (213) is non-degenerate (ii) the quadratic adjoint map, $\sharp \colon \mathfrak{J} \to \mathfrak{J}$, uniquely defined by $\mathrm{Tr}(X^\sharp, Y) = 3N(X,X,Y)$, satisfies $(X^\sharp)^\sharp = N(X)X$, $\forall X \in \mathfrak{J}$. The Jordan product is then given by

$$X \circ Y = \tfrac{1}{2}\big(X \times Y + \mathrm{Tr}(X)Y + \mathrm{Tr}(Y)X - S(X,Y)\mathbb{1}\big), \tag{214}$$

where, $X \times Y$ is the linearisation of the quadratic adjoint, $X \times Y = (X+Y)^\sharp - X^\sharp - Y^\sharp$.

7.2. *The Freudenthal triple system*

Given a cubic Jordan algebra $\mathfrak{J}$ defined over a field $\mathbb{F}$, one is able to construct an FTS by defining the vector space $\mathfrak{F}(\mathfrak{J})$,

$$\mathfrak{F}(\mathfrak{J}) = \mathbb{F} \oplus \mathbb{F} \oplus \mathfrak{J} \oplus \mathfrak{J}. \tag{215}$$

An arbitrary element $x \in \mathfrak{F}(\mathfrak{J})$ may be written as a "2×2 matrix",

$$x = \begin{pmatrix} \alpha & X \\ Y & \beta \end{pmatrix} \quad \text{where } \alpha, \beta \in \mathbb{F} \quad \text{and} \quad X, Y \in \mathfrak{J}. \tag{216}$$

The FTS comes equipped with a non-degenerate bilinear antisymmetric quadratic form, a quartic form and a trilinear triple product:[173,184–187]

(1) Quadratic form $\{x,y\} \colon \mathfrak{F} \times \mathfrak{F} \to \mathbb{F}$

$$\{x,y\} = \alpha\delta - \beta\gamma + \mathrm{Tr}(X,W) - \mathrm{Tr}(Y,Z), \tag{217a}$$

where

$$x = \begin{pmatrix} \alpha & X \\ Y & \beta \end{pmatrix}, \quad y = \begin{pmatrix} \gamma & Z \\ W & \delta \end{pmatrix}. \tag{217b}$$

(2) Quartic form $q : \mathfrak{F} \to \mathbb{F}$

$$q(x) = -2[\alpha\beta - \operatorname{Tr}(X,Y)]^2 - 8[\alpha N(X) + \beta N(Y) - \operatorname{Tr}(X^\sharp, Y^\sharp)]. \tag{217c}$$

(3) Triple product $T : \mathfrak{F} \times \mathfrak{F} \times \mathfrak{F} \to \mathfrak{F}$ which is uniquely defined by

$$\{T(x,y,w), z\} = q(x,y,w,z) \tag{217d}$$

where $q(x,y,w,z)$ is the full linearisation of $q(x)$ such that $q(x,x,x,x) = q(x)$.

The *automorphism* group is given by the set of invertible $\mathbb{F}$-linear transformations preserving the quadratic and quartic forms,[173,184]

$$\operatorname{Aut}(\mathfrak{F}) = \{\sigma \in \operatorname{Iso}_{\mathbb{F}}(\mathfrak{F}) | q(\sigma x) = q(x), \{\sigma x, \sigma y\} = \{x,y\}, \ \forall x, y \in \mathfrak{F}\}. \tag{218}$$

The automorphism group corresponds to the U-duality group of a variety 4-dimensional supergravities (see for example[188,189] and the references therein). The conventional concept of matrix rank may be generalised to Freudenthal triple systems in a natural and $\operatorname{Aut}(\mathfrak{F})$ invariant manner. The rank of an arbitrary element $x \in \mathfrak{F}$ is uniquely defined by:[186,187]

$$\begin{aligned}
&\operatorname{Rank} x = 1 \Leftrightarrow 3T(x,x,y) + x\{x,y\}x = 0 \ \forall y; \\
&\operatorname{Rank} x = 2 \Leftrightarrow \exists y \text{ s.t. } 3T(x,x,y) + x\{x,y\}x \neq 0, \ T(x,x,x) = 0; \\
&\operatorname{Rank} x = 3 \Leftrightarrow T(x,x,x) \neq 0, \ q(x) = 0; \\
&\operatorname{Rank} x = 4 \Leftrightarrow q(x) \neq 0.
\end{aligned} \tag{219}$$

8. The 3-qubit Freudenthal triple system

Definition 8.1 (*STU* **cubic Jordan algebra).** *We define the STU cubic Jordan algebra, denoted* $\mathfrak{J}_{STU}$*, as the real vector space* $\mathbb{R} \oplus \mathbb{R} \oplus \mathbb{R}$ *with elements,*

$$A = (A_1, A_2, A_3), \tag{220}$$

and cubic norm,

$$N_3(A) = A_1 A_2 A_3. \tag{221}$$

Using the cubic Jordan algebra construction (213), one finds

$$\mathrm{Tr}(A,B) = A_1B_1 + A_2B_2 + A_3B_3, \tag{222}$$

Then, using $\mathrm{Tr}(A^\sharp, B) = 3N(A,A,B)$, the quadratic adjoint is given by

$$A^\sharp = (A_2A_3, A_1A_3, A_1A_2), \tag{223}$$

and therefore

$$\begin{aligned}(A^\sharp)^\sharp &= (A_1A_2A_3A_1, A_1A_2A_3A_2, A_1A_2A_3A_3)\\ &= N(A)A.\end{aligned} \tag{224}$$

It is not hard to check $\mathrm{Tr}(A,B)$ is non-degenerate and so N_3 is Jordan cubic. Hence, we have a *bona fide* cubic Jordan algebra $\mathfrak{J}_{STU} = \mathbb{R} \oplus \mathbb{R} \oplus \mathbb{R}$ with product given by

$$A \circ B = (A_1B_1, A_2B_2, A_3B_3). \tag{225}$$

The structure and reduced structure groups are given by $[\mathrm{SO}(2,\mathbb{R})]^3$ and $[\mathrm{SO}(2,\mathbb{R})]^2$ respectively.

The 3-qubit cubic Jordan algebra is defined by simply promoting $\mathbb{R}$ to $\mathbb{C}$,

Definition 8.2 (3-qubit cubic Jordan algebra). *We define the 3-qubit cubic Jordan algebra, denoted $\mathfrak{J}_{ABC}$, as the complex vector space $\mathbb{C} \oplus \mathbb{C} \oplus \mathbb{C}$ with elements:*

$$A = (A_1, A_2, A_3), \tag{226}$$

and cubic norm,

$$N_3(A) = A_1A_2A_3. \tag{227}$$

Definition 8.3 (3-qubit Freudenthal triple system). *We define the 3-qubit Freudenthal triple system, denoted $\mathfrak{F}_{ABC}$, as the complex vector space,*

$$\mathfrak{F}_{ABC} := \mathbb{C} \oplus \mathbb{C} \oplus \mathfrak{J}_{ABC} \oplus \mathfrak{J}_{ABC}, \tag{228}$$

with elements

$$\begin{pmatrix} \alpha & A = (A_1, A_2, A_3) \\ B = (B_1, B_2, B_3) & \beta \end{pmatrix}. \tag{229}$$

We identify the eight complex components of $\mathfrak{F}_{ABC}$ with the three qubit wavefunction $|\psi\rangle = a_{ABC}|ABC\rangle$,

$$\begin{pmatrix} \alpha & (A_1, A_2, A_3) \\ (B_1, B_2, B_3) & \beta \end{pmatrix} \leftrightarrow \begin{pmatrix} a_{111} & (a_{001}, a_{010}, a_{100}) \\ (a_{110}, a_{101}, a_{011}) & a_{000} \end{pmatrix} \tag{230}$$

so that

$$|\Psi\rangle = a_{ABC}|ABC\rangle \quad \leftrightarrow \quad \Psi := \begin{pmatrix} a_{111} & (a_{001}, a_{010}, a_{100}) \\ (a_{110}, a_{101}, a_{011}) & a_{000} \end{pmatrix}. \tag{231}$$

Using (217c) one finds that the quartic norm $\Delta(\Psi)$ is related to Cayley's hyperdeterminant by

$$\begin{aligned} \Delta(\Psi) &= \{T(\Psi, \Psi, \Psi), \Psi\} \\ &= 2\det\gamma^A = 2\det\gamma^B = 2\det\gamma^C \\ &= -2\,\mathrm{Det}\,a_{ABC}, \end{aligned} \tag{232}$$

The triple product maps a state Ψ, which transforms as a $(\mathbf{2}, \mathbf{2}, \mathbf{2})$ of $[\mathrm{SL}(2, \mathbb{C})]^3$, to another state $T(\Psi, \Psi, \Psi)$, cubic in the state vector coefficients, also transforming as a $(\mathbf{2}, \mathbf{2}, \mathbf{2})$. Explicitly, $T(\Psi, \Psi, \Psi)$ may be written as

$$T(\Psi, \Psi, \Psi) = T_{ABC}|ABC\rangle \tag{233}$$

where T_{ABC} takes one of three equivalent forms

$$\begin{aligned} T_{A_3B_1C_1} &= \epsilon^{A_1A_2}a_{A_1B_1C_1}(\gamma^A)_{A_2A_3} \\ T_{A_1B_3C_1} &= \epsilon^{B_1B_2}a_{A_1B_1C_1}(\gamma^B)_{B_2B_3} \\ T_{A_1B_1C_3} &= \epsilon^{C_1C_2}a_{A_1B_1C_1}(\gamma^C)_{C_2C_3}. \end{aligned} \tag{234}$$

The γ's are related to the local entropies of subsection 1.5 by

$$S_A = 4\Big[\operatorname{tr}\gamma^{B\dagger}\gamma^B + \operatorname{tr}\gamma^{C\dagger}\gamma^C\Big], \quad \operatorname{tr}\gamma^{A\dagger}\gamma^A = \tfrac{1}{8}\big[S_B + S_C - S_A\big] \tag{235}$$

and their cyclic permutations. This permits us to link T to the norm, local entropies and the Kempe invariant of subsection 1.5:

$$\langle T|T\rangle = \tfrac{2}{3}(K - |\psi|^6) + \tfrac{1}{16}|\psi|^2(S_A + S_B + S_C). \tag{236}$$

Having couched the 3-qubit system within the FTS framework we may assign an abstract FTS rank (219) to an arbitrary state Ψ:

$$\begin{aligned} \mathrm{Rank}\Psi &= 1 \Leftrightarrow \Upsilon(\Psi, \Psi, \Phi) = 0,\ \Psi \neq 0; \\ \mathrm{Rank}\Psi &= 2 \Leftrightarrow T(\Psi) = 0,\ \Upsilon(\Psi, \Psi, \Phi) \neq 0; \\ \mathrm{Rank}\Psi &= 3 \Leftrightarrow \Delta(\Psi) = 0,\ T(\Psi) \neq 0; \\ \mathrm{Rank}\Psi &= 4 \Leftrightarrow \Delta(\Psi) \neq 0. \end{aligned} \tag{237}$$

Strictly speaking, the automorphism group $\text{Aut}(\mathfrak{F}_{ABC})$ is not simply $\text{SL}(2,\mathbb{C})\times\text{SL}(2,\mathbb{C})\times\text{SL}(2,\mathbb{C})$ but includes a semi-direct product with the interchange triality $A\leftrightarrow B\leftrightarrow C$. The rank conditions are invariant under this triality. Hence the ranks naturally provide an SLOCC* classification. However, as we shall demonstrate, the set of rank 2 states may be subdivided into three distinct classes which are inter-related by this triality. In the next section we show that these rank conditions give the correct entanglement classification of three qubits as in Table 11.

Table 11: The entanglement classification of three qubits as according to the FTS rank system.

Class	Rank	FTS rank condition	
		vanishing	non-vanishing
Null	0	Ψ	–
A-*B*-*C*	1	$3T(\Psi,\Psi,\Phi)+\{\Psi,\Phi\}\Psi$	Ψ
A-*BC*	2a	$T(\Psi,\Psi,\Psi)$	γ^A
B-*CA*	2b	$T(\Psi,\Psi,\Psi)$	γ^B
C-*AB*	2c	$T(\Psi,\Psi,\Psi)$	γ^C
W	3	$\Delta(\Psi)$	$T(\Psi,\Psi,\Psi)$
GHZ	4	–	$\Delta(\Psi)$

8.1. *The FTS rank entanglement classes*

8.1.1. *Rank 1 and the class of separable states*

A non-zero state Ψ is rank 1 if and only if

$$\Upsilon(\Psi,\Psi,\Phi):=3T(\Psi,\Psi,\Phi)+\{\Psi,\Phi\}\Psi=0,\quad\forall\,\Phi. \tag{238}$$

The weaker condition $T(\Psi,\Psi,\Psi)=0$ implies that there is at most one non-vanishing γ since,

$$\begin{aligned}(\gamma^A)_{A_1A_2}(\gamma^C)_{C_1C_2} &= \epsilon^{B_1B_2}\epsilon^{Z_1Z_2}a_{A_1B_1Z_1}a_{A_2B_2Z_2}(\gamma^C)_{C_1C_2}\\ &= \epsilon^{B_2B_1}a_{A_1B_1C_1}T_{A_2B_2C_2}+\epsilon^{B_1B_2}a_{A_2B_2C_1}T_{A_1B_1C_2},\end{aligned} \tag{239}$$

and similarly for $(\gamma^B)_{B_1B_2}(\gamma^A)_{A_1A_2}$ and $(\gamma^C)_{C_1C_2}(\gamma^B)_{B_1B_2}$. In component form Υ is given by,

$$\begin{aligned}-\Upsilon_{A_1B_1C_1} =&\epsilon^{A_2A_3}b_{A_3B_1C_1}(\gamma^A)_{A_1A_2}\\ &+\epsilon^{B_2B_3}b_{A_1B_3C_1}(\gamma^B)_{B_1B_2}\\ &+\epsilon^{C_2C_3}b_{A_1B_1C_3}(\gamma^C)_{C_1C_2}\end{aligned} \tag{240}$$

where

$$|\phi\rangle = b_{ABC}|ABC\rangle \quad \leftrightarrow \quad \Phi = \begin{pmatrix} b_{111} & (b_{001}, b_{010}, b_{100}) \\ (b_{110}, b_{101}, b_{011}) & b_{000} \end{pmatrix}. \quad (241)$$

Hence, (238) implies all three gammas must vanish. Using (235) it is then clear that all three local entropies vanish.

Conversely, $S_A = S_B = S_C = 0$ implies that each of the three γ's vanish and the rank 1 condition is satisfied. Hence FTS rank 1 is equivalent to the class of separable states as in Table 11.

8.1.2. *Rank 2 and the class of biseparable states*

A nonzero state Ψ is rank 2 or less if and only if $T(\Psi, \Psi, \Psi) = 0$, which implies there is at most one non-vanishing γ. To be rank > 1 there must exist some Φ such that $3T(\Psi, \Psi, \Phi) + \{\Psi, \Phi\}\Psi \neq 0$, which implies there is at least one non-vanishing γ. Hence, rank 2 states have precisely one nonzero γ.

Using (235) it is clear that the choices $\gamma^A \neq 0$ or $\gamma^B \neq 0$ or $\gamma^C \neq 0$ give $S_A = 0, S_{B,C} \neq 0$ or $S_B = 0, S_{C,A} \neq 0$ or $S_C = 0, S_{A,B,} \neq 0$, respectively. These are precisely the conditions for the biseparable class A-BC or B-CA or C-AB presented in Table 2.

Conversely, using (235) and the fact that the local entropies and $\mathrm{tr}(\gamma^\dagger\gamma)$ are positive semidefinite, we find that all states in the biseparable class are rank 2, the particular subdivision being given by the corresponding nonzero γ. Hence FTS rank 2 is equivalent to the class of biseparable states as in Table 11.

8.1.3. *Rank 3 and the class of W-states*

A non-zero state Ψ is rank 3 if $\Delta(\Psi) = -2\,\mathrm{Det}\, a = 0$ but $T(\Psi, \Psi, \Psi) \neq 0$. From (234) all three γ's are then non-zero but from (232) all have vanishing determinant. In this case (235) implies that all three local entropies are non-zero but $\mathrm{Det}\, a = 0$. So all rank 3 Ψ belong to the W-class.

Conversely, from (235) it is clear that no two γ's may simultaneously vanish when all three S's > 0. We saw in subsubsection 8.1.1 that $T(\Psi, \Psi, \Psi) = 0$ implied at least two of the γ's vanish. Consequently, for all W-states $T(\Psi, \Psi, \Psi) \neq 0$ and, therefore, all W-states are rank 3. Hence FTS rank 3 is equivalent to the class of W-states as in Table 11.

8.1.4. *Rank 4 and the class of GHZ-states*

The rank 4 condition is given by $\Delta(\Psi) \neq 0$ and, since for the 3-qubit FTS $\Delta(\Psi) = -2\operatorname{Det} a$, we immediately see that the set of rank 4 states is equivalent to the GHZ class of genuine tripartite entanglement as in Table 11.

Note, $\operatorname{Aut}(\mathfrak{F}_{ABC})$ acts transitively only on rank 4 states with the same value of $\Delta(\Psi)$ as in the standard treatment. The GHZ class really corresponds to a 1-dimensional space of orbits parametrised by Δ.

In summary, we have demonstrated that each rank corresponds to one of the entanglement classes described in subsection 1.3 subsection 1.5.

8.2. *SLOCC orbits*

We now turn our attention to the coset parametrisation of the entanglement classes. The coset space of each orbit ($i = 1, 2, 3, 4$) is given by G/H_i where $G = [\mathrm{SL}(2,\mathbb{C})]^3$ is the SLOCC group and $H_i \subset [\mathrm{SL}(2,\mathbb{C})]^3$ is the stability subgroup leaving the representative state of the ith orbit invariant. We proceed by considering the infinitesimal action of $\operatorname{Aut}(\mathfrak{F})$ on the representative states of each class. The subalgebra annihilating the representative state gives, upon exponentiation, the stability group H. Note, $\mathfrak{der}(\mathfrak{J}_{ABC})$ is empty due to the associativity of $\mathfrak{J}_{ABC}$. Consequently, $\mathfrak{Str}(\mathfrak{J}_{ABC}) = L_{\mathbb{1}}\mathbb{C} \oplus \mathfrak{Str}_0(\mathfrak{J}_{ABC})$ has complex dimension 3, while $\mathfrak{Str}_0(\mathfrak{J}_{ABC})$ is now simply $L_{\mathfrak{J}'}$ and has complex dimension 2. Recall, $\mathfrak{Str}(\mathfrak{J}_{ABC})$ and $\mathfrak{Str}_0(\mathfrak{J}_{ABC})$ generate $[\mathrm{SO}(2,\mathbb{C})]^3$ and $[\mathrm{SO}(2,\mathbb{C})]^2$, respectively the structure and reduced structure groups of $\mathfrak{J}_{\mathbb{C}}$.

The results are summarised in Table 12. To be clear, in the preceding analysis we have regarded the three-qubit state as a point in $\mathbb{C}^2 \otimes \mathbb{C}^2 \otimes \mathbb{C}^2$, the philosophy adopted in, for example,[60,69,72] We could have equally well considered the projective Hilbert space regarding states as rays in $\mathbb{C}^2 \otimes \mathbb{C}^2 \otimes \mathbb{C}^2$, that is, identifying states related by a global complex scalar factor, as was done in.[4,74,190] The coset spaces obtained in this case are also presented in Table 12, the dimensions of which agree with the results of.[4,76] Note that the three-qubit separable projective coset is just a direct product of three individual qubit cosets $\mathrm{SL}(2,\mathbb{C})/\,\mathrm{SO}(2,\mathbb{C}) \ltimes \mathbb{C}$. Furthermore, the biseparable projective coset is just the direct product of the two entangled qubits coset $[\mathrm{SL}(2,\mathbb{C})]^2/\,\mathrm{O}(3,\mathbb{C})$ and an individual qubit coset.

As noted in,[5,71] the case of real qubits or "rebits" is qualitatively different from the complex case. An interesting observation is that on restricting to real states the GHZ class actually has two distinct orbits, characterised

Table 12: Coset spaces of the orbits of the 3-qubit state space $\mathbb{C}^2 \otimes \mathbb{C}^2 \otimes \mathbb{C}^2$ under the action of the SLOCC group $[\mathrm{SL}(2,\mathbb{C})]^3$.

Class	Orbits	dim	Projective orbits	dim
Separable	$\frac{[\mathrm{SL}(2,\mathbb{C})]^3}{[\mathrm{SO}(2,\mathbb{C})]^2 \ltimes \mathbb{C}^3}$	4	$\frac{[\mathrm{SL}(2,\mathbb{C})]^3}{[\mathrm{SO}(2,\mathbb{C}) \ltimes \mathbb{C}]^3}$	3
Biseparable	$\frac{[\mathrm{SL}(2,\mathbb{C})]^3}{O(3,\mathbb{C}) \times \mathbb{C}}$	5	$\frac{[\mathrm{SL}(2,\mathbb{C})]^3}{O(3,\mathbb{C}) \times (\mathrm{SO}(2,\mathbb{C}) \ltimes \mathbb{C})}$	4
W	$\frac{[\mathrm{SL}(2,\mathbb{C})]^3}{\mathbb{C}^2}$	7	$\frac{[\mathrm{SL}(2,\mathbb{C})]^3}{\mathrm{SO}(2,\mathbb{C}) \ltimes \mathbb{C}^2}$	6
GHZ	$\frac{[\mathrm{SL}(2,\mathbb{C})]^3}{[\mathrm{SO}(2,\mathbb{C})]^2}$	7	$\frac{[\mathrm{SL}(2,\mathbb{C})]^3}{[\mathrm{SO}(2,\mathbb{C})]^2}$	7

Table 13: Coset spaces of the orbits of the real case $\mathfrak{J}_{\mathbb{R}} = \mathbb{R} \oplus \mathbb{R} \oplus \mathbb{R}$ under $[\mathrm{SL}(2,\mathbb{R})]^3$.

Class	FTS Rank	$\Delta(\Psi)$	Orbits	dim
Separable	1	$= 0$	$\frac{[\mathrm{SL}(2,\mathbb{R})]^3}{[\mathrm{SO}(1,1)]^2 \ltimes \mathbb{R}^3}$	4
Biseparable	2	$= 0$	$\frac{[\mathrm{SL}(2,\mathbb{R})]^3}{O(2,1) \times \mathbb{R}}$	5
W	3	$= 0$	$\frac{[\mathrm{SL}(2,\mathbb{R})]^3}{\mathbb{R}^2}$	7
GHZ	4	< 0	$\frac{[\mathrm{SL}(2,\mathbb{R})]^3}{[\mathrm{SO}(1,1)]^2}$	7
GHZ	4	> 0	$\frac{[\mathrm{SL}(2,\mathbb{R})]^3}{[U(1)]^2}$	7
GHZ	4	> 0	$\frac{[\mathrm{SL}(2,\mathbb{R})]^3}{[U(1)]^2}$	7

by the sign of $\Delta(\Psi)$. This difference shows up in the cosets in the different possible real forms of $[\mathrm{SO}(2,\mathbb{C})]^2$. For positive $\Delta(\Psi)$ there are two disconnected orbits, both with $[\mathrm{SL}(2,\mathbb{R})]^3/[\mathrm{U}(1)]^2$ cosets, while for negative $\Delta(\Psi)$ there is one orbit $[\mathrm{SL}(2,\mathbb{R})]^3/[\mathrm{SO}(1,1,\mathbb{R})]^2$. In which of the two positive $\Delta(\Psi)$ orbits a given state lies is determined by the sign of the eigenvalues of the three γ's, as shown in Table 13. This phenomenon also has its counterpart in the black-hole context,[13,136,163,167,188,191] where the two disconnected $\Delta(\Psi) > 0$ orbits are given by 1/2-BPS black holes and non-BPS black holes with vanishing central charge respectively.[188]

Lecture IV: *Superqubits*

9. Supersymmetric quantum information

In this final lecture we consider another cross-fertilization of ideas from the world of M-theory to quantum information. While this topic is actually logically independent of the black-hole/qubit correspondence, it has nonetheless been both inspired and informed by it.

What QIT has shown us is that since information is encoded and transformed by physical systems its nature depends fundamentally on the underlying physical principles. The shift from a classical to a quantum world has precipitated a change in information theory. Hence, one might anticipate that if supersymmetry is realized in nature, whether it be at a fundamental or emergent level, there may be still further implications for information theory.

Here we describe some initial steps exploring this possibility. These ideas were inspired by the fascinating observation made by Castellani *et al*[192] that Cayley's hyperdeterminant may be supersymmetrized. More precisely there is a superhyperdeterminant built from a superhypermatrix, which has 14 "bosonic" commuting components and 13 "fermionic" anticommuting components. This superhyperdeterminant is invariant under the supergroup $[\mathrm{OSp}(1|2)]^3$, where $\mathrm{OSp}(1|2)$ is the minimal extension of $\mathrm{SL}(2)$ to a supergroup.

Taking its cues from this construction the superqubit is defined as a 3-dimensional representation of $\mathrm{OSp}(1|2)$ consisting of two commuting "bosonic" components and one anticommuting "fermionic" component. The n-qubit SLOCC equivalence group $[\mathrm{SL}(2,\mathbb{C})]^n$ and the LOCC equivalence group $[\mathrm{SU}(2)]^n$ are promoted to the supergroups $[\mathrm{OSp}(1|2)]^n$ and $[\mathrm{uOSp}(1|2)]^n$, respectively, where $\mathrm{uOSp}(1|2)$ is a "compact real form" of $\mathrm{OSp}(1|2)$. For $n = 2$ and $n = 3$ one can build the appropriate supersymmetric generalizations of the conventional entanglement measures. In particular, super-Bell and super-GHZ states are characterized, respectively, by nonvanishing superdeterminant (distinct from the Berezinian) and superhyperdeterminant.

This mathematical construction seems a very natural one. Moreover, from a physical point of view, it makes contact with various condensed-matter systems. For example, the three-dimensional representation of $\mathrm{Osp}(1|2)$ is encountered in the supersymmetric t-J model where it describes spinons and holons on a one-dimensional lattice.[193–197] It also shows up in the quantum Hall effect[198] and Affleck-Kennedy-Lieb-Tasaki models of superconductivity[199]

We should mention that a very interesting paper has recently appeared,[200] which analyzes quantum computing with superqubits, but within a different framework to the one discussed here.

10. Supergroups

10.1. *Grassmann numbers*

Grassmann numbers are the 2^n-dimensional vectors populating the Grassmann algebra Λ_n, which is generated by n mutually anticommuting elements $\{\theta^i\}_{i=1}^n$.

Any Grassmann number z may be decomposed into "body" $z_\mathcal{B} \in \mathbb{C}$ and "soul" $z_\mathcal{S}$ viz.

$$\begin{aligned} z &= z_\mathcal{B} + z_\mathcal{S} \\ z_\mathcal{S} &= \textstyle\sum_{k=1}^{\infty} \frac{1}{k!} c_{a_1 \cdots a_k} \theta^{a_1} \cdots \theta^{a_k}, \end{aligned} \tag{242}$$

where $c_{a_1 \cdots a_k} \in \mathbb{C}$ are totally antisymmetric. For finite dimension n the sum terminates at $k = 2^n$ and the soul is nilpotent $z_\mathcal{S}^{n+1} = 0$.

One may also decompose z into even and odd parts u and v

$$\begin{aligned} u &= z_\mathcal{B} + \textstyle\sum_{k=1}^{\infty} \frac{1}{(2k)!} c_{a_1 \cdots a_{2k}} \theta^{a_1} \cdots \theta^{a_{2k}} \\ v &= \textstyle\sum_{k=0}^{\infty} \frac{1}{(2k+1)!} c_{a_1 \cdots a_{2k+1}} \theta^{a_1} \cdots \theta^{a_{2k+1}}, \end{aligned} \tag{243}$$

which may also be expressed as the direct sum decomposition $\Lambda_n = \Lambda_n^0 \oplus \Lambda_n^1$. Furthermore, analytic functions f of Grassmann numbers are defined via

$$f(z) := \sum_{k=0}^{\infty} \tfrac{1}{k!} f^{(k)}(z_\mathcal{B}) z_\mathcal{S}^k, \tag{244}$$

where $f^{(k)}(z_\mathcal{B})$ is the kth derivative of f evaluated at $z_\mathcal{B}$ and is well defined if f is nonsingular at $z_\mathcal{B}$.[201]

One defines the *grade of a Grassmann number* as

$$\deg x := \begin{cases} 0 & x \in \Lambda_n^0 \\ 1 & x \in \Lambda_n^1, \end{cases} \tag{245}$$

where the grades 0 and 1 are referred to as even and odd, respectively.

Define the star * and superstar $^\#$ operators[202–204] satisfying the following properties:

$$
\begin{gathered}
(\Lambda_n^0)^\star = \Lambda_n^0, \quad (\Lambda_n^1)^\star = \Lambda_n^1, \\
(\Lambda_n^0)^\# = \Lambda_n^0, \quad (\Lambda_n^1)^\# = \Lambda_n^1, \\
(x\theta_i)^\star = x^*\theta_i^\star, \qquad \theta_i^{\star\star} = \theta_i, \qquad (\theta_i\theta_j)^\star = \theta_j^\star\theta_i^\star, \\
(x\theta_i)^\# = x^*\theta_i^\#, \quad \theta_i^{\#\#} = -\theta_i, \quad (\theta_i\theta_j)^\# = \theta_i^\#\theta_j^\#,
\end{gathered}
\tag{246}
$$

where $x \in \mathbb{C}$ and * is ordinary complex conjugation, which means

$$
\alpha^{\star\star} = \alpha, \qquad\qquad \alpha^{\#\#} = (-)^{\deg\alpha}\alpha \tag{247}
$$

for pure even/odd Grassmann α. The impure case follows by linearity.

Following[201] one may, if so desired, take the formal limit $n \to \infty$ defining the infinite dimensional vector space Λ_∞. Elements of Λ_∞ are called *supernumbers*. Our results are independent of the dimension of the underlying Grassmann algebra and one can use supernumbers throughout, but for the sake of simplicity we restrict to finite dimensional algebra by assigning just one Grassmann generator θ and its superconjugate $\theta^\#$ to every superqubit.

10.2. *Super linear algebra*

The grade definition applies to the components $T_{X_1\cdots X_k}$ of any k-index array of Grassmann numbers T, but one may also define $\deg X_i$, the *grade of an index*, for such an array by specifying a characteristic function from the range of the index X_i to the set $\{0,1\}$. In general the indices can have different ranges and the characteristic functions can be arbitrary for each index. It is then possible to define $\deg T$, the *grade of an array*, as long as the compatibility condition

$$
\deg T \equiv \deg(T_{X_1\cdots X_k}) + \sum_{i=1}^{k} \deg X_i \mod 2 \quad \forall\ X_i \tag{248}
$$

is satisfied. In precisely such cases the entries of T satisfy

$$
\begin{gathered}
\deg(T_{X_1\cdots X_k}) = \deg T + \sum_{i=1}^{k} \deg X_i \mod 2, \\
\implies \deg T = \deg(T_{\underbrace{1\ldots 1}_{k}}), \\
\deg(T_1T_2) = \deg T_1 + \deg T_2 \mod 2,
\end{gathered}
\tag{249}
$$

so that in other words T is partitioned into blocks with definite grade such that the nearest neighbors of any block are of the opposite grade to that block. The array grade simply distinguishes the two distinct ways of accomplishing such a partition (i.e. the two possible grades of the first element $T_{1\cdots 1}$). Grassmann numbers and the Grassmann number grade may be viewed as special cases of arrays and the array grade.

Special care must be taken not to confuse this notion of array grade with whether the array entries at even/odd index positions vanish. An array T may be decomposed as

$$T = T_E + T_O, \tag{250}$$

where the pure even part T_E is obtained from T by setting to zero all entries satisfying $\deg(T_{X_1\cdots X_k}) = 1$, and similarly *mutatis mutandis* for T_O. The property of being pure even or pure odd is therefore independent of the array grade as defined above.

The various grades commonly appear in formulae as powers of -1 and the shorthand

$$(-)^X := (-1)^{\deg X} \tag{251}$$

is often used. The indices of superarrays may be supersymmetrized as follows:

$$\begin{gathered} T_{X_1\cdots [\![X_i|\cdots|X_j]\!]\cdots X_k} := \\ \tfrac{1}{2}[T_{X_1\cdots X_i\cdots X_j\cdots X_k} + (-)^{X_i X_j} T_{X_1\cdots X_j\cdots X_i\cdots X_k}]. \end{gathered} \tag{252}$$

While we require these definitions for some of our considerations, one typically only uses arrays with 0, 1, or 2 indices where the characteristic functions are monotonic: supernumbers, supervectors, and supermatrices, respectively. Functions of grades extend to mixed superarrays (with nonzero even *and* odd parts) by linearity.

A $(p|q)\times(r|s)$ supermatrix is just an $(p+q)\times(r+s)$-dimensional block partitioned matrix

$$M = \begin{array}{c} \\ p \\ q \end{array}\!\!\begin{array}{c} \begin{array}{c|c} r & s \end{array} \\ \left[\begin{array}{c|c} A & B \\ \hline C & D \end{array}\right] \end{array} \tag{253}$$

where entries in the A and D blocks are grade $\deg M$, and those in the B and C blocks are grade $\deg M + 1 \mod 2$. The special cases $s = 0$ or $q = 0$ can be permitted to make the definition encapsulate row and column supervectors. Supermatrix multiplication is defined as for ordinary matrices;

however, the trace, transpose, adjoint, and determinant have distinct super versions.[204,205]

The supertrace str M of a supermatrix is M defined as

$$\operatorname{str} M := \sum_X (-)^{(X+M)X} M_{XX} \tag{254}$$

and is linear, cyclic modulo sign, and insensitive to the supertranspose

$$\begin{aligned} \operatorname{str}(M+N) &= \operatorname{str}(M) + \operatorname{str}(N) \\ \operatorname{str}(MN) &= (-)^{MN} \operatorname{str}(NM) \\ \operatorname{str} M^{st} &= \operatorname{str} M. \end{aligned} \tag{255}$$

The supertranspose M^{st} of a supermatrix M is defined componentwise as

$$M^{st}{}_{X_1X_2} := (-)^{(X_2+M)(X_1+X_2)} M_{X_2X_1}. \tag{256}$$

Unlike the transpose the supertranspose is not idempotent; instead,

$$\begin{aligned} M^{st\,st}{}_{X_1X_2} &= (-)^{(X_1+X_2)} M_{X_1X_2}, \\ M^{st\,st\,st}{}_{X_1X_2} &= (-)^{(X_1+M)(X_1+X_2)} M_{X_2X_1}, \\ M^{st\,st\,st\,st}{}_{X_1X_2} &= M_{X_1X_2}, \end{aligned} \tag{257}$$

so that it is of order 4. The supertranspose also satisfies

$$(MN)^{st} = (-)^{MN} N^{st} M^{st}. \tag{258}$$

The adjoint † and superadjoint ‡ of a supermatrix are defined as

$$\begin{aligned} M^\dagger &:= M^{\star t} \\ M^\ddagger &:= M^{\# st}, \end{aligned} \tag{259}$$

and satisfy

$$\begin{aligned} M^{\dagger\dagger} &= M, & M^{\ddagger\ddagger} &= (-)^M M, \\ (MN)^\dagger &= N^\dagger M^\dagger, & (MN)^\ddagger &= (-)^{MN} N^\ddagger M^\ddagger. \end{aligned} \tag{260}$$

The preservation of anti-super-Hermiticity, $M^\ddagger = -M$, under scalar multiplication by Grassmann numbers, as required for the proper definition of $\mathfrak{uosp}(1|2)$,[206] necessitates the left/right multiplication rules:

$$\begin{aligned} (\alpha M)_{X_1X_2} &= (-)^{X_1\alpha} \alpha M_{X_1X_2}, \\ (M\alpha)_{X_1X_2} &= (-)^{X_2\alpha} M_{X_1X_2}\alpha. \end{aligned} \tag{261}$$

The Berezinian is defined as

$$\begin{aligned} \operatorname{Ber} M &:= \det(A - BD^{-1}C)/\det(D) \\ &= \det(A)/\det(D - CA^{-1}B) \end{aligned} \tag{262}$$

and is multiplicative, insensitive to the supertranspose, and generalizes the relationship between trace and determinant

$$\begin{aligned}\operatorname{Ber}(MN) &= \operatorname{Ber}(M)\operatorname{Ber}(N)\\ \operatorname{Ber} M^{st} &= \operatorname{Ber} M\\ \operatorname{Ber} e^{M} &= e^{\operatorname{str} M}.\end{aligned}\tag{263}$$

The direct sum and super tensor product are unchanged from their ordinary versions. In analogy with the ordinary case we have

$$\begin{aligned}(M\otimes N)^{t} &= M^{t}\otimes N^{t}\\ (M\otimes N)^{st} &= M^{st}\otimes N^{st}\\ \operatorname{str}(M\otimes N) &= \operatorname{str} M \operatorname{str} N.\end{aligned}\tag{264}$$

These definitions are manifestly compatible with Hermiticity and super-Hermiticity.

In supermatrix representations of superalgebras, one may represent the superbracket of generators M and N as

$$[\![M,N]\!] := MN - N_E M - N_O(M_E - M_O).\tag{265}$$

One may also consider supermatrices M and N whose components are themselves supermatrices. Provided the component supermatrices are pure even (odd) at even (odd) index positions (e.g. M_{11} is a pure even supermatrix for even M), one may write the superbracket of such supermatrices as

$$\begin{gathered}[\![M_{X_1X_2}, N_{X_3X_4}]\!] =\\ M_{X_1X_2}N_{X_3X_4} - (-)^{(X_1+X_2)(X_3+X_4)}N_{X_3X_4}M_{X_1X_2},\end{gathered}\tag{266}$$

where the final two indices are suppressed. This grouping of supermatrices into supermatrices is useful for summarizing the superbrackets of superalgebras.

10.3. *Orthosymplectic superalgebras*

Supermatrix representations of the orthosymplectic supergroup $\mathrm{OSp}(p|2q)$ consist of supermatrices $M \in \mathrm{GL}(p|2q)$ satisfying

$$M^{st}EM = E,\tag{267}$$

but for convenience we choose instead to use supermatrices $M \in \mathrm{GL}(2q|p)$ satisfying (267). In this convention, the invariant supermatrix E is defined by

$$E := \left(\begin{array}{c|c} \mathbb{J}_{2q} & 0 \\ \hline 0 & \mathbb{1}_p \end{array}\right), \qquad \mathbb{J}_{2q} := \begin{pmatrix} 0 & \mathbb{1}_q \\ -\mathbb{1}_q & 0 \end{pmatrix}. \tag{268}$$

Writing a generic supermatrix $\mathfrak{M}$ of the super Lie algebra $\mathfrak{osp}(p|2q)$ as

$$\mathfrak{M} = \left(\begin{array}{c|c} A & B \\ \hline C & D \end{array}\right) \tag{269}$$

permits (267) to be rewritten as the following conditions on the blocks of the algebra supermatrices:

$$A^t\mathbb{J} = -\mathbb{J}A, \qquad C = B^t\mathbb{J}, \qquad D^t = -D. \tag{270}$$

Depending on the value of p, the superalgebra falls into one of three basic, "classical" families

$$\mathfrak{osp}(p|2q) = \begin{cases} B(r,q) & p = 2r+1,\ r \geq 0 \\ C(q+1) & p = 2 \\ D(r,q) & p = 2r,\ r \geq 2. \end{cases} \tag{271}$$

Clearly it is the first case that will concern us, in particular, with $r = 0, q = 1$. $B(r,q)$ has rank $q+r$, dimension $2(q+r)^2+3q+r$, and even part $\mathfrak{so}(p) \oplus \mathfrak{sp}(2q)$, which for $\mathfrak{osp}(1|2)$ are 1, 5, and $\mathfrak{sl}(2)$, respectively.

The $\mathfrak{osp}(1|2)$ generators are denoted $T_{X_1X_2}$, where $X = 0, 1, \bullet$. Note, $T_{\bullet\bullet} = 0$. The supercommutators are given by

$$\begin{aligned} [T_{A_1A_2}, T_{A_3A_4}] &= 4E_{(A_1(A_3}T_{A_2)A_4)} \\ [T_{A_1A_2}, T_{A_3\bullet}] &= 2E_{(A_1|A_3}T_{A_2)\bullet} \\ \{T_{A_1\bullet}, T_{A_2\bullet}\} &= T_{A_1A_2}, \end{aligned} \tag{272}$$

which are written out in Table 14 with $T_A \equiv T_{A\bullet} \equiv T_{\bullet A}$. Explicitly the

Table 14: $\mathfrak{osp}(1|2)$ superbrackets.

	T_{01}	T_{00}	T_{11}	T_0	T_1
T_{01}	0	$-2T_{00}$	$2T_{11}$	$-T_0$	T_1
T_{00}	$2T_{00}$	0	$4T_{01}$	0	$2T_0$
T_{11}	$-2T_{11}$	$-4T_{01}$	0	$-2T_1$	0
T_0	T_0	0	$2T_1$	T_{00}	T_{01}
T_1	$-T_1$	$-2T_0$	0	T_{01}	T_{11}

generators are

$$
\begin{gathered}
T_{01} = \left(\begin{array}{cc|c} -1 & 0 & 0 \\ 0 & 1 & 0 \\ \hline 0 & 0 & 0 \end{array}\right), \\
T_{00} = \left(\begin{array}{cc|c} 0 & 2 & 0 \\ 0 & 0 & 0 \\ \hline 0 & 0 & 0 \end{array}\right), \quad T_{11} = \left(\begin{array}{cc|c} 0 & 0 & 0 \\ -2 & 0 & 0 \\ \hline 0 & 0 & 0 \end{array}\right), \\
T_{0} = \left(\begin{array}{cc|c} 0 & 0 & 1 \\ 0 & 0 & 0 \\ \hline 0 & 1 & 0 \end{array}\right), \quad T_{1} = \left(\begin{array}{cc|c} 0 & 0 & 0 \\ 0 & 0 & 1 \\ \hline -1 & 0 & 0 \end{array}\right).
\end{gathered}
\tag{273}
$$

In order to make contact with,[192] we rescale the generators into a new supermatrix P

$$P_{X_1X_2} := \tfrac{1}{2}T_{X_1X_2} \equiv E_{[\![X_1|X_3}U_{X_3|X_2]\!]} \tag{274}$$

to yield the superbrackets

$$
\begin{aligned}
[P_{A_1A_2}, P_{A_3A_4}] &= 2\varepsilon_{(A_1(A_3}P_{A_2)A_4)} \\
[P_{A_1A_2}, Q_{A_3}] &= \varepsilon_{(A_1|A_3}Q_{A_2)} \\
\{Q_{A_1}, Q_{A_2}\} &= \tfrac{1}{2}P_{A_1A_2},
\end{aligned}
\tag{275}
$$

where $Q_A \equiv P_A$, which are summarized as

$$[\![P_{X_1X_2}, P_{X_3X_4}]\!] = 2E_{[\![X_1[\![X_3}P_{X_2]\!]X_4]\!]}. \tag{276}$$

The rescaled generators have an action on the 3-dimensional representation a_X given by

$$(P_{X_1X_2}a)_{X_3} = E_{[\![X_1|X_3}a_{X_2]\!]} \tag{277}$$

which naturally generalizes to the action on the N-fold exterior tensor product $a_{X_1\cdots X_N}$,

$$(P_{X_kY_k}a)_{Z_1\cdots Z_k\cdots Z_N} = (-)^{(X_k+Y_k)\sum_{i=1}^{k-1}Z_i}E_{[\![X_k|Z_k}a_{Z_1\cdots|Y_k]\!]\cdots Z_N}. \tag{278}$$

The three even elements $P_{A_1A_2}$ form an $\mathfrak{sl}(2)$ subalgebra generating the bosonic SLOCC equivalence group, under which Q_A transforms as a spinor.

The supersymmetric generalization of the conventional group of local unitaries is given by uOSp(1|2), a compact subgroup of OSp(1|2).[203,204] It has a supermatrix representation as the subset of OSp(1|2) supermatrices satisfying the additional superunitary condition

$$M^{\ddagger}M = \mathbb{1}, \tag{279}$$

where ‡ is the superadjoint given by

$$M^\ddagger = (M^{st})^\#. \tag{280}$$

The uOSp(1|2) algebra is given by

$$\mathfrak{uosp}(1|2) := \{X \in \mathfrak{osp}(1|2) | X^\ddagger = -X\}. \tag{281}$$

An arbitrary element $X \in \mathfrak{uosp}(1|2)$ may be written as

$$X = \xi_i A_i + \eta^\# Q_0 + \eta Q_1, \tag{282}$$

where ξ_i and η are pure even/odd Grassmann numbers respectively and

$$\begin{gathered} A_1 = \tfrac{i}{2}(P_{00} - P_{11}), \qquad A_2 = \tfrac{1}{2}(P_{00} + P_{11}), \\ A_3 = iP_{01}, \\ Q_A^\ddagger = \varepsilon_{AA'} Q_{A'}, \qquad A_i^\ddagger = -A_i. \end{gathered} \tag{283}$$

11. Super Hilbert space and uOSp(1|2)

With one important difference, explained below, our definition of a super Hilbert space follows that of DeWitt.[201] We define a super Hilbert space to be a supervector space $\mathcal{H}$ equipped with an injection to its dual space $\mathcal{H}^\ddagger$,

$$\begin{aligned} ^\ddagger : \mathcal{H} &\to \mathcal{H}^\ddagger, \\ |\psi\rangle &\mapsto (|\psi\rangle)^\ddagger := \langle\psi|. \end{aligned} \tag{284}$$

A basis in which all basis vectors are pure even or odd is said to be pure. Such a basis may always be found.[201]

The map $^\ddagger : \mathcal{H} \to \mathcal{H}^\ddagger$ defines an inner product $\langle\psi|\phi\rangle$ and satisfies the following axioms:

(1) ‡ sends pure bosonic (fermionic) supervectors in $\mathcal{H}$ into bosonic (fermionic) supervectors in $\mathcal{H}^\ddagger$.

(2) ‡ is linear

$$(|\psi\rangle + |\phi\rangle)^\ddagger = \langle\psi| + \langle\phi|. \tag{285}$$

(3) For pure even/odd α and $|\psi\rangle$

$$(|\psi\rangle\alpha)^\ddagger = (-)^{\alpha\psi}\alpha^\#\langle\psi| \tag{286}$$

and

$$(\alpha\langle\psi|)^\ddagger = (-)^{\psi+\alpha\psi}|\psi\rangle\alpha^\#. \tag{287}$$

In particular,

$$|\psi\rangle^{\ddagger\ddagger} = (-)^\psi|\psi\rangle. \tag{288}$$

In a pure even/odd orthonormal basis $\{|i\rangle\}$ we adopt the following convention:

$$|\psi\rangle = |i\rangle\psi_i \tag{289}$$

so that for pure even/odd ψ (286) and (287) imply

$$\begin{aligned}(|i\rangle\psi_i)^\ddagger &= (-)^{\psi_i i}\psi_i^\# \langle i| = (-)^{i+i\psi}\psi_i^\# \langle i| \\ ((-)^{i+i\psi}\psi_i^\# \langle i|)^\ddagger &= (-)^{\psi}|i\rangle\psi_i\end{aligned} \tag{290}$$

where we have used $\deg(\psi_i) = \deg(i) + \deg(\psi)$.

For all pure even/odd $|\psi\rangle, |\phi\rangle \in \mathcal{H}$ the inner product $\langle\psi|\phi\rangle$ satisfies

$$\langle\psi|\phi\rangle^\# = (-)^{\psi+\psi\phi}\langle\phi|\psi\rangle. \tag{291}$$

Consequently,

$$\langle\psi|\phi\rangle^{\#\#} = (-)^{\psi+\phi}\langle\phi|\psi\rangle, \tag{292}$$

as would be expected of a pure even/odd Grassmann number since $\deg(\langle\phi|\psi\rangle) = \deg(\psi) + \deg(\phi)$. In a pure even/odd orthonormal basis we find

$$\langle\phi|\psi\rangle = (-)^{i+i\phi}\phi_i^\# \psi_i. \tag{293}$$

In using the superstar we depart from the formalism presented in,[201] which uses the ordinary star. The use of the superstar is necessary for the use of the super LOCC group.

11.0.1. *Physical states*

A Grassmann number has an inverse iff it has a nonvanishing body. Consequently, a state $|\psi\rangle$ is normalizable iff $\langle\psi|\psi\rangle_{\mathcal{B}} > 0$. The state may then be normalized,

$$|\hat{\psi}\rangle = N_\psi|\psi\rangle, \quad N_\psi = \langle\psi|\psi\rangle^{-1/2}, \tag{294}$$

where N_ψ is given by the general definition of an analytic function f on the space of Grassmann numbers (244). Explicitly,

$$\langle\psi|\psi\rangle^{-1/2} = \sum_{k=0}^{\infty} \frac{1}{k!2^k} \prod_{j=0}^{k} (1-2j)\langle\psi|\psi\rangle_{\mathcal{B}}^{-\frac{2k+1}{2}} \langle\psi|\psi\rangle_{\mathcal{S}}^k. \tag{295}$$

Motivated by the above considerations a state $|\psi\rangle$ is said to be *physical* iff $\langle\psi|\psi\rangle_{\mathcal{B}} > 0$. We restrict our attention to physical states throughout.

11.1. *The superqubit*

The 1-superqubit system (Alice) is described by the state

$$|\Psi\rangle = |A\rangle a_A + |\bullet\rangle a_\bullet, \tag{296}$$

where a_A is commuting with $A = 0, 1$ and $a_\bullet$ is anticommuting. That is to say, the state vector is promoted to a supervector. The super Hilbert space has dimension 3, two "bosons", and one "fermion". In more compact notation we may write

$$|\Psi\rangle = |X\rangle a_X, \tag{297}$$

where $X = (A, \bullet)$.

The super SLOCC equivalence group for a single qubit is $OSp(1|2)_A$. Under the $SL(2)_A$ subgroup a_A transforms as a **2** while $a_\bullet$ is a singlet as shown in Table 15. The super LOCC entanglement equivalence group, i.e. the group of local unitaries, is given by $uOSp(1|2)_A$, the unitary subgroup of $OSp(1|2)_A$.

Table 15: The action of the $\mathfrak{osp}(1|2)$ generators on the superqubit fields.

Generator	Field acted upon a_{A_3}	Field acted upon $a_\bullet$
$P_{A_1A_2}$	$\varepsilon_{(A_1\|A_3}a_{\|A_2)}$	0
$2Q_{A_1}$	$\varepsilon_{A_1A_3}a_\bullet$	a_{A_1}

The norm squared $\langle\Psi|\Psi\rangle$ is given by

$$\langle\Psi|\Psi\rangle = \delta^{A_1A_2}a^{\#}_{A_1}a_{A_2} - a^{\#}_\bullet a_\bullet, \tag{298}$$

where $\langle\Psi| = (|\Psi\rangle)^\ddagger$ and $\langle\Psi|\Psi\rangle$ is the conventional inner product that is manifestly $uOSp(1|2)$ invariant. The one-superqubit state may then be normalized. For one superqubit this gives

$$\begin{aligned}\langle\Psi|\Psi\rangle^{-1/2} = & \; (\delta^{A_1A_2}a^*_{A_1}a_{A_2})^{-1/2} \\ & +\tfrac{1}{2}(\delta^{A_1A_2}a^*_{A_1}a_{A_2})^{-3/2}a^{\#}_\bullet a_\bullet\end{aligned} \tag{299}$$

so the normalized wave function $|\hat{\Psi}\rangle$, for which $\langle\hat{\Psi}|\hat{\Psi}\rangle = 1$, is

$$|\hat{\Psi}\rangle = |A\rangle\hat{a}_A + |\bullet\rangle\hat{a}_\bullet \tag{300}$$

where

$$\begin{aligned}\hat{a}_A &= a_A[(\delta^{A_1A_2}a^*_{A_1}a_{A_2})^{-1/2} \\ &\quad + \tfrac{1}{2}(\delta^{A_1A_2}a^*_{A_1}a_{A_2})^{-3/2}a^\#_\bullet a_\bullet], \\ \hat{a}_\bullet &= a_\bullet(\delta^{A_1A_2}a^*_{A_1}a_{A_2})^{-1/2}.\end{aligned} \tag{301}$$

12. Super entanglement

12.1. *Two superqubits*

In seeking a supersymmetric generalization of the 2-tangle one might be tempted to replace the determinant of a_{AB} by the Berezinian of a_{XY}

$$\operatorname{Ber} a_{XY} = \det(a_{AB} - a_{A\bullet}a_{\bullet\bullet}^{-1}a_{\bullet B})a_{\bullet\bullet}^{-1}. \tag{302}$$

However, although the Berezinian is the natural supersymmetric extension of the determinant, it is not defined for vanishing $a_{\bullet\bullet}$, making it unsuitable as an entanglement measure.

A better candidate follows from writing

$$\begin{aligned}\det a_{AB} = \tfrac{1}{2}a^{AB}a_{AB} &= \tfrac{1}{2}\operatorname{tr}(a^t\varepsilon a\varepsilon^t) \\ &= \tfrac{1}{2}\operatorname{tr}[(a\varepsilon)^t\varepsilon a],\end{aligned} \tag{303}$$

This expression may be generalized by a straightforward promotion of the trace and transpose to the supertrace and supertranspose and replacing the SL(2) invariant tensor ε with the OSp(1|2) invariant tensor E. This yields a quadratic polynomial, which we refer to as the superdeterminant, denoted sdet:

$$\begin{aligned}\operatorname{sdet} a_{XY} &= \tfrac{1}{2}\operatorname{str}[(aE)^{st}Ea] \\ &= \tfrac{1}{2}(a^{AB}a_{AB} - a^{A\bullet}a_{A\bullet} - a^{\bullet B}a_{\bullet B} - a^{\bullet\bullet}a_{\bullet\bullet}) \\ &= (a_{00}a_{11} - a_{01}a_{10} + a_{0\bullet}a_{1\bullet} + a_{\bullet 0}a_{\bullet 1}) - \tfrac{1}{2}{a_{\bullet\bullet}}^2,\end{aligned} \tag{304}$$

which is clearly not equal to the Berezinian, but is nevertheless supersymmetric since Q_A annihilates $a^{AB}a_{AB} - a^{\bullet B}a_{\bullet B}$ and $a^{A\bullet}a_{A\bullet} + a^{\bullet\bullet}a_{\bullet\bullet}$, while Q_B annihilates $a^{AB}a_{AB} - a^{A\bullet}a_{A\bullet}$ and $a^{\bullet B}a_{\bullet B} + a^{\bullet\bullet}a_{\bullet\bullet}$. Satisfyingly, (304) reduces to $\det a_{AB}$ when $a_{A\bullet}$, $a_{\bullet B}$, and $a_{\bullet\bullet}$ are set to zero. We then define the super 2-tangle as

$$\tau_{XY} = 4 \operatorname{sdet} a_{XY}(\operatorname{sdet} a_{XY})^\#. \tag{305}$$

In summary, 2-superqubit entanglement seems to have the same two entanglement classes as 2-qubits with the invariant $\det a_{AB}$ replaced by its supersymmetric counterpart $\operatorname{sdet} a_{XY}$.

12.2. *Three superqubits*

In seeking to generalize the 3-tangle, invariant under $[\mathrm{SL}(2)]^3$, to a supersymmetric object, invariant under $[\mathrm{OSp}(1|2)]^3$, we need to find a quartic polynomial that reduces to Cayley's hyperdeterminant when $a_{AB\bullet}$, $a_{A\bullet C}$, $a_{\bullet BC}$, $a_{A\bullet\bullet}$, $a_{\bullet B\bullet}$, $a_{\bullet\bullet C}$, and $a_{\bullet\bullet\bullet}$ are set to zero. We do this by generalizing the γ matrices:

$$\begin{aligned}\gamma_{A_1A_2} :=\ & a_{A_1}{}^{BC}a_{A_2BC} - a_{A_1}{}^{B\bullet}a_{A_2B\bullet} \\ & - a_{A_1}{}^{\bullet C}a_{A_2\bullet C} - a_{A_1}{}^{\bullet\bullet}a_{A_2\bullet\bullet},\end{aligned} \tag{306a}$$

$$\begin{aligned}\gamma_{A_1\bullet} :=\ & a_{A_1}{}^{BC}a_{\bullet BC} + a_{A_1}{}^{B\bullet}a_{\bullet B\bullet} \\ & + a_{A_1}{}^{\bullet C}a_{\bullet\bullet C} - a_{A_1}{}^{\bullet\bullet}a_{\bullet\bullet\bullet},\end{aligned} \tag{306b}$$

$$\begin{aligned}\gamma_{\bullet A_2} :=\ & a_{\bullet}{}^{BC}a_{A_2BC} - a_{\bullet}{}^{B\bullet}a_{A_2B\bullet} \\ & - a_{\bullet}{}^{\bullet C}a_{A_2\bullet C} - a_{\bullet}{}^{\bullet\bullet}a_{A_2\bullet\bullet},\end{aligned} \tag{306c}$$

together with their B and C counterparts; notice that the building blocks with two indices are bosonic and those with one index are fermionic. The final bosonic possibility, $\gamma_{(\bullet\bullet)}$, vanishes identically. The simple supersymmetry relations are given by

$$\begin{aligned} Q_{A_1}\gamma_{A_2A_3} &= \varepsilon_{A_1(A_2}\gamma_{A_3)\bullet} \\ Q_{A_1}\gamma_{A_2\bullet} &= \tfrac{1}{2}\gamma_{A_1A_2} \\ Q_B\gamma_{A_1A_2} &= 0 = Q_C\gamma_{A_1A_2} \\ Q_B\gamma_{A\bullet} &= 0 = Q_C\gamma_{A\bullet}. \end{aligned} \tag{307}$$

Using these expressions we define the superhyperdeterminant, denoted $\mathrm{sDet}\, a$:

$$\mathrm{sDet}\, a_{XYZ} = \tfrac{1}{2}(\gamma^{A_1A_2}\gamma_{A_1A_2} - \gamma^{A\bullet}\gamma_{A\bullet} - \gamma^{\bullet A}\gamma_{\bullet A}) \tag{308}$$

which is invariant under the action of the superalgebra. The corresponding expressions singling out superqubits B and C are also invariant and equal to (308). $\mathrm{sDet}\, a_{XYZ}$ can be seen as the definition of the super-Cayley determinant of the cubic superhypermatrix a_{XYZ}.

Writing

$$\Gamma^A := \left(\begin{array}{c|c} \gamma_{A_1A_2} & \gamma_{A_1\bullet} \\ \hline \gamma_{\bullet A_2} & \gamma_{\bullet\bullet} \end{array}\right) = \left(\begin{array}{c|c} \gamma_{A_1A_2} & \gamma_{A_1\bullet} \\ \hline \gamma_{A_2\bullet} & 0 \end{array}\right), \tag{309}$$

we obtain an invariant analogous to (304)

$$\mathrm{sDet}\, a_{XYZ} = \tfrac{1}{2}\,\mathrm{str}[(\Gamma^A E)^{st} E\Gamma^A] \tag{310}$$

so that

$$\text{sDet}\, a_{XYZ} = -\,\text{sdet}\, \Gamma^A \tag{311}$$

in direct analogy to the conventional 3-qubit identity construction. This result for sDet agrees with that of.[192]

Finally, using Γ^A we are able to define the supersymmetric generalization T_{XYZ} of the 3-qubit tensor T_{ABC} as defined in (234),

$$T_{XYZ} = \Gamma^A_{XX'} a^{X'}{}_{YZ}. \tag{312}$$

It is not difficult to verify that T_{XYZ} transforms in precisely the same way as a_{XYZ} under $\mathfrak{osp}(1|2) \oplus \mathfrak{osp}(1|2) \oplus \mathfrak{osp}(1|2)$. The superhyperdeterminant may then also be written as

$$\begin{aligned} \text{sDet}\, a_{XYZ} = {} & T_{ABC}a^{ABC} + T_{\bullet BC}a^{\bullet BC} \\ & - T_{A\bullet C}a^{A\bullet C} - T_{AB\bullet}a^{AB\bullet} \\ & - T_{A\bullet\bullet}a^{A\bullet\bullet} + T_{\bullet B\bullet}a^{\bullet B\bullet} \\ & + T_{\bullet\bullet C}a^{\bullet\bullet C} - T_{\bullet\bullet\bullet}a^{\bullet\bullet\bullet}. \end{aligned} \tag{313}$$

In this sense $\text{sDet}\, a_{XYZ}$, $(\Gamma^A)_{X_1X_2}$, and T_{XYZ} are the natural supersymmetric generalizations of the hyperdeterminant, $\text{Det}\, a_{ABC}$, and the covariant tensors, $(\gamma^A)_{A_1A_2}$ and T_{ABC}, of the conventional 3-qubit treatment in section 8. Finally we are in a position to define the super 3-tangle:

$$\tau_{XYZ} = 4\sqrt{\text{sDet}\, a_{XYZ}(\text{sDet}\, a_{XYZ})^{\#}}. \tag{314}$$

In summary 3-superqubit entanglement seems to have the same five entanglement classes as that of 3-qubits shown in Table 11, with the covariants $a_{ABC}, \gamma^A, \gamma^B, \gamma^C, T_{ABC}$ and $\text{Det}\, a_{ABC}$ replaced by their supersymmetric counterparts $a_{XYZ}, \Gamma^A, \Gamma^B, \Gamma^C, T_{XYZ}$ and $\text{sDet}\, a_{ABC}$.

Completely separable nonsuperentangled states are given by product states for which $a_{ABC} = a_A b_B c_C, a_{AB\bullet} = a_A b_B c_\bullet, a_{A\bullet C} = a_A b_\bullet c_C, a_{\bullet BC} = a_\bullet b_B c_C, a_{A\bullet\bullet} = a_A b_\bullet c_\bullet, a_{\bullet B\bullet} = a_\bullet b_B c_\bullet, a_{\bullet\bullet C} = a_\bullet b_\bullet c_C, a_{\bullet\bullet\bullet} = a_\bullet b_\bullet c_\bullet$, and $\text{sDet}\, a_{XYZ}$ vanishes. This provides a nontrivial consistency check.

Acknowledgments

The work of LB is supported by an Imperial College Junior Research Fellowship. The work of MJD is supported by the STFC under rolling grant ST/G000743/1.

References

1. A. Cayley, On the theory of linear transformations, *Camb. Math. J.* **4** (1845) 193–209.
2. M. J. Duff, *Phys. Rev.* **D76**, p. 025017 (2007).
3. V. Coffman, J. Kundu and W. K. Wootters, *Phys. Rev.* **A61**, p. 052306 (2000).
4. A. Miyake and M. Wadati, *Quant. Info. Comp.* **2 (Special)**, 540 (2002).
5. P. Lévay, *Phys. Rev.* **D74**, p. 024030 (2006).
6. M. J. Duff and S. Ferrara, *Phys. Rev.* **D76**, p. 025018 (2007).
7. P. Lévay, *Phys. Rev.* **D75**, p. 024024 (2007).
8. M. J. Duff and S. Ferrara, *Phys. Rev.* **D76**, p. 124023 (2007).
9. P. Lévay, *Phys. Rev.* **D76**, p. 106011 (2007).
10. L. Borsten, D. Dahanayake, M. J. Duff, W. Rubens and H. Ebrahim, *Phys. Rev. Lett.* **100**, p. 251602 (2008).
11. L. Borsten, *Fortschr. Phys.* **56**, 842 (2008).
12. P. Lévay, M. Saniga and P. Vrana, *Phys. Rev.* **D78**, p. 124022 (2008).
13. L. Borsten, D. Dahanayake, M. J. Duff, H. Ebrahim and W. Rubens, *Phys. Rep.* **471**, 113 (2009).
14. P. Levay, M. Saniga, P. Vrana and P. Pracna, *Phys. Rev.* **D79**, p. 084036 (2009).
15. L. Borsten, D. Dahanayake, M. J. Duff and W. Rubens, *Phys. Rev.* **D81**, p. 105023 (2010).
16. P. Levay and S. Szalay, *Phys.Rev.* **D82**, p. 026002 (2010).
17. P. Levay, *Phys.Rev.* **D82**, p. 026003 (2010).
18. L. Borsten, D. Dahanayake, M. J. Duff, A. Marrani and W. Rubens, *Phys. Rev. Lett.* **105**, p. 100507 (2010).
19. L. Borsten, M. Duff, A. Marrani and W. Rubens, *Eur.Phys.J.Plus* **126**, p. 37 (2011).
20. M. Rios (2011), 19 pages, 1 figure.
21. P. Levay, *Phys.Rev.* **D84**, p. 025023 (2011).
22. P. Levay, *Phys.Rev.* **D84**, p. 125020 (2011), 32 pages Latex.
23. M. Redhead, *Incompleteness, nonlocality, and realism: a prolegomenon to the philosophy of quantum mechanics* (Oxford University Press, 1989).
24. C. J. Isham, *LECTURES ON QUANTUM THEORY: Mathematical and Structural Foundations* (Imperial College Press, London, 1995).
25. C. Isham, *Some reflections on the status of conventional quantum gravity.*, in *The future of theoretical Physics and Cosmology*, ed. Gibbons, G. W., Shellard, E. P. S., & Rankin, S. J. November 2003, pp. 384–408.
26. A. Doring and C. Isham, 'What is a Thing?': Topos Theory in the Foundations of Physics, in *New Structures for Physics*, ed. B. Coecke (Springer Berlin / Heidelberg, 2011) pp. 753–934.
27. M. Born and I. Born, *The Born-Einstein letters* (Walker and Co., London, 1971).
28. S. Kochen and E. P. Specker, *Journal of Mathematics and Mechanics* **17**, 59 (1967).
29. A. Einstein, B. Podolsky and N. Rosen, *Phys. Rev.* **47**, 777 (1935).

30. D. Bohm, *Quantum Theory* (Prentice-Hall, Englewood Cliffs, N.J., 1951).
31. J. S. Bell, On the Einstein-Podolsky-Rosen paradox, *Physics* **1** (1964) no. 3, 195.
32. A. Einstein, *Autobiographical Notes in Albert Einstein: Philosopher-Scientist P. A. Schilpp (ed.), Library of Living Philosophers* (Cambridge Unicersity Press, 1970).
33. N. D. Mermin, *Am. J. Phys.* **58**, 731 (1990).
34. N. D. Mermin, *Quantum Information Processing* **5**, 239 (2006).
35. A. Aspect, P. Grangier and G. Roger, *Phys. Rev. Lett.* **49**, 91(Jul 1982).
36. M. A. Nielsen and I. L. Chuang, *Quantum Computation and Quantum Information* (Cambridge University Press, New York, NY, USA, 2000).
37. J. Audretsch (ed.), *Entangled World: The Fascination of Quantum Information and Computation*, second edn. (Wiley VCH, 2005).
38. G. Jaeger, *Quantum Information: An Overview* (Springer, December 2006).
39. M. B. Plenio and S. Virmani, *Quant. Inf. Comp.* **7**, p. 1 (2007).
40. R. Horodecki, P. Horodecki, M. Horodecki and K. Horodecki, *Rev. Mod. Phys.* **81**, 865(Jun 2009).
41. L. Amico, R. Fazio, A. Osterloh and V. Vedral, *Rev. Mod. Phys.* **80**, 517 (2008).
42. G. Jaeger, *Entanglement, Information, and the Interpretation of Quantum Mechanics* (Springer-Verlag Berlin and Heidelberg GmbH & Co. KG, 2009).
43. A. K. Ekert, *Phys. Rev. Lett.* **67**, 661(Aug 1991).
44. J. F. Clauser, M. A. Horne, A. Shimony and R. A. Holt, *Phys. Rev. Lett.* **23**, 880(Oct 1969).
45. W. K. Wootters and W. H. Zurek, *Nature* **299**, 802 (1982).
46. P. W. Shor, *SIAM J. Sci. Statist. Comput.* **26**, p. 1484 (1997).
47. D. S. Naik, C. G. Peterson, A. G. White, A. J. Berglund and P. G. Kwiat, *Phys. Rev. Lett.* **84**, 4733(May 2000).
48. W. Tittel, J. Brendel, H. Zbinden and N. Gisin, *Phys. Rev. Lett.* **84**, 4737(May 2000).
49. C. H. Bennett, G. Brassard, C. Crépeau, R. Jozsa, A. Peres and W. K. Wootters, *Phys. Rev. Lett.* **70**, 1895(Mar 1993).
50. D. Bouwmeester, J. Pan, K. Mattle, M. Eibl, H. Weinfurter and A. Zeilinger, *Nature* **390**, 575 (1997).
51. D. M. Greenberger, M. Horne and A. Zeilinger, *Bell's Theorem, Quantum Theory and Conceptions of the Universe* (Kluwer Academic, Dordrecht, 1989).
52. R. K. Clifton, M. L. G. Redhead and J. N. Butterfield, *Foundations of Physics* **21**, 149(Feb 1990).
53. J. Pan, D. Bouwmeester, M. Daniell, H. Weinfurter and A. Zeilinger, *Nature* **403**, 515(Feb 2000).
54. C. H. Bennett, D. P. DiVincenzo, J. A. Smolin and W. K. Wootters, *Phys. Rev. A* **54**, 3824(Nov 1996).
55. C. H. Bennett, S. Popescu, D. Rohrlich, J. A. Smolin and A. V. Thapliyal, *Phys. Rev.* **A63**, p. 012307 (2000).
56. O. Ghne and G. Tth, *Physics Reports* **474**, 1 (2009).

57. V. Vedral, M. B. Plenio, M. A. Rippin and P. L. Knight, *Phys. Rev. Lett.* **78**, 2275(Mar 1997).
58. G. Vidal, *J. Mod. Opt.* **47**, p. 355 (2000).
59. S. Albeverio and S.-M. Fei, *J. Opt. B Quant. Semiclass. Opt.* **B3**, 223 (2001).
60. H. Carteret and A. Sudbery, *J. Phys.* **A33**, 4981 (2000).
61. H. Carteret, A. Higuchi and A. Sudbery, *J. Math. Phys.* **41**, 7932 (2000).
62. J. Kempe, *Phys. Rev.* **A60**, 910 (1999).
63. F. Verstraete, J. Dehaene and B. De Moor, *Phys. Rev.* **A68**, p. 012103 (2003).
64. I. S. Abascal and G. Björk, *Phys. Rev.* **A75**, p. 062317 (2007).
65. M. Grassl, M. Rotteler and T. Beth, *Phys. Rev.* **A58**, 1833 (1998).
66. P. Lévay, *J. Phys.* **A37**, 1821 (2004).
67. F. Toumazet, J. Luque and J. Thibon, *Math. Struct. Comp. Sci.* **17**, 1133 (2006).
68. W. Dür, G. Vidal and J. I. Cirac, *Phys. Rev.* **A62**, p. 062314 (2000).
69. N. Linden and S. Popescu, *Fortschr. Phys.* **46**, 567 (1998).
70. W. K. Wootters, *Phys. Rev. Lett.* **80**, 2245 (1998).
71. A. Acin, A. Andrianov, L. Costa, E. Jané, J. I. Latorre and R. Tarrach, *Phys. Rev. Lett.* **85**, 1560 (2000).
72. A. Sudbery, *J. Phys.* **A34**, 643 (2001).
73. P. Lévay, *Phys. Rev.* **A71**, p. 012334 (2005).
74. D. C. Brody, A. C. T. Gustavsson and L. P. Hughston, *J. Phys.: Conf. Ser.* **67**, p. 012044 (2007).
75. A. Acin, A. Andrianov, E. Jane and R. Tarrach, *J. Phys.* **A34**, 6725 (2001).
76. I. M. Gelfand, M. M. Kapranov and A. V. Zelevinsky, *Discriminants, Resultants and Multidimensional Determinants* (Birkhäuser, Boston, 1994).
77. P. Gibbs (2001).
78. K. V. Fernando (2006), Oxford preprint.
79. S. Lee, J. Joo and J. Kim, *Phys. Rev.* **A72**, p. 024302 (2005).
80. S. Lee, J. Joo and J. Kim, *Phys. Rev.* **A76**, p. 012311 (2007).
81. P. A. M. Dirac, *Proc. R. Soc. A* **268** (1962).
82. G. Veneziano, *Nuovo Cim.* **A57** (1968).
83. M. B. Green, J. H. Schwarz and E. Witten, *Superstring Theory vol. 2: Loop Amplitudes, Anomalies and Phenomenology*Cambridge Monographs on Mathematical Physics, Cambridge Monographs on Mathematical Physics (Cambridge University Press, Cambridge, UK, 1987). 596 p.
84. R. Haag, J. T. Lopuszanski and M. Sohnius, *Nucl. Phys.* **B88**, p. 257 (1975).
85. P. Candelas, G. T. Horowitz, A. Strominger and E. Witten, *Nucl. Phys.* **B258**, 46 (1985).
86. W. Nahm, *Nucl. Phys.* **B135**, p. 149 (1978).
87. E. Cremmer, B. Julia and J. Scherk, *Phys. Lett.* **B76**, 409 (1978).
88. M. J. Duff, (ed.) Bristol, UK: IOP (1999) 513 p.
89. E. Cremmer and B. Julia, *Nucl. Phys.* **B159**, p. 141 (1979).
90. E. Witten, *Nucl. Phys.* **B186**, p. 412 (1981).
91. M. J. Duff, B. E. W. Nilsson and C. N. Pope, *Phys. Rept.* **130**, 1 (1986).
92. J. Hughes, J. Liu and J. Polchinski, *Phys. Lett.* **B180**, p. 370 (1986).

93. E. Bergshoeff, E. Sezgin and P. K. Townsend, *Phys. Lett.* **B189**, 75 (1987).
94. M. J. Duff and K. S. Stelle, *Phys. Lett.* **B253**, 113 (1991).
95. R. Gueven, *Phys. Lett.* **B276**, 49 (1992).
96. E. Witten Lecture given at Shelter Island II Conf., Shelter Island, N.Y., 1-2 Jun 1983.
97. M. J. Duff, *Sci. Am.* **278**, 64 (1998).
98. M. J. Duff, P. S. Howe, T. Inami and K. S. Stelle, *Phys. Lett.* **B191**, p. 70 (1987).
99. M. J. Duff and J. Rahmfeld, *Phys. Lett.* **B345**, 441 (1995).
100. P. K. Townsend, *Phys. Lett.* **B350**, 184 (1995).
101. A. Sen, *Mod. Phys. Lett.* **A10**, 2081 (1995).
102. J. H. Schwarz, *Phys. Lett.* **B367**, 97 (1996).
103. P. Horava and E. Witten, *Nucl. Phys.* **B460**, 506 (1996).
104. M. J. Duff and J. X. Lu, *Nucl. Phys.* **B347**, 394 (1990).
105. M. J. Duff and R. R. Khuri, *Nucl. Phys.* **B411**, 473 (1994).
106. J. H. Schwarz and A. Sen, *Phys. Lett.* **B312**, 105 (1993).
107. C. M. Hull and P. K. Townsend, *Nucl. Phys.* **B438**, 109 (1995).
108. M. J. Duff, J. T. Liu and J. Rahmfeld, *Nucl. Phys.* **B459**, 125 (1996).
109. M. J. Duff, *Nucl. Phys.* **B442**, 47 (1995).
110. E. Witten, *Nucl. Phys.* **B443**, 85 (1995).
111. J. Polchinski, *Phys. Rev. Lett.* **75**, 4724 (1995).
112. A. Sen, *Nucl. Phys.* **B440**, 421 (1995).
113. M. J. Duff, R. R. Khuri and J. X. Lu, *Phys. Rept.* **259**, 213 (1995).
114. M. Cvetic and D. Youm, *Phys. Rev.* **D53**, 584 (1996).
115. J. Rahmfeld, *Phys. Lett.* **B372**, 198 (1996).
116. M. Cvetic and D. Youm, *Nucl. Phys.* **B472**, 249 (1996).
117. I. R. Klebanov and A. A. Tseytlin, *Nucl. Phys.* **B475**, 179 (1996).
118. J. P. Gauntlett, D. A. Kastor and J. H. Traschen, *Nucl. Phys.* **B478**, 544 (1996).
119. V. Balasubramanian and F. Larsen, *Nucl. Phys.* **B478**, 199 (1996).
120. A. Strominger and C. Vafa, *Phys. Lett.* **B379**, 99 (1996).
121. G. 't Hooft, *in Proceedings of the International School of Subnuclear Physics, Erice* **40**, 249 (2002).
122. J. D. Bekenstein, *Phys. Rev.* **D7**, 2333 (1973).
123. J. M. Bardeen, B. Carter and S. W. Hawking, *Commun. Math. Phys.* **31**, 161 (1973).
124. S. W. Hawking, *Commun. Math. Phys.* **43**, 199 (1975).
125. A. Salam and E. Sezgin (eds.), *Supergravities in diverse dimensions* (Netherlands: North-Holland, Amsterdam, 1989, 1499 p. Singapore: World Scientific, Singapore, 1989, 1499 p.
126. S. Ferrara, R. Kallosh and A. Strominger, *Phys. Rev.* **D52**, 5412 (1995).
127. A. Strominger, *Phys. Lett.* **B383**, 39 (1996).
128. S. Ferrara and R. Kallosh, *Phys. Rev.* **D54**, 1514 (1996).
129. S. Ferrara and R. Kallosh, *Phys. Rev.* **D54**, 1525 (1996).
130. S. Ferrara, G. W. Gibbons and R. Kallosh, *Nucl. Phys.* **B500**, 75 (1997).
131. M. K. Gaillard and B. Zumino, *Nucl. Phys.* **B193**, p. 221 (1981).

132. R. Kallosh and B. Kol, *Phys. Rev.* **D53**, 5344 (1996).
133. L. Andrianopoli, R. D'Auria, S. Ferrara and M. Trigiante, *Lect. Notes Phys.* **737**, 661 (2008).
134. A. Sen and C. Vafa, *Nucl. Phys.* **B455**, 165 (1995).
135. A. Gregori, C. Kounnas and P. M. Petropoulos, *Nucl. Phys.* **B553**, 108 (1999).
136. S. Bellucci, S. Ferrara, A. Marrani and A. Yeranyan, *Entropy* **10**, 507 (2008).
137. K. Behrndt, R. Kallosh, J. Rahmfeld, M. Shmakova and W. K. Wong, *Phys. Rev.* **D54**, 6293 (1996).
138. A. Ceresole, R. D'Auria, S. Ferrara and A. Van Proeyen, On electromagnetic duality in locally supersymmetric $\mathcal{N} = 2$ Yang-Mills theory, in *Proceedings of the Workshop on Physics from the Planck Scale to Electromagnetic Scale*, eds. P. Nath, T. Taylor and S. Pokorski (World Scientific, 1994).
139. R. Kallosh and A. Linde, *Phys. Rev.* **D73**, p. 104033 (2006).
140. B. A. Bernevig and H.-D. Chen, *J. Phys.* **A36**, 8325 (2003).
141. S. Hawking, J. M. Maldacena and A. Strominger, *JHEP* **05**, p. 001 (2001).
142. R. Britto-Pacumio, J. Michelson, A. Strominger and A. Volovich, Lectures on superconformal quantum mechanics and multi-black hole moduli spaces, in *Cargese 1999, Progress in string theory and M-theory*, 1999. Contributed to NATO Advanced Study Institute on Quantum Geometry, Akureyri, Iceland, 10-20 Aug 1999.
143. J. Batle, A. R. Plastino, M. Casas and A. Plastino, *Optics and Spectroscopy* **94**, 700 (2003).
144. C. M. Caves, C. A. Fuchs and C. A. Rungta, *Found. Phys. Lett.* **14**, 199 (2001).
145. M. J. Duff and J. Rahmfeld, *Nucl. Phys.* **B481**, 332 (1996).
146. R. Kallosh (2006).
147. E. G. Gimon, F. Larsen and J. Simon, *JHEP* **01**, p. 040 (2008).
148. V. Balasubramanian, How to count the states of extremal black holes in $\mathcal{N} = 8$ supergravity, in *Cargese 1997, Strings, branes and dualities*, 1997. Published in the proceedings of NATO Advanced Study Institute on Strings, Branes and Dualities, Cargese, France, 26 May - 14 Jun 1997.
149. M. Bertolini and M. Trigiante, *Nucl. Phys.* **B582**, 393 (2000).
150. D. Dahanayake (2010), PhD thesis.
151. P. Breitenlohner, D. Maison and G. W. Gibbons, *Commun. Math. Phys.* **120**, p. 295 (1988).
152. M. Gunaydin, A. Neitzke, B. Pioline and A. Waldron, *JHEP* **09**, p. 056 (2007).
153. E. Bergshoeff, W. Chemissany, A. Ploegh, M. Trigiante and T. Van Riet, *Nucl. Phys.* **B812**, 343 (2009).
154. G. Bossard, Y. Michel and B. Pioline, *JHEP* **01**, p. 038 (2010).
155. G. Bossard, H. Nicolai and K. Stelle, *JHEP* **0907**, p. 003 (2009).
156. D. H. Collingwood and W. M. McGovern, *Nilpotent orbits in semisimple Lie algebras*Van Nostrand Reinhold mathematics series, Van Nostrand Reinhold mathematics series (CRC Press, 1993).

157. F. Verstraete, J. Dehaene, B. De Moor and H. Verschelde, *Phys. Rev.* **A65**, p. 052112 (2002).
158. O. Chterental and D. Ž. Djoković, Normal forms and tensor ranks of pure states of four qubits, in *Linear Algebra Research Advances*, ed. G. D. Ling (Nova Science Publishers Inc, 2007) pp. 133–167.
159. E. Cartan, Œeuvres complètes Editions du Centre National de la Recherche Scientifique, (1984).
160. S. Ferrara and R. Kallosh, *Phys. Rev.* **D73**, p. 125005 (2006).
161. V. Balasubramanian, F. Larsen and R. G. Leigh, *Phys. Rev.* **D57**, 3509 (1998).
162. M. Günaydin, K. Koepsell and H. Nicolai, *Commun. Math. Phys.* **221**, 57 (2001).
163. S. Ferrara and J. M. Maldacena, *Class. Quant. Grav.* **15**, 749 (1998).
164. M. Cvetic and A. A. Tseytlin, *Phys. Rev.* **D53**, 5619 (1996).
165. M. Cvetic and C. M. Hull, *Nucl. Phys.* **B480**, 296 (1996).
166. M. Bertolini, P. Fre and M. Trigiante, *Class. Quant. Grav.* **16**, 2987 (1999).
167. S. Ferrara and M. Günaydin, *Int. J. Mod. Phys.* **A13**, 2075 (1998).
168. J. C. Baez, *Bull. Amer. Math. Soc.* **39**, 145 (2001).
169. N. Jacobson, *Rend. Circ. Mat. Palermo* **7**, 58 (1958).
170. A. Hurwitz, *Nachr. Ges. Wiss. Gottingen* , 309 (1898).
171. R. B. Brown, On generalized Cayley-Dickson algebras, *Pacific J. Math.* **20** (1967) no. 3, 415–422.
172. M. Günaydin and S. V. Ketov, *Nucl. Phys.* **B467**, 215 (1996).
173. R. B. Brown, *J. Reine Angew. Math.* **236**, 79 (1969).
174. M. Günaydin, G. Sierra and P. K. Townsend, *Nucl. Phys.* **B242**, p. 244 (1984).
175. M. Günaydin, G. Sierra and P. K. Townsend, *Phys. Lett.* **B133**, p. 72 (1983).
176. L. Borsten, D. Dahanayake, M. J. Duff and W. Rubens, *Phys. Rev.* **D80**, p. 026003 (2009).
177. S. Ferrara, A. Marrani and A. Yeranyan, *Phys.Lett.* **B701**, 640 (2011).
178. L. Borsten, M. Duff, S. Ferrara, A. Marrani and W. Rubens, *Phys.Rev. D (to appear)* (2011).
179. L. Borsten, M. Duff, S. Ferrara, A. Marrani and W. Rubens (2011).
180. S. Ferrara and R. Kallosh (2011).
181. S. Ferrara and A. Marrani (2011).
182. L. Borsten Ph.D. thesis, Imperial College (2010).
183. K. McCrimmon, *A Taste of Jordan Algebras* (Springer-Verlag New York Inc., New York, 2004).
184. H. Freudenthal, *Nederl. Akad. Wetensch. Proc. Ser.* **57**, 218 (1954).
185. J. R. Faulkner, A Construction of Lie Algebras from a Class of Ternary Algebras, *Trans. Amer. Math. Soc.* **155** (1971) no. 2, 397–408.
186. C. J. Ferrar, Strictly Regular Elements in Freudenthal Triple Systems, *Trans. Amer. Math. Soc.* **174** (1972) 313–331.
187. S. Krutelevich, *J. Algebra* **314**, 924 (2007).
188. S. Bellucci, S. Ferrara, M. Günaydin and A. Marrani, *Int. J. Mod. Phys.* **A21**, 5043 (2006).

189. L. Borsten, D. Dahanayake, M. Duff, S. Ferrara, A. Marrani *et al.*, *Class.Quant.Grav.* **27**, p. 185003 (2010).
190. A. Miyake, *Phys. Rev.* **A67**, p. 012108 (2003).
191. S. Bellucci, S. Ferrara and A. Marrani (2007), Contribution to the Proceedings of the XVII SIGRAV Conference, Turin, Italy, 4–7 Sep 2006.
192. L. Castellani, P. A. Grassi and L. Sommovigo, *Phys. Lett.* **B678**, 308 (2009).
193. P. B. Wiegmann, *Phys. Rev. Lett.* **60**, p. 2445(Jun 1988).
194. S. Sarkar, *J. Phys.* **A24**, 1137 (1991).
195. A. Foerster and M. Karowski, *Phys. Rev. B* **46**, 9234(Oct 1992).
196. F. H. L. Essler and V. E. Korepin (1992).
197. N. E. Mavromatos and S. Sarkar, *Phys. Rev.* **B62**, p. 3438 (2000).
198. K. Hasebe, *Phys. Rev. Lett.* **94**, p. 206802 (2005).
199. D. P. Arovas, K. Hasebe, X.-L. Qi and S.-C. Zhang, *Phys. Rev.* **B79**, p. 224404 (2009).
200. L. Castellani, P. A. Grassi and L. Sommovigo (2010).
201. B. DeWitt, *Supermanifolds*Cambridge Monographs on Mathematical Physics, Cambridge Monographs on Mathematical Physics, second edn. (Cambridge University Press, 1984).
202. M. Scheunert, W. Nahm and V. Rittenberg, *J. Math. Phys.* **18**, 155 (1977).
203. F. A. Berezin and V. N. Tolstoy, *Commun. Math. Phys.* **78**, 409 (1981).
204. L. Frappat, A. Sciarrino and P. Sorba, *Dictionary on Lie Algebras and Superalgebras* (Academic Press, 2000).
205. V. S. Varadarajan, *Supersymmetry for mathematicians: an introduction* (American Mathematical Society, 2004).
206. A. F. Schunck and C. Wainwright, *J. Math. Phys.* **46**, p. 033511 (2005).

WEAK VALUE WITH DECOHERENCE

AKIO HOSOYA

Department of Physics, Tokyo Institute of Technology,
2-12-1Oh-Okayama, Meguro, Tokyo, 152-8551, Japan
E-mail: ahosoya@th.phys.titech.ac.jp

As Aharonov *et al.*proposed there is a way to measure a physical quantity of a system only slightly disturbing the quantum state of the system. Recently many weak value experiments have been carried out. In the present talk we present a framework to describe the decoherence effect on the weak value in a parallel way to the Kraus representation of the CP-map of the density operator. As an application we give an expression for the geometric phase for a mixed state, which can be experimentally verified.

1. Introduction

The concept of measurement has been a longstanding problem in quantum mechanics. It was discussed by von Neumann[1] and further developed by Kraus,[2] Davies and Lewis,[3] Ozawa,[4] and other people (e. g., see[5,6])in a mathematically rigorous way. The positive operator valued measure (POVM) M_i gives the probability distribution $P_i = \mathrm{Tr}[\rho M_i]$ in quantum measurement with ρ being the density operator corresponding to a prepared state. Measurement changes the quantum state as a positive map from a density operator $\rho(> 0)$ to another density operator $\mathcal{E}(\rho)(> 0)$ because of the probabilistic interpretation. The map $\mathcal{E}$ is called a completely positive map (CP map) if the map remains positive when the map is trivially extended to any larger Hilbert space. That is, $(\mathcal{E} \otimes I)(\rho_{ex}) > 0$ for any state ρ_{ex} of an arbitrarily extended system when the map $\mathcal{E}$ is completely positive. Physically, this is a very reasonable requirement because there should always exist the outside of an experimental set up which is inactive during the experiment procedures.[4] It is known that the seemingly humble requirement of the complete positivity of the quantum operation $\mathcal{E}$ together with the trace preservation and the positive convexity for the density operator ρ implies an explicit representation of the physical operation in the Kraus form: $\mathcal{E}(\rho) = \sum_i E_i \rho E_i^\dagger$, where E_i is called the Kraus operator and $M_i := E_i^\dagger E_i$ is the POVM, with the property of the decomposition of unity, $\sum_i E_i^\dagger E_i = \sum_i M_i = 1$. [7] It is also shown that there always exists a measurement model with an interaction between the system and detector corresponding to the Kraus representation.[4]

However, the probability distribution is not the only thing that is experimentally accessible in quantum mechanics. The phase is also an essential ingredient, e.g., the geometric phase.[8] The general experimentally accessible quantity which contains complete information of the probability and the phase seems to be the *weak value* advocated by Aharonov and his collaborators.[9,10] They proposed a model of weakly coupled system and probe (see Sec. 3.1.) to obtain information of a physical quantity as a "weak value" almost without disturbing the state. Below, we briefly review the formal aspects of the weak value.

2. Weak Value

Weak Value: For an observable A, the weak value $\langle A\rangle_w$ is defined as

$$\langle A\rangle_w = \frac{\langle f|U(t_f,t)AU(t,t_i)|i\rangle}{\langle f|U(t_f,t_i)|i\rangle} \in \mathbb{C}, \tag{1}$$

where $|i\rangle$ and $\langle f|$ are normalized pre-selected ket and post-selected bra state vectors, respectively. Here, $U(t_2,t_1)$ is an evolution operator from the time t_1 to t_2. The weak value $\langle A\rangle_w$ actually depends on the pre- and post-selected states $|i\rangle$ and $\langle f|$ but we omit them for notational simplicity in the case that we fix them. Otherwise, we write them explicitly as ${}_f\langle A\rangle_i^w$ instead for $\langle A\rangle_w$. The denominator is assumed to be non-vanishing.

The physical intuition may be enhanced by looking at the identity for the expectation value of an observable A,

$$\begin{aligned}
&\langle i|U^\dagger(t,t_i)AU(t,t_i)|i\rangle \\
&= \sum_f |\langle f|U(t_f,t_i)|i\rangle|^2 \frac{\langle f|U(t_f,t)AU(t,t_i)|i\rangle}{\langle f|U(t_f,t_i)|i\rangle} \\
&= \sum_f p_f \cdot {}_f\langle A\rangle_i^w,
\end{aligned} \tag{2}$$

where $p_f = |\langle f|U(t_f,t_i)|i\rangle|^2$ is the probability to obtain the final state $\langle f|$ given the initial state $|i\rangle$.[11] Comparing with the standard probability theory,[12] one may interpret the weak value as a complex random variable with the probability measure p_f. The statistical average of the weak value coincides with the expectation value in quantum mechanics. Further, if an operator A is a projection operator $A = |a\rangle\langle a|$, the above identity becomes an analog of the Bayesian formula,

$$|\langle a|U(t,t_i)|i\rangle|^2 = \sum_f p_f \cdot {}_f\,\langle|a\rangle\langle a|\rangle_i^w. \tag{3}$$

The left hand side is the probability to obtain the state $|a\rangle$ given the initial state $|i\rangle$. From this, one may get some intuition by interpreting the weak value ${}_f\langle|a\rangle\langle a|\rangle_i^w$ as the (complex!) conditional probability of obtaining the result $|a\rangle$ under an initial condition $|i\rangle$ and a final condition $|f\rangle$ in the process $|i\rangle \to |a\rangle \to |f\rangle$.[13] We believe that the concept of a quantum trajectory can be formulated in the framework of weak values.[14] This interpretation of the weak values gives many possible examples of strange phenomena like a negative probability,[15] a negative kinetic energy,[16] a spin $100\hbar$ for an electron[9] and a superluminal propagation of light.[17] Of course, we should not take the strange weak values too literally but the remarkable consistency of the framework of the weak values due to Eq. (3) and a consequence of the completeness relation,

$$\sum_a \langle|a\rangle\langle a|\rangle_w = 1, \tag{4}$$

may give a useful concept to further push theoretical consideration by intuition. The most important fact is that the weak value is experimentally accessible by the weak measurement (e.g., see[18–24]) so that the intuitive argument based on the weak values can be either verified or falsified by experiments.

3. Weak Measurement with Decoherence

In this section, we would like to study the effect of environment in the course of the weak measurement[9] and see how the shift of the probe position is affected by the environment.

3.1. *Weak Measurement—Review*

First, we recapitulate the idea of the weak measurement.[9,25–31] Consider a target system and a probe defined in the Hilbert space $\mathcal{H}_s \otimes \mathcal{H}_p$. The interaction of the target system and the probe is assumed to be weak and instantaneous,

$$H_{int}(t) = g\delta(t - t_0)(A \otimes P), \tag{5}$$

where an observable A is defined in $\mathcal{H}_s$, while P is the momentum operator of the probe. The time evolution operator becomes

$$e^{-ig(A\otimes P)}. \tag{6}$$

Suppose the probe state is initially $\xi(q) \in \mathbb{R}$ in the coordinate representation with the probe position q. For the transition from the pre-selected state $|i\rangle$ to the post-selected state $|f\rangle$, the probe wave function becomes

$$\langle f|Ve^{-ig(A\otimes P)}U|i\rangle\xi(q), \tag{7}$$

which is in the weak coupling case,[32]

$$\begin{aligned}
&\langle f|V[1 - ig(A\otimes P)]U|i\rangle\xi(q) \\
&= \langle f|VU|i\rangle\xi(q) - g\langle f|VAU|i\rangle\xi'(q) \\
&\approx \langle f|VU|i\rangle\xi\left(q - g\frac{\langle f|VAU|i\rangle}{\langle f|VU|i\rangle}\right).
\end{aligned} \tag{8}$$

In the previous notation, the argument of the wave function is shifted by

$$g\frac{\langle f|VAU|i\rangle}{\langle f|VU|i\rangle} = g\langle A\rangle_w \tag{9}$$

so that the shift of the expectation value is the real part of the weak value, $g \cdot \mathrm{Re}[\langle A\rangle_w]$. The shift of the momentum distribution can be similarly calculated to give $2g \cdot Var(p) \cdot \mathrm{Im}[\langle A\rangle_w]$, where $Var(p)$ is the variance of the probe momentum before the interaction. Putting together, we can measure the weak value $\langle A\rangle_w$ by observing the shift of the expectation value of the probe both in the coordinate and momentum representations.

3.2. *Weak Measurement and Environment*

Let us consider a target system coupled with an environment and a general weak measurement for the compound of the target system and the environment. We assume that there is no interaction between the probe and the environment. The Hamiltonian for the target system and the environment is given by

$$H = H_0 \otimes I_e + H_1, \tag{10}$$

where H_0 acts on the target system $\mathcal{H}_s$ and the identity operator I_e is for the environment $\mathcal{H}_e$, while H_1 acts on $\mathcal{H}_s \otimes \mathcal{H}_e$. The evolution operators $U = U(t, t_i)$ and $V = U(t_f, t)$ can be expressed by

$$\begin{aligned}
U &= U_0 K(t_0, t_i), \\
V &= K(t_f, t_0)V_0,
\end{aligned} \tag{11}$$

where U_0 and V_0 are the evolution operators forward in time and backward in time, respectively, by the target Hamiltonian H_0. K's are the evolution operators in the interaction picture,

$$K(t_0, t_i) = \mathcal{T} e^{-i\int_{t_i}^{t_0} dt U_0^\dagger H_1 U_0},$$
$$K(t_f, t_0) = \overline{\mathcal{T}} e^{-i\int_{t_0}^{t_f} dt V_0 H_1 V_0^\dagger}, \tag{12}$$

where $\mathcal{T}$ and $\overline{\mathcal{T}}$ stand for the time-ordering and anti time-ordering products.

Let the initial and final environmental states be $|e_i\rangle$ and $|e_f\rangle$, respectively. The probe state now becomes

$$\langle f|\langle e_f|VU|e_i\rangle|i\rangle\xi\left(q - g\frac{\langle f|\langle e_f|VAU|e_i\rangle|i\rangle}{\langle f|\langle e_f|VU|e_i\rangle|i\rangle}\right). \tag{13}$$

Plugging the expressions for U and V into the above, we obtain the probe state as

$$N\xi\left(q - g\frac{\langle f|\langle e_f|K(t_f,t_0)V_0AU_0K(t_0,t_i)|e_i\rangle|i\rangle}{N}\right), \tag{14}$$

where $N = \langle f|\langle e_f|K(t_f,t_0)V_0U_0K(t_0,t_i)|e_i\rangle|i\rangle$ is the normalization factor. We define the dual quantum operation of A as

$$\mathcal{E}^*(A) := \langle e_f|K(t_f,t_0)V_0AU_0K(t_0,t_i)|e_i\rangle$$
$$= \sum_m V_0 F_m^\dagger A E_m U_0, \tag{15}$$

where

$$F_m^\dagger := V_0^\dagger\langle e_f|K(t_f,t_0)|e_m\rangle V_0,$$
$$E_m := U_0\langle e_m|K(t_0,t_i)|e_i\rangle U_0^\dagger. \tag{16}$$

Here, we have inserted the completeness relation $\sum_m |e_m\rangle\langle e_m| = 1$. The basis $|e_i\rangle$ and $|e_f\rangle$ are the initial and final environmental states, respectively.

We can compute

$$\sum_m F_m^\dagger E_m = \sum_m {}_e\langle e_f|V|e_m\rangle_e\langle e_m|U|e_i\rangle_e$$
$$= {}_e\langle e_f|VU|e_i\rangle_e. \tag{17}$$

The above equality (17) may be interpreted as a "decomposition of the history" in analogy to the decomposition of unity because

$${}_e\langle e_f|VU|e_i\rangle_e = {}_e\langle e_f|S|e_i\rangle_e = S_{fi} \tag{18}$$

is the S-matrix element.

We obtain the wave function of the probe as

$$\xi\left(q - g\frac{\langle f|\mathcal{E}^*(A)|i\rangle}{N}\right) = \xi\left(q - g\frac{\sum_m \langle f|V_0 F_m^\dagger A E_m U_0|i\rangle}{\sum_m \langle f|V_0 F_m^\dagger E_m U_0|i\rangle}\right)$$
$$= \xi\left(q - g\frac{\mathrm{Tr}\left[A\sum_m E_m U_0|i\rangle\langle f|V_0 F_m^\dagger\right]}{\mathrm{Tr}\left[\sum_m E_m U_0|i\rangle\langle f|V_0 F_m^\dagger\right]}\right) \tag{19}$$

with $N = \langle f|\mathcal{E}^*(I)|i\rangle$ up to the overall normalization factor. To simplify the exposition we introduce the "weak operator" W as

$$W := U_0|i\rangle\langle f|V_0 \tag{20}$$

and its quantum operation $\mathcal{E}(W)$ as

$$\mathcal{E}(W) := \sum_m E_m W F_m^\dagger. \tag{21}$$

This together with (17) and (18) can be viewed as an extension of the Kraus representation for the density operator to the one for the weak operator. It is shown in [33] that this result can also be formally obtained assuming the complete positivity.[34]

Then the right hand side of (19) becomes[a]

$$\xi\left(q - g\frac{\mathrm{Tr}[\mathcal{E}(W)A]}{\mathrm{Tr}[\mathcal{E}(W)]}\right) = \xi(q - g\langle A\rangle_{\mathcal{E}(W)}).^1 \tag{22}$$

This is the main result of this talk. The shift of the expectation value of the position operator on the probe is

$$\delta q = g \cdot \mathrm{Re}[\langle A\rangle_{\mathcal{E}(W)}]. \tag{23}$$

From an analogous discussion, we obtain the shift of the expectation value of the momentum operator on the probe as

$$\delta p = 2g \cdot Var(p) \cdot \mathrm{Im}[\langle A\rangle_{\mathcal{E}(W)}]. \tag{24}$$

Thus, we have shown that the probe shift in the weak measurement is exactly given by the weak value defined by the quantum operation of the weak operator due to the environment.

[a]Note that if there were no decoherence, i.e., $\mathcal{E}(W) = W$, the result would reduce to the previous one (8).

4. Geometric Phase

We present a simple application of our framework of the physical operation of the weak operators to the geometric phase. There have been many works on the "geometric phase of mixed states" but the very definition seems under controversy.[35] We would like to start with the geometric phase γ in a pure state[35] which is well-defined and can be expressed in terms of the weak value,

$$\gamma := \arg\left[\langle a|\tilde{i}\rangle\langle\tilde{i}|\tilde{f}\rangle\langle\tilde{f}|a\rangle\right] = \arg\left[\frac{\langle a|\tilde{i}\rangle\langle\tilde{f}|a\rangle}{\langle\tilde{f}|\tilde{i}\rangle}\right] = \arg\left[\frac{\mathrm{Tr}(WP)}{\mathrm{Tr}(W)}\right], \qquad (25)$$

where $P = |P\rangle\langle P|$ is a projector to a pure state and $W = |\tilde{i}\rangle\langle\tilde{f}|$. Here, we simplify the notations $|\tilde{i}\rangle := U(t, t_i)|i\rangle$ and $\langle\tilde{f}| := \langle f|U(t_f, t)$ only in this section. The geometric phase γ corresponds to the quantum path $|\tilde{i}\rangle \to |P\rangle \to |\tilde{f}\rangle$. By the physical operation $\mathcal{E}$, the weak operator and the density operator are mapped to $\mathcal{E}(W)$ and $\mathcal{E}(\rho)$, respectively. The new state $\mathcal{E}(\rho)$ is in general a mixed state. The new geometric phase γ_g is correspondingly given by

$$\gamma_g = \arg\left[\frac{\mathrm{Tr}(\mathcal{E}(W)P)}{\mathrm{Tr}\,\mathcal{E}(W)}\right], \qquad (26)$$

which might be called the geometric phase of the mixed state $\mathcal{E}(\rho)$ by a slight abuse of words. This operational definition fits well to experimental situation. That is, an experimentalist starts with a pure state $|\tilde{i}\rangle$ and then makes a trip $|\tilde{i}\rangle \to |P\rangle \to |\tilde{f}\rangle$ by manipulating the external field. If there were no decoherence during that process, one would get the geometric phase defined above (25). Otherwise, one would instead get the value (26) for the geometric phase, while one can presume that the state is $\mathcal{E}(\rho) = \mathcal{E}(WW^\dagger)$. Furthermore, we would like to point out that this definition (26) coincides with the formal definition in the Uhlmann approach.[36] In the generalized Kraus representation, the geometric phase for that path, $|\tilde{i}\rangle \to |P\rangle \to |\tilde{f}\rangle$ can be written as

$$\gamma_g = \arg\left[\frac{\sum_m \langle\tilde{f}|F_m^\dagger|P\rangle\langle P|E_m|\tilde{i}\rangle}{\sum_m \langle\tilde{f}|F_m^\dagger E_m|\tilde{i}\rangle}\right]. \qquad (27)$$

5. Summary

The weak measurement proposed by Aharonov *et al.* is a device to obtain the information of a quantum system as the weak value so that the quantum state of the system is only weakly disturbed. In the present work we have developed the framework to study the effect of decoherence to the weak value. The result is parallel to the Kraus representation for the density operator.

This talk is based on the collaboration with Yutaka Shikano.[33]

Bibliography

1. J. von Neumann, *Mathematische Grundlagen der Quantumechanik* (Springer, Berlin, 1932), [*Mathematical foundations of quantum mechanics* (Princeton University Press, Princeton, 1955).]
2. K. Kraus, Ann. Phys. **64**, 311-335 (1971).
3. E. B. Davies and J. T. Lewis, Commun. Math. Phys. **17**, 239 (1970).
4. M. Ozawa, J. Math. Phys. **25**, 79-87 (1984).
5. P. Busch, P. Mittelstaedt, and P. J. Lahti, *Quantum Theory of Measurement* (Springer-Verlag, Berlin, 1991).
6. V. B. Braginsky and F. Y. Khalili, *Quantum Measurement* (Cambridge University Press, Cambridge, 1992).
7. M. A. Nielsen and I. L. Chuang, *Quantum Computation and Quantum Information* (Cambridge University Press, Cambridge, 2000).
8. A. Shapere and F. Wilczek, *Geometric Phases in Physics* (World Scientific, Singapore, 1988).
9. Y. Aharonov, D. Z. Albert, and L. Vaidman, Phys. Rev. Lett. **60**, 1351 (1988).
10. Y. Aharonov and D. Rohrlich, *Quantum Paradoxes* (Wiley-VCH, Weibheim, 2005).
11. Y. Aharonov and A. Botero, Phys. Rev. A **72**, 052111 (2005).
12. In the standard probability theory,[37] the expectation value of an observable A is given as a probabilistic average, $\langle A \rangle = \int dp h_A(p)$, with $h_A(p)$ and dp being the characteristic function associated with A and probability measure, which is independent of A. The standard expression of the expectation value, $\langle A \rangle = \sum_n a_n |\langle a_n | i \rangle|^2$, is given by the Born rule. However, this does not fit the standard probability theory because the probability measure depends on A.
13. The interpretation of the weak value as a complex probability is suggested in the literature.[38]
14. H. Wiseman, Phys. Rev. A **65**, 032111 (2002).
15. The concept of complex probability in quantum mechanics is not new, e. g., see.[39,40]
16. Y. Aharonov, S. Popescu, D. Rohrlich, and L. Vaidman, Phys. Rev. A **48**, 4084 (1993).
17. D. Rohrlich and Y. Aharonov, Phys. Rev. A **66**, 042102 (2002).
18. N. W. M. Ritchie, J. G. Story, and R. G. Hulet, Phys. Rev. Lett. **66**, 1107 (1991).
19. K. J. Resch, J. S. Lundeen, and A. M. Steinberg, Phys. Lett. A **324**, 125 (2004).
20. K. J. Resch, Science **319**, 733 (2008).

21. A. D. Parks, D. W. Cullin, and D. C. Stoudt, Proc. R. Soc. Lond. A **454**, 2997 (1998).
22. O. Hosten and P. Kwiat, Science **319**, 787 (2008).
23. G. L. Pryde, J. L. O'Brien, A. G. White, T. C. Ralph, and H. M. Wiseman, Phys. Rev. Lett. **94**, 220405 (2005).
24. K. Yokota, T. Yamamoto, M. Koashi, and N. Imoto, arXiv:0811.1625.
25. Y. Aharonov and L. Vaidman, Phys. Rev. A **41**, 11 (1990).
26. Y. Aharonov, P. G. Bergmann, and J. L. Lebowitz, Phys. Rev. **134**, B1410 (1964).
27. Y. Aharonov and L. Vaidman, J. Phys. A **24**, 2315 (1991).
28. Y. Aharonov and L. Vaidman, in *Time in Quantum Mechanics*, Vol. 1, edited by J. G. Muga, R. Sala Mayato, and I. L. Egusquiza (Springer, Berlin Heidelberg, 2008) pp. 399-447.
29. Y. Aharonov and J. Tollaksen, in *Visions of Discovery: New Light on Physics Cosmology and Consciousness*, edited by R. Y. Chiao, M. L. Cohen, A. J. Leggett, W. D. Phillips, and C. L. Harper, Jr. (Cambridge University Press, Cambridge, in press), arXiv:0706.1232.
30. B. Reznik and Y. Aharonov, Phys. Rev. A **52** 2358 (1995).
31. R. Jozsa, Phys. Rev. A **76**, 044103 (2007).
32. $\xi\left(q - g\frac{\langle f|VAU|i\rangle}{\langle f|VU|i\rangle}\right)$ stands for $\xi(q)|_{q \to q - g\frac{\langle f|VAU|i\rangle}{\langle f|VU|i\rangle}}$.
33. Y.Shikano and A. Hosoya, "Weak Valus with Decoherence", arxiv 0812.4502
34. Of course, if we introduce a probe Hilbert space and then consider the probability distribution of the measurement outcome, we can extract information of the phase by an interference pattern. The virtue of the CP map is that it is defined solely by operations of the target Hilbert space. We would like to find out a representation analogous to the Kraus representation for the phase related object in a way defined only in the target system.
35. E. Sjöqvist, Phys. Lett. A **359**, 187 (2006); Acta Phys. Hung. B **26**, 195 (2006).
36. A. Uhlmann, Rep. Math. Phys. **24**, 229 (1986).
37. K. Itô, *An Introduction to Probability Theory* (Cambridge University Press, Cambridge, 1978).
38. G. Mitchison, R. Jozsa, and S. Popescue, Phys. Rev. A **76**, 060302(R) (2007).
39. R. P. Feynman, Int. J. of Theor. Phys. **21**, 467 (1982).
40. H. F. Hofmann, arXiv:0805.0029.

LECTURES ON MATRIX PRODUCT REPRESENTATION OF STATES*

VAHID KARIMIPOUR†

Department of Physics, Sharif University of Technology,
P.O. Box 11365-9161, Tehran, Iran
E-mail: vahid@sharif.edu

M. ASOUDEH

Department of Physics, Islamic Azad University,
Northern Branch, Tehran, Iran
E-mail: m_asoudeh@sbu.ac.ir

We present in a pedagogical way, the basics of matrix product representation of states, both for the ground states of quantum spin chains and for the steady states of classical one-dimensional non-equilibrium systems. While there are conceptual similarities between the two forms of representations, the mathematical and physical frameworks are quite different. The two subjects comprise the two parts of these lecture notes and in each part, after explaining the method, we provide illustrative examples.

1. Introduction

In these lectures we want to study two kinds of states which are of interest in studying one-dimensional many-body systems. These kinds of states are

i) ground states of quantum many body systems or quantum spin chains and

ii) steady states of classical non-equilibrium processes on in one dimensions.

These two areas which have formal similarities are rather different, in terms of basic concepts, motivations and objectives. In quantum spin chains, the ground state of a given Hamiltonian, encompasses the properties of the system at zero or very low temperatures. On the other hand, in stochastic systems, when formulated

*Based on the lectures given by the first author in the Summer School, "Decoherence, Entanglement and Entropy", August 2009. Oxford Kobe Institute (Kobe, Japan).

†Corresponding author.

in a proper language, a state very akin to the ground state of a (generally non-Hermitian) Hamiltonian, describes the steady state of the stochastic process, that is, the state that the system relaxes into after a very long time. It is important to note that the second problem is entirely classical, that is, it describes the probabilistic processes (i.e. diffusion and reaction) between classical particles with no quantum behavior involved, while the former is indeed a quantum mechanical process, i.e. a quantum spin chain.

While a plethora of techniques have been accumulated over the decades for attacking these problems, here we want to concentrate on a rather recent formalism, which is called the Matrix Product Formalism. The states thus constructed are called Matrix Product States, or MPS for short, and have the interesting property which allows us to find exact solutions for both of the above problems, in a rather wide range of cases. This of course does not mean that we can find exact solutions for free, but as expected and in accord with "The principle of conservation of difficulty", the difficulty of finding the exact solution is transferred to another corner, hopefully a corner where we are in more control and command.

I will divide the lectures into two parts: in the first part, I discuss the MPS for quantum spin chains and in the second part, I discuss the stochastic processes in one dimension. In each part I try to explain in detail the very basic formalism and then review a few illustrative examples. These lectures are meant to be as pedagogical as possible, so I will pay attention to details of the concepts and calculations, rather than introducing an exhaustive list of references. Instead I quote a few papers and review articles which may guide the reader through the vast literature on the subject.

Part I: Matrix Product States in Quantum Spin Chains

2. Quantum Spin Chains

In quantum many body systems, one is given a Hamiltonian describing the interactions of an array of atoms, ions, or spins and the objective is to find the ground state or the low-lying excited states of such a Hamiltonian. All the physical properties and characteristics of such a system, like the magnetization in the case of a spin system, its susceptibility, its correlations, and so on are determined once we have a good knowledge of the ground state. Of particular interest is the phenomena of quantum phase transition,[1] in which a slight change in the coupling constants of the Hamiltonian, may result in discontinuous changes in the properties of the ground state and hence in observable macroscopic properties of the system.

To find if a system allows a quantum phase transition or not and if it does, how such a transition occurs, is a highly nontrivial problem and usually requires the determination of the ground state for arbitrary sizes of the system and for arbitrary values of the coupling constants. A simple example in this category is provided by the Ising model in transverse field, with the Hamiltonian given by

$$H = -J \sum_{k=1}^{N} \sigma_{z,k}\sigma_{z,k+1} + h\sigma_{x,k}, \tag{1}$$

where $J > 0$ and h denotes the magnetic field.

To see why a quantum phase transition should occur in this system, we can look at two extreme regimes, namely when $h = 0$, and when $h \longrightarrow +\infty$. In the case $h = 0$, the system is an Ising chain with doubly degenerate ground states, where all spins are aligned in the $+z$ or $-z$ directions,

$$|\phi_+\rangle = |z,+\rangle|z,+\rangle \cdots |z,+\rangle, \qquad |\phi_-\rangle = |z,-\rangle|z,-\rangle \cdots |z,-\rangle, \tag{2}$$

while in the case $h \longrightarrow +\infty$, the Hamiltonian describes a set of non-interacting spins in a magnetic field (in the x direction) whose unique ground state is one in which all spins are aligned in the negative x direction

$$|\psi\rangle = |x,-\rangle|x,-\rangle \cdots |x,-\rangle. \tag{3}$$

When we gradually increase h from zero to very large values, the number of ground states, changes from two to one, but this change in a discrete quantity cannot be continuous, it should happen sharply at a particular point. That point is where phase transition occurs. Detailed calculation shows that this transition happens at $h = J$.

There are many other examples of quantum spin systems and quantum phase transitions and all of them have one feature in common: drastic changes of the ground state (in the thermodynamic limit) as a result of slight changes of the coupling constants of the Hamiltonian. Exactly what drastic change depends on the model and the type of transition. For example we know that in classical phase transitions, thermal fluctuations become correlated in all length scales and hence the correlation length diverges at the critical point. Whether or not quantum correlations become long-ranged also at quantum phase transition, is a subject of current study.

3. Ground States of Spin Chains as Matrix Product States

The problem of determining the ground state of a quantum spin chain , is an important problem in condensed matter and mathematical physics. The exponential rise in the dimension of Hilbert space of such a system, as the number of particles increase, turns this problem into a computationally formidable one, beyond the capability of any classical computer. While the original problem is quite difficult, we may be able to solve the converse problem much more easily. The main theme is that one starts with a state with prescribed symmetries and properties, and then construct the family of Hamiltonians for which this state is an exact ground state. It is obvious that for any given state ψ, the equation $H|\psi\rangle = 0$ has always many solutions for the unknown H, since the number of equations is much less than the number of unknowns. In fact such an equation can always be solved by taking H to be the sum of projectors onto subspaces orthogonal to $|\psi\rangle$, that is by taking

$$H = \sum_i g_i |\psi_i\rangle\langle\psi_i|, \tag{4}$$

where each of the states $|\psi_i\rangle$ is orthogonal to $|\psi\rangle$ and $g_i > 0$ for all i. However the problem becomes interesting and quite non-trivial when we put physical constraints on the Hamiltonian. That is we demand that

i) H be be a sum of local terms, i.e. $H = \sum_k h_{k,k+l}$, where $h_{k,k+l}$ acts on a block of $l+1$ spins, and

ii) that both the state and the Hamiltonian have some reasonable physical symmetries, like parity, spin-flip, and at times rotational symmetries.

It is for these two conditions that the simple solution (142) should be dismissed as a naive and non-physical solution. As we will see the MPS provides a beautiful framework for modifying the simple solution (4) to a physically interesting solution.

3.1. *Construction of the MPS*

Let us review the basics of matrix product states.[2–4] Consider a homogeneous lattice of N sites, where each site describes a $d-$ level state. The Hilbert space of each site is spanned by the basis vectors $|i\rangle, i = 0, \cdots d-1$. A state

$$|\Psi\rangle = \sum_{i_1,i_2,\cdots i_N} \psi_{i_1 i_2 \cdots i_N} |i_1, i_2, \cdots, i_N\rangle \tag{5}$$

is called a matrix product state if there exists D dimensional complex matrices such that

$$\psi_{i_1,i_2,\cdots i_N} = \frac{1}{\sqrt{Z}} tr(A_{i_1} A_{i_2} \cdots A_{i_N}), \tag{6}$$

where Z is a normalization constant. Here we have assumed periodic boundary conditions, hence the trace. In case of open boundary conditions, we should replace (6) with

$$\psi_{i_1,i_2,\cdots i_N} = \frac{1}{\sqrt{Z}} \langle W|A_{i_1} A_{i_2} \cdots A_{i_N}|V\rangle, \tag{7}$$

where $|V\rangle$ and $\langle W|$ are vectors to be determined. For concreteness we continue our discussion for the case of period boundary conditions.

First we note that Z is given by

$$Z = tr(E^N), \tag{8}$$

where

$$E := \sum_{i=0}^{d-1} A_i^* \otimes A_i. \tag{9}$$

We verify the above formula for the simple case when N = 2, which shows the basic idea of the proof. It also contains the basic idea of many other calculations of the correlation functions in the sequel. We have

$$\begin{aligned} 1 = Z^{-1}\langle\psi|\psi\rangle &= \sum_{i,j} \psi_{i,j}^* \psi_{i,j} \\ &= Z^{-1} \sum_{i,j} tr(A_i^* A_j^*) tr(A_i A_j) \\ &= Z^{-1} \sum_{i,j} tr((A_i^* \otimes A_i)(A_j^* \otimes A_j)) \\ &= Z^{-1} tr(E^2). \end{aligned} \tag{10}$$

In writing (6) and by taking the matrices to be site independent, we are restricting ourselves to translationally invariant states. The most general matrix product state should be written as

$$\psi_{i_1,i_2,\cdots i_N} = \frac{1}{\sqrt{Z}} tr(A_{i_1}^{[1]} A_{i_2}^{[2]} \cdots A_{i_N}^{[N]}), \tag{11}$$

where the matrices $A_i^{[l]}$ pertain to the site l. In such a case the normalization will be

$$Z = tr(E^{[1]}E^{[2]}\cdots E^{[N]}). \tag{12}$$

It is worth mentioning that it has been shown that any arbitrary state, can be MPS-represented, provided that we allow site-dependent matrices of arbitrary dimensions. The matrices need not even be square matrices and the only requirement is that their dimension is compatible for multiplication.

Let us now go back to our homogeneous system. By defining the vector valued matrix

$$\mathcal{A} := \sum_i A_i|i\rangle, \tag{13}$$

one can write the MPS in a more concise way as

$$|\psi\rangle = tr(\mathcal{A}\otimes\mathcal{A}\otimes\cdots\mathcal{A}), \tag{14}$$

where we use the convention

$$tr(\mathcal{A}\otimes\mathcal{A}) = tr(A_iA_j)|i\rangle\otimes|j\rangle. \tag{15}$$

It is important to note that the MPS representation (6) is not unique and a transformation such as $A_i \longrightarrow A'_i = \mu U A_i U^{-1}$ where U is an invertible matrix, and μ is a constant, leaves the state invariant.

Exercise 1: Let $A_0 = \begin{pmatrix} 1 & 0 \\ g & 0 \end{pmatrix}$ and $A_1 = \begin{pmatrix} 1 & g \\ 0 & 0 \end{pmatrix}$. Explicitly determine the normalized MPS state for a chain of length $N = 3$.

3.2. *Correlation Functions*

The simple form of a matrix product state allows us to easily calculate any correlation function. In fact the complexity of this calculation reduces considerably from an exponential one to a polynomial one. First consider a one-point function. Let O be a local observable with corresponding operator $\hat{O}$ (i.e. spin, density, etc) for which we want to find the average $\langle O\rangle_k := \langle\psi|\hat{O}_k|\psi\rangle$. Here $\hat{O}_k$ is the embedding of the operator $\hat{O}$ in site k, that is

$$\hat{O}_k := I^{\otimes(k-1)}\otimes O\otimes I^{\otimes(N-k)}, \tag{16}$$

We will then have

$$\begin{aligned}\langle O\rangle_k &= \langle\psi|I^{\otimes(k-1)}\otimes O\otimes I^{\otimes(N-k)}|\psi\rangle \\ &= \sum \psi^*_{i_1,\cdots i_{k-1},j,i_{k+1},\cdots i_N} O_{jl}\psi_{i_1,\cdots i_{k-1},l,i_{k+1},\cdots i_N}.\end{aligned} \tag{17}$$

If we now use the matrix product representation and proceed in a way similar to what we did in Eq. (10) for dealing with traces, we find that the one-point function finds the very simple form

$$\langle\psi|\hat{O}_k|\psi\rangle = \frac{tr(E^{k-1}E_O E^{N-k})}{tr(E^N)}, \tag{18}$$

where

$$E_O := \sum_{i,j=0}^{d-1} \langle i|O|j\rangle A_i^* \otimes A_j. \tag{19}$$

In the thermodynamic limit $N \longrightarrow \infty$, equation (18) gives

$$\langle\psi|\hat{O}_k|\psi\rangle = \frac{\langle\lambda_{max}|E_O|\lambda_{max}\rangle}{\lambda_{max}}, \tag{20}$$

where we have used the translation invariance of the model and λ_{max} is the eigenvalue of E with the largest absolute value and $|\lambda_{max}\rangle$ and $\langle\lambda_{max}|$ are the right and left eigenvectors corresponding to this eigenvalue, normalized such that $\langle\lambda_{max}|\lambda_{max}\rangle = 1$. Here we are assuming that the largest eigenvalue of E is non-degenerate. In case λ_{max} is degenerate with degree equal to g, then equation (20)) will be modified to

$$\langle\psi|\hat{O}_k|\psi\rangle = \frac{\sum_{i=1}^{g}\langle\lambda_{max,i}|E_O|\lambda_{max,i}\rangle}{\lambda_{max}}. \tag{21}$$

The n-point functions can be obtained in a similar way. For example, the two-point function $\langle\psi|O(k)O(l)|\psi\rangle$ can be obtained as

$$\langle\psi|O(k)O(l)|\psi\rangle = \frac{tr(E_O(k)E_O(l)E^N)}{tr(E^N)}, \tag{22}$$

where $E_O(k) := E^{k-1}E_O E^{-k}$. Note that this is a formal notation which allows us to write the n-point functions in a uniform way, it does not require that E is an invertible matrix. In the thermodynamic limit the two point function turns out to be

$$\langle\psi|O(1)O(r)|\psi\rangle = \frac{1}{\lambda_{max}^r}\sum_i \lambda_i^{r-2}\langle\lambda_{max}|E_O|\lambda_i\rangle\langle\lambda_i|E_O|\lambda_{max}\rangle. \tag{23}$$

For large distances $r >> 1$, this formula reduces to

$$\langle\psi|O(1)O(r)|\psi\rangle - \langle\psi|O|\psi\rangle^2 = \frac{\lambda_1^{r-2}}{\lambda_{max}^r}\langle\lambda_{max}|E_O|\lambda_1\rangle\langle\lambda_1|E_O|\lambda_{max}\rangle, \tag{24}$$

where λ_1 is the second largest eigenvalue of E (in absolute value) for which the matrix element $\langle\lambda_1|E_O|\lambda_{max}\rangle$ is non-vanishing and we have assumed that the eigenvectors of E have been normalized, i.e. $\langle\lambda_i|\lambda_j\rangle = \delta_{i,j}$. Writing the correlation (24) in the form

$$\langle\psi|O(1)O(r)|\psi\rangle - \langle\psi|O|\psi\rangle^2 \sim e^{-\frac{r}{\xi}},$$

where by $\sim$ we mean that the form of the pre-factor is not of interest to us, we obtain the correlation length ξ as

$$\xi = \frac{1}{ln\frac{\lambda_{max}}{\lambda_1}}. \tag{25}$$

Any level crossing in the largest eigenvalue of the matrix E signals a possible quantum phase transition. Also, due to (25), any level crossing in the second largest eigenvalue of E implies the correlation length of the system has undergone a discontinuous change. Here we are using a broader definition of quantum phase transition, that is we call any non-analytical behavior of a macroscopic property, a quantum phase transition.[4]

Exercise 2: For the MPS in exercise 1, determine the one point functions $\langle\sigma_x\rangle$, $\langle\sigma_y\rangle$ and $\langle\sigma_z\rangle$ for a chain of arbitrary length N. Compare your results for what you find from explicit calculations for $N = 3$.

3.3. *Symmetries of the Ground State*

It is readily verified from the basic definition of MPS, that if $A_i = A_i^T$, where T stands for the transpose, then

$$\psi_{i_1,i_2,\ldots i_N} = \psi_{i_N,i_{N-1},\ldots i_1}, \tag{26}$$

so that the state has left-right symmetry and is invariant under parity. In fact it is not necessary that the matrices A_i be symmetric, but they should be conjugate to their transpose, that is they should satisfy $A_i^T = \Pi^{-1}A_i\Pi$. In this case, if a term like

$$|000...101\rangle \tag{27}$$

exists in the expansion, a state of the form

$$|101...000\rangle \tag{28}$$

will also be in the expansion with the same amplitude. Also, if there is a matrix X, such that $X^{-1}A_iX = A_{\bar{i}}$, where $\bar{i}$, means $1 - i$(or in general $d - i$), then the state will be invariant under spin flip, and we will have

$$\psi_{i_1,i_2,\ldots i_N} = \psi_{\bar{i}_1\bar{i}_2,\ldots \bar{i}_{N-1}}. \tag{29}$$

With such matrices,if a state of the form

$$|000...101\rangle, \tag{30}$$

exists in the expansion, a state of the form

$$|111...010\rangle, \tag{31}$$

will also be in the expansion with the same amplitude. More generally, consider a local symmetry operator R acting on a site as $R|i\rangle = R_{ij}|j\rangle$ where summation convention is being used. R is a $d-$dimensional unitary representation of the symmetry. A global symmetry operator $\mathcal{R} := R^{\otimes N}$ will then change this state to another matrix product state

$$\psi_{i_1 i_2 ... i_N} \rightarrow \psi'_{i_1 i_2 ... i_N} := tr(A'_{i_1} A'_{i_2} ... A'_{i_N}), \tag{32}$$

where

$$A'_i := R_{ij} A_j. \tag{33}$$

The state $|\psi$ is invariant (i.e. a singlet) under this symmetry if there exist an operator $U(R)$ such that

$$R_{ij} A_j = U^{-1}(R) A_i U(R). \tag{34}$$

Repeating this transformation leads to the relation

$$U_{R'} U_R = U_{R'R}. \tag{35}$$

Thus $U(R)$ is a $D-$ dimensional representation of the symmetry R. In case that R is a continuous symmetry with generators T_a, i.e. if

$$R = I + \theta_a T_a + O(\theta^2), \tag{36}$$

then

$$U(R) = I + \theta_a \mathcal{T}_a + O(\theta^2), \tag{37}$$

where θ_a are parameters of the transformations. Equation (142), leads to

$$(T_a)_{ij} A_j = [\mathcal{T}_a, A_i], \tag{38}$$

where T_a and $\mathcal{T}_a$ are the d and D dimensional representations of the Lie algebra of the symmetry. We will see examples of symmetry in the sequel.

Exercise 3: Determine the most general 2×2 matrices which can represent an MPS state of spin $1/2$ chain, allowing spin-flip symmetry and /or left-right inversion symmetry.

Exercise 4: Determine the most general 2×2 matrices which can represent an MPS state of spin $1/2$ chain, allowing rotation symmetry only around the $z-$ axis. Can you extend this symmetry to full rotational symmetry with these dimensions?

4. The Parent Hamiltonian

Now we can ask if there is a Hamiltonian or family of Hamiltonians for which the given matrix product state is a ground state. In fact we can construct these Hamiltonians in a straightforward way. The first step is to calculate the reduced density matrices corresponding to this state, a calculation which is very simple given the sample calculation (10). For the density matrix of site k, we find

$$\begin{aligned}\rho^{(k)}_{i,j} &= \sum_{l_1,\ldots l_N} \psi_{l_1,\ldots l_{k-1},i,l_{k+1},\ldots l_N} {\psi^*}_{l_1,\ldots l_{k-1},j,l_{k+1},\ldots l_N} \\ &= \frac{1}{Z_N}\sum_{l_1,\ldots l_N} tr(A_{l_1}\ldots A_{l_{k-1}} A_i A_{l_{k+1}}\ldots A_{l_N}) tr({A^*}_{l_1}\ldots {A^*}_{l_{k-1}} {A^*}_i {A^*}_{l_{k+1}}\ldots {A^*}_{l_N}) \\ &= \frac{1}{tr(E^N)} tr(E^{k-1}(A_i \otimes {A^*}_j)E^{N-k}) = \frac{tr(E^{N-1}(A_i \otimes {A^*}_j))}{tr(E^N)}. \qquad (39)\end{aligned}$$

In a similar way the reduced density matrix of two consecutive sites turns out to be

$$\rho_{ij;kl} = \frac{tr((A_iA_j \otimes {A^*}_k{A^*}_l)E^{N-k})}{tr(E^N)}. \qquad (40)$$

With similar relations for the density matrices of k consecutive sites. Consider the density matrix of k consecutive sites. The null space of this matrix includes the solutions of the following system of equations

$$\sum_{j_1,\ldots,j_k=0}^{d-1} c_{j_1\cdots j_k} A_{j_1}\ldots A_{j_k} = 0. \qquad (41)$$

Given that the matrices A_i are of size $D \times D$, there are D^2 equations with d^k unknowns. Since there can be at most D^2 independent equations, there are at least $d^k - D^2$ solutions for this system of equations. Thus for the density matrix of k sites to have null space it is sufficient that the following inequality holds

$$d^k > D^2. \qquad (42)$$

Let the null space of the reduced density matrix be spanned by the orthogonal vectors $|e_\alpha\rangle$, ($\alpha = 1, \ldots s, \geq d^k - D^2$). Then we can construct the local Hamiltonian acting on k consecutive sites as

$$h := \sum_{\alpha=1}^{s} \mu_\alpha |e_\alpha\rangle\langle e_\alpha|, \qquad (43)$$

where $\mu_\alpha^{,}s$ are positive constants. These parameters together with the parameters of the vectors $|e_i\rangle$ inherited from those of the original matrices A_i, determine the total number of coupling constants of the Hamiltonian. If we call the embedding of this local Hamiltonian into the sites l to $l+k$ by $h_{l,l+k}$ then the full Hamiltonian on the chain is written as

$$H = \sum_{l=1}^{N} h_{l,l+k}. \tag{44}$$

The state $|\psi\rangle$ is then a ground state of this Hamiltonian with vanishing energy. The reason is as follows:

$$\langle\psi|H|\psi\rangle = tr(H|\psi\rangle\langle\psi|) = \sum_{l=1}^{N} tr(h_{l,l+k}\rho_{l,l+k}) = 0, \tag{45}$$

where$\rho_{l,k+l}$ is the reduced density matrix of sites l to $k+l$ and in the last line we have used the fact that h is constructed from the null eigenvectors of ρ for k consecutive sites. Given that H is a positive operator, this proves the assertion.

4.1. *Symmetries of the Hamiltonian*

A Hamiltonian derived as above does not have any particular symmetry. Indeed the above class include all types of Hamiltonians which have the matrix product state as their ground state. A subclass of these Hamiltonians do have the symmetry of the ground state. Consider equation (41). Multiplying both sides of this equation by $U^{-1}(R)$ from left and $U(R)$ from right, and using (34), we find that if $c_{i...j}$ is a solution of (41), then $R_{ii'}...R_{jj'}c_{i'...j'}$ is also a solution of the same equation, that is:

$$R_{ii'}...R_{jj'}c_{i'...j'} = \lambda(R)c_{i...j}. \tag{46}$$

This means that the null space of the reduced density matrix is an invariant subspace under the action of the symmetry group $R^{\otimes k}$. Thus the null vectors $|e_\alpha$ transform into each other under the action of the reducible representation $R^{\otimes}$. Such vectors can be classified into multiplets such that each multiplet transforms under one irreducible representation of the group R. Let the states transforming under the irreducible representation $D^\mu$of the group, be denoted by $|e^\mu_\beta\rangle$. Then the operators $h^\mu = \sum_\beta |e^\mu_\beta\rangle\langle e^\beta_\mu|$, are scalar under the action of the group, that is

$$[D^\mu, h^\mu] = 0. \tag{47}$$

Hence to ensure the symmetry of the local Hamiltonian we write it as

$$h_{i,i+k} = \sum_\mu \lambda_\mu h_\mu = \sum_{\beta,\mu} \lambda_\mu |e^\mu_\beta\rangle\langle e^\beta_\mu|, \tag{48}$$

where the number of free couplings λ_μ is equal to the number of multiplets which span the null space Δ_k.

4.2. *Some Examples*

Within the matrix product formalism, and in its various formulations given by different authors, many interesting models have been introduced and solved.[3,5–18] Here we cannot give a comprehensive list of these models and our aim is only to substantiate the formalism in the previous sections with a few examples so that the reader can follow the literature and also construct his or her own models.

4.2.1. *The GHZ state*

The GHZ states are of the form

$$|GHZ\rangle = \frac{1}{\sqrt{2}}(|000...0\rangle + |111...1\rangle). \tag{49}$$

Consider the following matrices

$$A_0 = \begin{pmatrix} 1 & 0 \\ 0 & 0 \end{pmatrix}, \quad A_1 = \begin{pmatrix} 0 & 0 \\ 0 & 1 \end{pmatrix}. \tag{50}$$

It is now easily verified that $A_0^N = A_0$, $A_1^N = A_1$ and any other combination of products will vanish. These relations indicate that the MPS constructed from these two matrices is a Greenberger-Horne-Zeilinger or GHZ state, namely

$$|GHZ\rangle = \frac{1}{\sqrt{2}}(|000...0\rangle + |111...1\rangle). \tag{51}$$

Exercise 5: Determine the MPS representation of a GHZ state in arbitrary dimension d, i.e.

$$|GHZ\rangle_d = \frac{1}{\sqrt{2}}(|000...0\rangle + |111...1\rangle + \cdots |d-1, d-1, d-1...d-1\rangle). \tag{52}$$

4.2.2. *The W state*

The so-called W states are of the form

$$|W\rangle = \frac{1}{\sqrt{N}}(|100...0\rangle + |010...0\rangle + |000...1\rangle). \tag{53}$$

Let us take the matrices as follows:

$$A_0 = I, \quad A_1 = \sigma^+. \tag{54}$$

The reason for this choice of matrices is that any power of A_1 except power 1, will immediately give a zero result, so all the amplitudes $\psi_{0...010...010...0}$ with

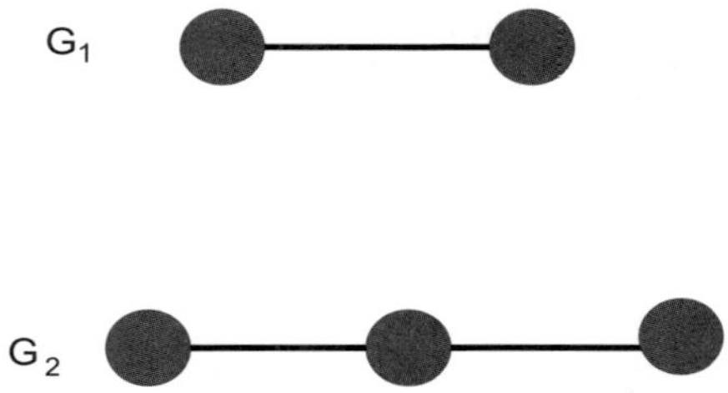

Fig. 1. Two simple cluster states. Each link means application of a CZ operation between the qubits which are already in the $|+\rangle$ state.

more than one 1 will vanish. However this choice of matrices will also give a non-zero amplitude for $\psi_{000\ldots0}$ which is absent in the W state. To complete the representation, we have to choose different matrices for the last site. This should be such that the amplitudes for $\psi_{000\ldots0}$ will also vanish. Thus we choose the N-th matrices as follows:

$$A_0^{[N]} = A_1^{[N]} = \sigma_- = \begin{pmatrix} 0 & 0 \\ 1 & 0 \end{pmatrix}. \tag{55}$$

4.3. *Cluster States*

Cluster states or more generally graph states, have recently found widespread interest in the quantum information community. This interest comes mainly from Measurement-Based Quantum Computation or MBQC, which is a different model of quantum computation than the circuit model. In MBQC, computation (implementation of specific unitary gates) is given to successive controlled measurements on an initially large cluster state having sufficient entanglement. Given a graph $G = (V, E)$ where V and E respectively represent the set of vertices and links of the graph, a graph state $|G\rangle$ is defined as follows: First we form the product state

$$|\Omega\rangle := \bigotimes_{j \in V} |+\rangle_j, \tag{56}$$

where $|+\rangle_j$ is a spin-up state in the x direction on site j. ($|+\rangle := |0\rangle + |1\rangle$.) Then we glue these separate states by the action of CZ operators along the link to make a large entangled state. That is, we form

$$|G\rangle := \bigotimes_{l \in E} (CZ)_l |\Omega\rangle. \tag{57}$$

Note that CZ or controlled-Z operation is defined in the computational basis $|0\rangle, |1\rangle$ as follows:

$$CZ|i, j\rangle := (-1)^{i,j} |i, j\rangle_j. \tag{58}$$

It is useful to note that CZ is a symmetric operator, i.e. its effect is the same if we swap the first and the second particles. Moreover it is useful to bear in mind the following equations:

$$\begin{aligned} CZ|0,\pm\rangle &= |0,\pm\rangle, \\ CZ|1,\pm\rangle &= |1,\mp\rangle. \end{aligned} \tag{59}$$

It is now easily verified that the explicit form of the simple graph states of figure (1) are as follows:

$$\begin{aligned} |G_1\rangle &= CZ|++\rangle = |0,+\rangle + |1,-\rangle, \\ |G_2\rangle &= (CZ)_{12}(CZ)_{23}|+++\rangle \\ &= |0,0,+\rangle + |0,1,-\rangle + |1,0,+\rangle - |1,1,-\rangle. \end{aligned} \tag{60}$$

The full expansion of these states in the computational basis will be

$$\begin{aligned} |G_1\rangle &= |0,0\rangle + |0,1\rangle + |1,0\rangle - |1,1\rangle, \\ |G_2\rangle &= |0,0,0\rangle + |0,0,1\rangle + |0,1,0\rangle - |0,1,1\rangle \\ &\quad + |1,0,0\rangle + |1,0,1\rangle - |1,1,0\rangle + |1,1,1\rangle. \end{aligned} \tag{61}$$

Given a cluster state corresponding to a linear graph L, we can ask if it can be represented as a matrix product state. The answer is positive and one can check that the matrices are

$$A_0 = |0\rangle\langle+| = \begin{pmatrix} 1\,1 \\ 0\,0 \end{pmatrix}, \quad A_1 = |1\rangle\langle-| = \begin{pmatrix} 0 & 0 \\ 1 & -1 \end{pmatrix}. \tag{62}$$

Let us see if this cluster state is a ground state of a Hamiltonian with nearest neighbor interaction. To verify this we form all the products A_iA_j and find that

$$A_0A_0 == |0\rangle\langle+|, \quad A_0A_1 = |0\rangle\langle-|, \quad A_1A_0 = |1\rangle\langle+|, \quad A_1A_1 = -|1\rangle\langle-|. \tag{63}$$

Therefore all these terms are independent and the system of equations $\sum_{ij} c_{ij}A_iA_j = 0$ has no non-trivial solution. Therefore there is no Hamiltonian with nearest-neighbor interaction, annihilating this cluster state. However if we set the range of interaction to be $k = 3$, then we find a non-trivial solution. In fact we find

$$\begin{aligned} A_0A_0A_0 = |0\rangle\langle+|, &\quad A_0A_0A_1 = |0\rangle\langle-|, \\ A_0A_1A_0 = |0\rangle\langle+|, &\quad A_0A_1A_1 = -|0\rangle\langle-|, \end{aligned} \tag{64}$$

and

$$\begin{aligned} A_1A_0A_0 = |1\rangle\langle+|, &\quad A_1A_0A_0A_1 = |1\rangle\langle-|, \\ A_1A_1A_0 = -|1\rangle\langle+|, &\quad A_1A_1A_1 = |1\rangle\langle-|. \end{aligned} \tag{65}$$

Fig. 2. A linear cluster states. It's MPS representation is given by (62) and its parent Hamiltonian by (128).

These two equations indicate that the system of equations $\sum_{ijk} c_{ijk} A_i A_j A_k = 0$ has a simple solution, namely:

$$c_{000} = -c_{010}, \quad c_{001} = c_{011}, \quad c_{100} = c_{110}, \quad c_{101} = -c_{111}. \tag{66}$$

In other words the solution space is spanned by the vectors

$$\begin{aligned} |e_1\rangle &= |0,0,0\rangle - |0,1,0\rangle = |0,-,0\rangle, \\ |e_2\rangle &= |0,0,1\rangle + |0,1,1\rangle = |0,+,1\rangle, \\ |e_3\rangle &= |1,0,0\rangle + |1,1,0\rangle = |1,+,0\rangle, \\ |e_4\rangle &= |1,0,1\rangle - |1,1,1\rangle = |1,-,1\rangle. \end{aligned} \tag{67}$$

A spin-flip and parity invariant Hamiltonian is constructed by taking the new combination of vectors as

$$\begin{aligned} |e'_1\rangle &= |e_1\rangle + |e_4\rangle \\ |e'_2\rangle &= |e_1\rangle - |e_4\rangle \\ |e'_3\rangle &= |e_2\rangle + |e_3\rangle \\ |e'_4\rangle &= |e_2\rangle - |e_3\rangle. \end{aligned} \tag{68}$$

Exercise 6: Taking the local Hamiltonian as $h = \sum_{i=1}^{4} |e'_i\rangle\langle e'_i|$, show that the final Hamiltonian is of the form

$$H = \sum_{i=1}^{N} \sigma_{z,i-1}\sigma_{x,i}\sigma_{z,i+1}. \tag{69}$$

4.4. *A Simple Spin One-half Chain with Phase Transition*

Let us consider a simple case[4]

$$A_0 = \begin{pmatrix} 1 & 1 \\ 0 & 0 \end{pmatrix} \qquad A_1 = \begin{pmatrix} 0 & 0 \\ 1 & g \end{pmatrix}. \tag{70}$$

A simple calculation shows that this MPS is symmetric under parity, with $X = \begin{pmatrix} 0 & 1 \\ g & 0 \end{pmatrix}$. The matrix E is equal to

$$E = \begin{pmatrix} 1 & 1 & 1 & 1 \\ 0 & 0 & 0 & 0 \\ 0 & 0 & 0 & 0 \\ 1 & g & g & g^2 \end{pmatrix}, \tag{71}$$

and two of its largest eigenvalues are $1 + g$ and $1 - g$, which switch their role according to whether g is greater or less than 0. Therefore $g_c = 0$ is a point of phase transition. A straightforward calculation shows a discontinuity in the first derivative of $\langle \sigma_i^\alpha \sigma_{i+1}^\alpha \rangle$ for $\alpha = (x, y, z)$, whereas all two-point expectation values are continuous. The magnetization in the $-$direction can serve as an order-parameter since $\langle \sigma_i^x \rangle = \frac{4g}{(1+g)^2}$ for $g > 0$, whereas it vanishes for $g < 0$. At the quantum phase transition point g_c the state is a Greenberger-Horne-Zeilinger. For $g = -1$ (disordered phase) it is equal to the cluster state, see below, and for $g = 1$ (ordered phase) all spins point in the x-direction. The parent Hamiltonian turns out to be

$$H = \sum_i 2(g^2 - 1)\sigma_i^z \sigma_{i+1}^z - (1 + g)^2 \sigma_i^x + (g - 1)^2 \sigma_i^z \sigma_{i+1}^x \sigma_{i+2}^z. \tag{72}$$

Exercise 7: Derive this Hamiltonian. Show that it has spin-flip symmetry. Also derive the Hamiltonian for the MPS state of exercise 1.

Note that when $g = -1$, the above Hamiltonian reduces to the one of the previous model.

4.5. *The Majumdar-Ghosh Models*

Up to now we have not taken into account rotational symmetry. Our Hamiltonians were not scalars and the ground states were not singlets. Consider now the matrices,

$$A_0 = \begin{pmatrix} 0 & 1 & 0 \\ 0 & 0 & 1 \\ 0 & 0 & 0 \end{pmatrix}, \qquad A_1 = \begin{pmatrix} 0 & 0 & 0 \\ -1 & 0 & 0 \\ 0 & 1 & 0 \end{pmatrix}. \tag{73}$$

Exercise 8: Check that these matrices satisfy the condition of rotational invariance. **Hint:** Take your representation of rotation algebra as follows:

$$L_z = \begin{pmatrix} \frac{1}{2} & 0 & 0 \\ 0 & 0 & 0 \\ 0 & 0 & -\frac{1}{2} \end{pmatrix}, \qquad L_+ = \begin{pmatrix} 0 & 0 & 1 \\ 0 & 0 & 0 \\ 0 & 0 & 0 \end{pmatrix}, \qquad L_- = \begin{pmatrix} 0 & 0 & 0 \\ 0 & 0 & 0 \\ 1 & 0 & 0 \end{pmatrix}. \tag{74}$$

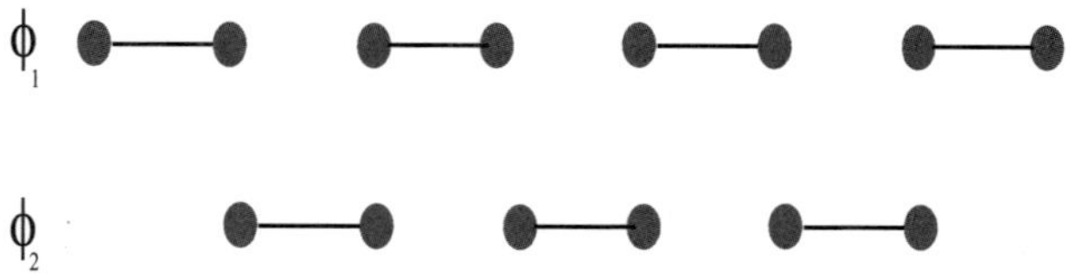

Fig. 3. The two ground states of the Majumdar-Ghosh model. The MPS representation is given in (73) and the parent Hamiltonian in (77).

Here we have $d = 2$ (spin 1/2) and $D = 3$. Although the condition $d^k > D^2$ is not satisfied for $k = 3$, we will see that the system of equations $\sum_{1i,j,k=0} c_{ijk} A_i A_j A_k = 0$ has indeed a solution. The solution space is spanned by the vectors

$$\begin{aligned} |e_1\rangle &= |000\rangle \\ |e_2\rangle &= |111\rangle \\ |e_3\rangle &= |001\rangle + |010\rangle + |100\rangle \\ |e_4\rangle &= |011\rangle + |101\rangle + |110\rangle. \end{aligned} \tag{75}$$

These vectors are parity invariant (invariant under left-right reflection) and are mapped to each other under spin-flip transformation ($|0\rangle \longrightarrow |1\rangle, |1\rangle \longrightarrow |0\rangle$). Furthermore they transform like a spin 3/2 multiplet under the $su(2)$ algebra. Therefore a Hamiltonian with all three symmetries is constructed as:

$$h = |e_1\rangle\langle e_1| + |e_2\rangle\langle e_2| + |e_3\rangle\langle e_3| + |e_4\rangle\langle e_4|. \tag{76}$$

Exercise 9: Let the local Hamiltonian be as $h = \sum_{i=1}^{4} |e_i'\rangle\langle e_i'|$. Show that the final Hamiltonian is of the form, modulo additive and multiplicative constants:

$$H = \sum_{i=1}^{N} \sigma_{z,i-1}\sigma_{x,i}\sigma_{z,i+1}. \tag{77}$$

This is the Majumdar-Ghosh Hamiltonian first found and studied before the advent of the matrix product formalism.[19,20]

What is the explicit form of the MPS in this case? If we define the states

$$|0\rangle = \begin{pmatrix} 1 \\ 0 \\ 0 \end{pmatrix}, \qquad |b\rangle = \begin{pmatrix} 0 \\ 1 \\ 0 \end{pmatrix}, \qquad |1\rangle = \begin{pmatrix} 0 \\ 0 \\ 1 \end{pmatrix}, \tag{78}$$

then the matrices have the simple form

$$A_m = |m\rangle\langle b| + (-1)^m |b\rangle\langle \overline{m}|, \tag{79}$$

where by $\overline{m}$ we mean $m+1, \mod 2$. So $|\overline{m}\rangle$ is the flipped state of $|m\rangle$ in the computational basis. Note that the states $|\rangle$, $|1\rangle$ and $|b\rangle$ are orthogonal. Moreover these states form the reducible representation of spin $\frac{1}{2} \oplus 0$, under the rotation group, that is why the matrices L_+, L_- and L_0 have this shape. Now it is straightforward to verify that

$$A_{m_1}A_{m_2} = (-1)^{m_2}|m_1\rangle\langle\overline{m_2}| + (-1)^{m_1}\delta_{\overline{m_1},m_2}|b\rangle\langle b|. \tag{80}$$

We can safely ignore the product of an odd number of A_m's, since they lead to zero when we take their trace. For even number of A's, say four, it is very simple to show that

$$tr(A_{m_1}A_{m_2}A_{m_3}A_{m_4}) = (-1)^{m_2+m_3}\delta_{\overline{m_2},m_3}\delta_{\overline{m_4},m_1} + (-1)^{m_1+m_4}\delta_{\overline{m_1},m_2}\delta_{\overline{m_3},m_4}. \tag{81}$$

If we now take the trace and note that the expression of a singlet state is

$$|S\rangle = |0,1\rangle - |1,0\rangle = \sum_{m_1}(-1)^{m_1}|m_1,\overline{m_1}\rangle, \tag{82}$$

then we find that the MPS defined by the above matrices is the sum of two states, namely

$$|\psi\rangle = |\phi_1\rangle + |\phi_2\rangle, \tag{83}$$

where

$$|\phi_1\rangle = |S_{1,2}\rangle|S_{3,4}\rangle\cdots|S_{2N-1,2N}\rangle, \tag{84}$$

and

$$|\phi_2\rangle = |S_{2,3}\rangle|S_{4,5}\rangle\cdots|S_{2N,1}\rangle. \tag{85}$$

That is, $|\phi_1\rangle$ and $|\phi_2\rangle$ are two dimerized states, i.e. states which are composed of juxtapositions of singlet states on adjacent sites, figure (3). The difference between $|\phi_1\rangle$ and $|\phi_2\rangle$ is that one has been shifted with respect to the other by one single site. In fact neither of them is translational-invariant, but their sum is. So it is natural that only their sum has found an MPS representation. Further inspection shows that each of these states alone is a ground state of the Hamiltonian and so each of them break the translational symmetry of the Hamiltonian.

4.5.1. *The AKLT states*

The matrix product representation of the spin-1 AKLT state, named after Affleck, Kennedy, Lieb and Tasaki[21] is constructed by taking the matrices to be as follows:

$$A_1 = \sqrt{2}\sigma_+ = \begin{pmatrix} 0 & \sqrt{2} \\ 0 & 0 \end{pmatrix}$$

$$A_0 = \sigma_z = \begin{pmatrix} 1 & 0 \\ 0 & -1 \end{pmatrix}$$

$$A_{\overline{1}} = -\sqrt{2}\sigma_- = \begin{pmatrix} 0 & 0 \\ -\sqrt{2} & 0 \end{pmatrix}, \tag{86}$$

where we have used and will use hereafter the symbol $\overline{1}$ for -1. These matrices satisfy the following relations:

$$\begin{aligned} [L_z, A_m] &= mA_m \\ [L_+, A_m] &= \sqrt{2 - m(m+1)}A_{m+1} \\ [L_-, A_m] &= \sqrt{2 - m(m-1)}A_{m-1}, \end{aligned} \tag{87}$$

where $L_z = \frac{1}{2}\sigma_z, L_+ = \frac{1}{2}\sigma_+$ and $L_- = \frac{1}{2}\sigma_-$, form a two dimensional representation of the $su(2)$ algebra. In order to find the parent Hamiltonian we have to solve the system of equations $\sum_{i,j} c_{i,j} A_i A_j = 0$ the solution space of which is spanned by the following vector:

$$|e\rangle = |1, \overline{1}\rangle - |0, 0\rangle + |\overline{1}, 1\rangle, \tag{88}$$

where $S_z|\overline{1}\rangle = -|\overline{1}\rangle, S_z|0\rangle = 0, S_z|1\rangle = |1\rangle$. This then leads to the following Hamiltonian

$$H = \sum_j S_j \cdot S_{j+1} + \frac{1}{3}(S_j \cdot S_{j+1})^2. \tag{89}$$

5. Conclusion of Part I

We have shown in this part that there are many interesting models which can be solved within the matrix product formalism. What we have focused on is the construction of models with low dimensional matrices, so that these models can be solved in closed analytical form. If we relax this condition and then use matrices with large sizes and also use numerical techniques for our calculations, then we can solve much more realistic models by suggesting MPS ansatz for the ground states of known Hamiltonians.

Part II: One Dimensional Reaction Diffusion Systems

We now move on to the second part of these lectures, classical stochastic systems on one dimensional lattices, where again matrix product states find a very wide application with a very reach literature. First we have to introduce the basic concepts of stochastic processes and their Hamiltonian formulation. The simplest stochastic process which we are all familiar with is a random walker on an infinite one dimensional lattice. The possible configurations of this random process is the set of integer numbers n, where $-\infty < n < \infty$, and the basic quantity of interest to us is $P(t,n)$, the probability that at time t the random walker be at position n. The time step t can be discrete and changes by units of $1(t \longrightarrow t+1)$ or it can be continuous. The random walker moves in the space of configurations according to a set of definite stochastic rules. For example in continuous time, the basic moves may be as follows:

$$\begin{aligned} n &\longrightarrow n+1 \qquad \text{with rate } \lambda \\ n &\longrightarrow n-1 \qquad \text{with rate } \mu. \end{aligned} \tag{90}$$

This means in an infinitesimal time interval dt, a particle at site n moves forward with probability λdt, moves backward with probability μdt and remains still with probability $1 - \lambda dt - \mu dt$. The elementary moves (90), allows us to find the dynamics of the probability function $P(t,n)$. The dynamics is encoded in an equation which is called the master equation

$$\frac{\partial P(t,n)}{\partial t} = \lambda P(t,n-1) + \mu P(t,n+1) - (\lambda+\mu)P(t,n). \tag{91}$$

The first two terms, the so called source terms indicate the ways that the probability $P(t,n)$ increases and the last terms, the so called sink terms, indicate the ways that lead to a decrease of $P(t,n)$. The simple example mentioned above, deals with one particle on a lattice. Consider now a system of identical particles doing random walks on a one dimensional lattice, and at the same time interacting with each other. There are many ways of interaction for two particles when they come close to each other. They may stop each other from hopping to the right or left, they may annihilate each other or may combine to form a single particle, among other possible types of interaction. For such a system, the set of configurations consists of the set of all $N-$ tuples, $(s_1, s_2, \cdots s_N)$, where $s_i = 0, 1$in which $s_i = 0$ means that the site i is empty and $s_i = 1$ means that site i is occupied with a particle. There are in total $2N$ different configurations. For more general interactions, i.e. for the case when there can be more than one particle, say up to $d-1$ particles on a single site, or for the case where there are $d-1$ types (colors) of particles, where each site can be either empty or occupied by one of them, the random variables s_i will then take d values, namely $s_i = 0, 1, 2, \cdots d-1$. In such a case, various types of particles, in addition to hopping can transform into each other upon meeting each other.

6. The Master Equation and its Hamiltonian Form

The above example can be generalized to a general setting. Let $\mathcal{H} = \{C, C', C", \cdots\}$ be the set of all accessible configurations for a system. A configuration can have any general form, i.e. the position of an arbitrary number of particles, the population of bacteria in a colony and so on. For such a system $P(t, C)$ denotes the probability that at time t, the system be in configuration C. The master equation takes the form

$$\frac{\partial}{\partial t}P(t,C) = \sum_{C' \neq C} P(t,C')\omega(C' \longrightarrow C) - \sum_{C' \neq C} P(t,C)\omega(C \longrightarrow C'), \quad (92)$$

where $\omega(C \longrightarrow C')$ is the rate of change of configuration C to C'. This equation is linear in $P(t, C)$ and hence can be set into a matrix form. One writes

$$\frac{\partial}{\partial t}P(t,C) = -\sum_{C'} H(C,C')P(t,C'), \quad (93)$$

where H is called the stochastic matrix. If a system has N configurations, then by using the abbreviations $\omega_{ij} := \omega((C_j \longrightarrow C_i)$, and $\omega_{ii} := -\sum_j \omega_{ji}$, the stochastic matrix H takes the form

$$H = -\begin{pmatrix} \omega_{11} & \omega_{12} & \omega_{13} & \omega_{14} & \cdots & \omega_{1N} \\ \omega_{21} & \omega_{22} & \omega_{23} & \omega_{24} & \cdots & \omega_{2N} \\ \omega_{31} & \omega_{32} & \omega_{33} & \omega_{34} & \cdots & \omega_{3N} \\ \cdot & \cdot & \cdot & \cdot & \cdots & \cdot \\ \cdot & \cdot & \cdot & \cdot & \cdots & \cdot \\ \omega_{N1} & \omega_{N2} & \omega_{N3} & \omega_{N4} & \cdots & \omega_{NN} \end{pmatrix}. \quad (94)$$

This matrix has two important characteristics.

i) Its off-diagonal terms are negative.

ii) The sum of all elements in each column is zero.

Any matrix with these two properties is called a stochastic matrix. The linear structure of the master equation, allows us to formalize the tools of stochastic processes in a way which is very similar to that of quantum mechanics. To this end, we first construct a Hilbert space $\mathcal{V}$ spanned by formal vectors $\{|C\rangle\}$ which are in one to one correspondence to the admissible configurations of the system. These vectors are orthonormal and complete:

$$\langle C|C'\rangle = \delta_{C,C'}, \qquad \sum_C |C\rangle\langle C| = I. \quad (95)$$

At each instant of time, the state of the system is denoted by the following vector in $\mathcal{V}$,

$$|P(t)\rangle = \sum_C P(t,C)|C\rangle, \tag{96}$$

from which the probabilities of different configurations are obtained by a simple projection

$$P(t,C) = \langle C|P(t)\rangle. \tag{97}$$

Note that contrary to quantum mechanics, the components of the state vector are probabilities and not probability amplitudes, hence the normalization of the state vector reads

$$\langle \Omega|P(t)\rangle = 1, \tag{98}$$

where $\langle\Omega|$ is a unique bra-vector defined as

$$\langle\Omega| = \sum_C \langle C|. \tag{99}$$

7. The Stationary State and it's Matrix Product Form

For a stationary stochastic process, i.e. one whose rates and hence its Hamiltonian are constant in time, the solution of the master equation is simply given by

$$|P(t)\rangle = e^{-Ht}|P(0)\rangle. \tag{100}$$

Therefore if we know the spectrum of H, i.e.

$$H|n\rangle = \lambda_n|n\rangle, \tag{101}$$

and the decomposition of the initial state in terms of the eigenvectors of H,

$$|P(0)\rangle = \sum_n c_n|n\rangle, \tag{102}$$

then the state at any arbitrary time t can be found simply from

$$|P(t)\rangle = \sum_n e^{-\lambda_n} c_n|n\rangle. \tag{103}$$

In most cases, what we are really interested in, is not the detailed dynamics of the system, but its stationary state, that is the state after an infinitely long time. We want to see if the system is left to itself, then after a very long time, what will be the probability of finding the system in different configurations. This stationary state is denoted by $|P^*\rangle$. To find this state, we need some information about the spectrum of the Hamiltonian. To this end we invoke a theorem which we state without proof. For the proofs, see.[22] Theorem: For every stochastic matrix H, there is an eigenvalue λ_0 which is exactly equal to zero. Furthermore the real part of each eigenvalue is non-negative. The eigenvalue λ_0 is non-degenerate if and only if, the space of configurations is connected, that is any two configurations C and C' can be connected by a sequence of elementary moves. For such a system, the stationary state is independent of the initial state and the system relaxes to the state which is defined by the eigenvector $|\lambda_0\rangle$, regardless of the initial state $P(0)$

$$|P*\rangle \equiv |P(\infty)\rangle = |\lambda_0\rangle. \tag{104}$$

The problem of determining the stationary state of a stochastic process, then boils down to finding the eigenstate $|\lambda_0\rangle$ of H,

$$H|\lambda_0\rangle = 0. \tag{105}$$

We are thus led to a problem which is very similar to the one we had already encountered in part I, namely finding the ground state of a Hamiltonian. There are however some differences: the Hamiltonian is not Hermition and it is not positive definite. However as we will see, with some modifications we can also use the technique of matrix product states for such a system, specially when the configurations are the positions of a set of particles on a one dimensional lattice. As indicated, for a set of identical particles hopping and interacting on a one dimensional lattice of N sites, the Hilbert space $\mathcal{V}$ becomes equal to $C_2^{\otimes N}$, where C_2 is the space spanned by the basic vectors $|0\rangle$ and $|1\rangle$. If there can be up to $d-1$ particles on each site, or if there are $d-1$ types of particles, then the Hilbert space V will be $C_d^{\otimes N}$, where C_d is spanned by the vectors $\{|0\rangle, |1\rangle, \cdots |d-1\rangle\}$. For such a system, the steady state takes the form $P^*(s_1, s_2, \cdots s_N)$ and the matrix product form of such a state will be written as

$$P^*(s_1, s_2, \cdots s_N) = \frac{1}{Z_N}\langle W|A_{s_1}A_{s_2}\cdots A_{s_N}|V\rangle, \tag{106}$$

where we have explicitly written this form for an open boundary condition which is the condition prevailing in a stochastic system, where particles can enter and or exit at the boundaries. With the definition $\mathcal{A} := \sum_i A_i|i\rangle$, this state can be written in the compact form

$$|P*\rangle = \frac{1}{Z_N}\langle W|\mathcal{A}\otimes\mathcal{A}\otimes\cdots\mathcal{A}|V\rangle, \tag{107}$$

where the normalization constant Z_N is called the partition function and is given by

$$Z_N = \langle W|\mathcal{E}^N|V\rangle, \tag{108}$$

in which

$$\mathcal{E} = \sum_s A_s. \tag{109}$$

Note the difference between the way normalization enters in matrix product states of stationary states of classical stochastic systems of part II and quantum systems of part I. These are due to the particular way that these states are normalized, i.e. $\langle\Omega|P(t)\rangle$, compared with $\langle P(t)|P(t)\rangle = 1$ which is what we have in quantum mechanics. If we have local hopping and interactions, in the sense that particles on adjacent sites interact with each other, then the Hamiltonian is of the following general form

$$H = h_L + \sum_{i=1}^{N-1} h_{i,i+1} + h_R. \tag{110}$$

where h_L and h_R act respectively on the leftmost and the rightmost sites of the lattice. The types of interactions in the bulk and the boundaries, will determine the explicit form of the Hamiltonian H. We will see a specific example of this in the following sections. For the present our objective is to find the ground state of this Hamiltonian, that is the state which is annihilated by this Hamiltonian,

$$H|P^*\rangle = 0. \tag{111}$$

Contrary to part I, we now have the Hamiltonian beforehand and we want to find the ground state. Actually the word ground state is not used here in its proper sense of having the lowest value of energy, since H is not Hermitian and its eigenvalues are not real. However we continue to use this term for ease of reference. To find the steady state, let us assume that there are matrices X_i (organized as usually in the vector-valued matrix $\mathcal{X} := \sum_s X_s|s\rangle$) such that

$$h\mathcal{A} \otimes \mathcal{A} = \mathcal{X} \otimes \mathcal{A} - \mathcal{A} \otimes \mathcal{X}. \tag{112}$$

If this local condition is satisfied then the action of the bulk Hamiltonian will be simplified due to cancelations of adjacent terms, and the action of the total Hamiltonian H on the steady state will be

$$H|P^*\rangle = \langle W|\mathcal{A}^{\otimes N-1} \otimes [h_R\mathcal{A} - \mathcal{X}]|V\rangle + \langle W|[h_L\mathcal{A} + \mathcal{X} \otimes \mathcal{A}^{\otimes N-1}] \otimes |V\rangle. \tag{113}$$

Therefore H annihilates $|P^*\rangle$, provided that the following relations are satisfied by the boundary terms

$$h_R\mathcal{A} - \mathcal{X}|V\rangle = 0, \qquad \langle W|h_L\mathcal{A} + \mathcal{X} = 0. \tag{114}$$

In this way a many-body problem reduces to the solutions of local problems, one in the bulk (112) and two in the boundaries (114). These equations are a set of matrix equations for the unknown matrices A_i and the boundary vectors $|V\rangle$ and $\langle W|$.

The dimension of the matrices A_i and the boundary vectors $|V\rangle$ and $\langle W|$ and also the form of the auxiliary matrices X_i are not known at this stage. These are determined by the solutions of equations (112) and (114). These equations are called the MPS algebra associated to the given Hamiltonian. We emphasize again the difference between the matrix product formalism for this non-equilibrium problem and the quantum many body problem. Note that here we cannot begin from a given stationary state in the form of MPS and look for a stochastic Hamiltonian, since we do not have any method for satisfying the conditions of stochasticity of such a Hamiltonian. This is in contrast with the quantum case, where the conditions of positivity and hermiticity are easily implemented. After these generalities, we can now study a specific example. This example clarifies many of the ideas explained above.

8. An Example: The Asymmetric Simple Exclusion Process

The prototype of stochastic processes on a one dimensional lattice, is the Asymmetric Simple Exclusion Process or simply called ASEP and was exactly solved by MPS techniques in.[23] Other processes are modifications and extensions of this model. The process describes a collection of particles which hop on a one dimensional lattice only to the right and interact with each other. The interaction is such that each site at most one particle so that when a site in front of a given particle is occupied, hence the name exclusion. A particle whose right hand neighboring site is empty will hop to that site with a rate μ and will remain in its place otherwise. Furthermore the system is open which means that particles are injected in the left hand side and extracted from the right hand side. The Hilbert space is $V = C_2^{\otimes N}$, where C_2 is spanned by the vectors $|0\rangle$ for empty site and $|1\rangle$ for occupied site. The rules of hopping and interactions are indicated in figure (4) and in the following formulas (where we have set the scale of time so that the rate of hopping μ_i is normalized to 1):

$$\begin{array}{rcll} (10)_{i,i+1} & \longrightarrow & (01)_{i,i+1} & \text{with rate } \mu \\ (0)_1 & \longrightarrow & (1)_1 & \text{with rate } \alpha \\ (1)_N & \longrightarrow & (0)_N & \text{with rate } \beta. \end{array} \tag{115}$$

Note that a rate normalized to unity does't mean that a particle hops with certainty (with probability 1), it only says that in a time interval dt, out of m particles, a fraction of mdt particles hop forward. Figure (4) also shows that when the leftmost site is empty a particle is injected into it with rate α and when the rightmost site is occupied, it is extracted with rate β. From equation (115), we can now derive the

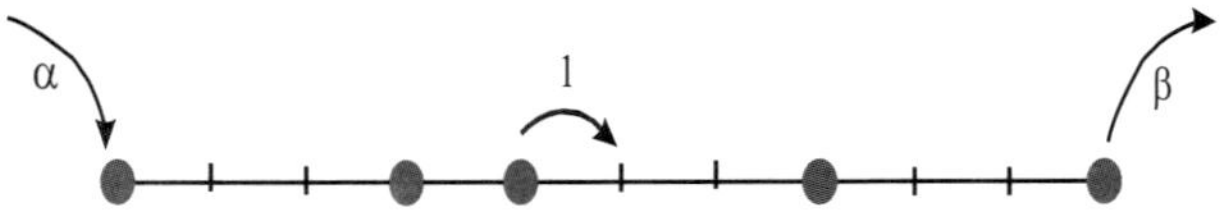

Fig. 4. The basic processes of the Asymmetric Simple Exclusion Process (ASEP).

form of the bulk local Hamiltonian in the form

$$h = \begin{pmatrix} 0 & 0 & 0 & 0 \\ 0 & 1 & 0 & 0 \\ 0 & -1 & 0 & 0 \\ 0 & 0 & 0 & 0 \end{pmatrix}. \tag{116}$$

Also the form of the local Hamiltonians h_L and h_R are

$$h_L = \begin{pmatrix} \alpha & 0 \\ -\alpha & 0 \end{pmatrix}, \qquad h_R = \begin{pmatrix} 0 & -\beta \\ 0 & \beta \end{pmatrix}. \tag{117}$$

With the introduction of the local operators

$$\hat{n} = \begin{pmatrix} 0 & 0 \\ 0 & 1 \end{pmatrix}, \qquad 1 - \hat{n} = \begin{pmatrix} 1 & 0 \\ 0 & 0 \end{pmatrix}, \tag{118}$$

and

$$\hat{\sigma}^+ = |1\rangle\langle 0| = \begin{pmatrix} 0 & 1 \\ 0 & 0 \end{pmatrix}, \qquad \hat{\sigma}^- = |0\rangle\langle 1| = \begin{pmatrix} 0 & 0 \\ 1 & 0 \end{pmatrix}, \tag{119}$$

the bulk local Hamiltonian will be

$$\hat{h}_{i,i+1} = -\hat{\sigma}_i^- \hat{\sigma}_{i+1}^+ + (1 - \hat{n}_i)\hat{n}_{i+1}, \tag{120}$$

and the boundary Hamiltonians will be

$$\hat{h}_L = -\alpha\hat{\sigma}_1^+ + \alpha(1 - \hat{n}_1), \qquad \hat{h}_R = -\beta\hat{\sigma}_N^- + \beta\hat{n}_N. \tag{121}$$

Therefore the total Hamiltonian will be

$$H = -\alpha\hat{\sigma}_1^+ + \alpha(1-\hat{n}_1) - \sum_{i=1}^{N-1} \left[\hat{\sigma}_i^- \hat{\sigma}_{i+1}^+ - (1 - \hat{n}_i)\hat{n}_{i+1}\right] - \beta\hat{\sigma}_N^- + \beta\hat{n}_N. \tag{122}$$

Once again we have ended up with finding the ground state of quantum spin chain. We now assume a matrix product form for the ground state and use equations (112) and (114) to derive the MPS algebra. For convenience and also for conformity with the usage in the literature, we re-label the matrices as $A_0 \longrightarrow E$ and $A_1 \longrightarrow D$. Equations (112) will then lead to

$$[X_0, D] = 0,$$
$$[X_1, E] = 0,$$

$$DE = X_0 E - DX_1. \tag{123}$$

Also the boundary equations (114) will lead to

$$\begin{pmatrix} \beta D - X_0 \\ \beta D + X_1 \end{pmatrix} |V\rangle = 0, \tag{124}$$

and

$$\langle W| \begin{pmatrix} \alpha E - X_0 \\ \alpha E + X_1 \end{pmatrix} = 0. \tag{125}$$

In order to make these equations consistent, we choose the matrices X_0 and X_1 proportional to identity as follows: $X_0 = I$, and $X_1 = -I$. With this choice, the MPS algebra takes the simple form

$$DE = D + E,$$
$$D|V\rangle = \frac{1}{\beta}|V\rangle, \qquad \langle W|E = \frac{1}{\alpha}\langle W|, \tag{126}$$

and the partition function will be

$$Z_N = \langle W|(D+E)^N|V\rangle =: \langle W|C^N|V\rangle. \tag{127}$$

Exercise 10: Show that if we allow a hopping rate p to the right and a hopping rate q to the left, then the bulk MPS algebra will be

$$pDE - qED = D + E. \tag{128}$$

8.1. *Correlation Functions*

In non-equilibrium system like ASEP, a very important quantity is the density profile. Such a density is easily calculated as

$$\langle n_k \rangle = \frac{\langle W|C^{k-1}DC^{N-k}|V\rangle}{\langle W|C^N|V\rangle}. \tag{129}$$

Another important characteristic of the process is the current which we denote by J. Note that in the steady state, the current should be constant along the chain and hence J has no site dependence. A little calculation, specially paying attention to the fact that J is nothing but the average rate of hopping gives the result that

$$J = \langle \hat{n}_k(1 - \hat{n}_{k+1})\rangle = \frac{\langle W|C^{k-1}DEC^{N-k-1}|V\rangle}{\langle W|C^N|V\rangle} = \frac{\langle W|C^{N-1}|V\rangle}{\langle W|C^N|V\rangle} = \frac{Z_{N-1}}{Z_N}. \tag{130}$$

Therefore to find the current, the partition function Z_N should be known for arbitrary number of sites N. This justifies the name partition function for Z_N, since many of the important characteristics of the process are obtained from this quantity.

Exercise 11: Use the MPS algebra (126) to calculate the density of particles at each site and also the current for chains of length 3 and 4. You do not need any representation of the algebra, just use the algebraic relations to pass matrices through each other.

Contrary to the quantum many body system, where symmetry considerations play a very important role in defining the matrices and then finding the Hamiltonian, here symmetry plays very little role and the Hamiltonian is determined from the outset by the process itself. Therefore the matrices should be found by solving the system of MPS equations. To solve (126), we rewrite the matrices as

$$D = I + \hat{d} \qquad E = I + \hat{e}, \tag{131}$$

which transform the first equation of (126) into the following

$$\hat{d}\hat{e} = I. \tag{132}$$

We now prove that the matrices $\hat{d}$ and $\hat{e}$ are either one or infinite dimensional. To see this, suppose that these matrices are finite dimensional. Then equation (132) means that $\hat{d}$ is the inverse of $\hat{e}$ which implies that $\hat{e}\hat{d} = I$, and hence, both matrices can be diagonalized and in view of the relations (126), both matrices can be taken to be one dimensional. However this puts a heavy constraint on the rates. To see this we note that

$$\langle W|V\rangle = \langle W|\hat{d}\hat{e}|V\rangle = \langle W|\hat{e}\hat{d}|V\rangle = \langle W|(\frac{1}{\alpha}-1)(\frac{1}{\beta}-1)|V\rangle = (\frac{1}{\alpha}-1)(\frac{1}{\beta}-1), \tag{133}$$

which implies

$$\alpha + \beta = 1. \tag{134}$$

This means that for one-dimensional matrices, we can solve the problem only for a particular line in the space of the control parameters α and β. On this line the steady state takes the factorized un-correlated form. We can say alternatively that on this line, the mean field approximation is exact, since there is no correlation in the steady state. Before going to the infinite dimensional matrices, which can explore the other parts of the phase diagram, let us see what are the properties of the system on this line. We have

$$E = \frac{1}{\alpha}, \qquad D = \frac{1}{\beta}, \tag{135}$$

and hence

$$C = \frac{1}{\alpha} + \frac{1}{\beta}. \tag{136}$$

Consequently the density will be constant along the chain and will be equal to

$$\langle n_k \rangle = \alpha = 1 - \beta, \tag{137}$$

and the current will be

$$J = \frac{1}{C} = \alpha(1-\alpha). \tag{138}$$

The current vanishes when $\alpha = 0$ (when no particle is injected and the chain is completely empty) or $\alpha = 1$ (when no particle is extracted and the chain is congested). Maximum current is obtained for $\alpha = \beta = \frac{1}{2}$, where particles are injected and extracted at reasonable rates into and out of the chain respectively. We now can study the state with infinite dimensional matrices.

We choose the following form for these matrices, where the states $\{|n\rangle, n = 0, 1, 2 \cdots\}$ are an infinite set of orthonormal vectors

$$\hat{d} = \sum_{n=0}^{\infty} |n\rangle\langle n+1|, \qquad \hat{e} = \sum_{n=0}^{\infty} |n+1\rangle\langle n|. \tag{139}$$

The following relations can be verified

$$\hat{d}\hat{e} = I, \qquad \hat{e}\hat{d} = I - |0\rangle\langle 0|, \tag{140}$$

so that while relation (132) is satisfied, the matrices do not commute any longer. Also we can find the boundary vectors as follows:

$$|V\rangle = v_0 \sum_{n=0}^{\infty} (\frac{1}{\beta} - 1)^n |n\rangle \tag{141}$$

$$\langle W| = w_0 \sum_{n=0}^{\infty} (\frac{1}{\alpha} - 1)^n \langle n|, \tag{142}$$

where v_0 and w_0 are constants which we choose so that

$$w_0 v_0 = \frac{\alpha + \beta - 1}{\alpha\beta}. \tag{143}$$

to make the boundary vectors satisfy the condition $\langle W|V\rangle = 1$.

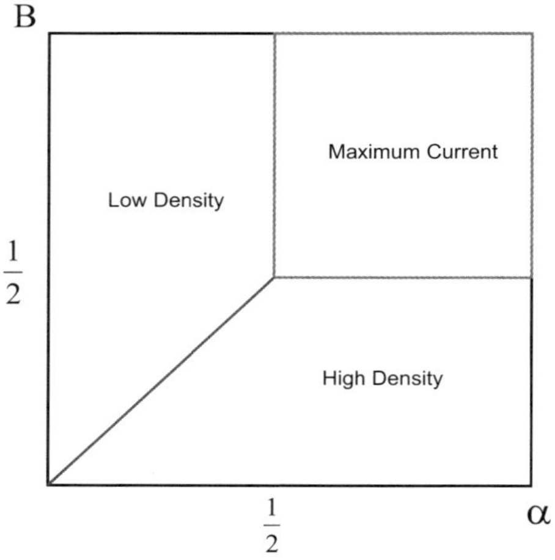

Fig. 5. The phase diagram of the ASEP.

9. The Phase Diagram

Having obtained the matrices and the boundary vectors, we can find many of the properties of the ASEP. The basic quantity to be calculated is the partition function ZN. There are many ways for obtaining Z_N. The interested reader can consult[24] for a comprehensive review. Here we mention only one method which is based on the exact diagonalization of the matrix $C = D + E$. The matrix C has a simple symmetric form, namely

$$C = \begin{pmatrix} 2 & 1 & \cdot & \cdot & \cdot & \cdot \\ 1 & 2 & 1 & \cdot & \cdot & \cdot \\ \cdot & 1 & 2 & 1 & \cdot & \cdot \\ \cdot & \cdot & 1 & 2 & 1 & \cdot \\ \cdot & \cdot & \cdot & \cdot & \cdot & \cdot \\ \cdot & \cdot & \cdot & \cdot & \cdot & \cdot \end{pmatrix}, \tag{144}$$

where· stands for zero. The spectrum of this matrix is found in a straightforward way

$$C|\theta\rangle = 2(1 + \cos\theta)|\theta\rangle, \tag{145}$$

where

$$|\theta\rangle = \frac{1}{\sin\theta} \sum_{n=0}^{\infty} \sin n\theta |n\rangle. \tag{146}$$

We can then show that

$$I = \frac{1}{\pi} \int d\theta \sin\theta |\theta\rangle\langle\theta|, \tag{147}$$

and with similar calculations

$$\langle W|\theta\rangle = \frac{\kappa}{(1 - ae^{i\theta})(1 - ae^{-i\theta})}, \tag{148}$$

and

$$\langle\theta|V\rangle = \frac{\kappa}{(1 - be^{i\theta})(1 - be^{-i\theta})}, \tag{149}$$

where

$$\kappa^2 := \frac{\alpha + \beta - 1}{\alpha\beta}, \qquad a := \frac{1-\alpha}{\alpha}, \qquad b := \frac{1-\beta}{\beta}. \tag{150}$$

Finally we obtain the integral representation of the partition function

$$Z_N = \frac{\kappa^2}{\pi} \int d\theta \frac{\sin^2\theta[2(1+\cos\theta)]^N}{(1 - ae^{i\theta})(1 - ae^{-i\theta})(1 - be^{i\theta})(1 - be^{-i\theta})}. \tag{151}$$

10. Extensions of the ASEP

The simple exclusion process has been extended and studied in many different directions. The most immediate extension is a model in which particles hop asymmetrically to the right and the left, the other ones are the multi-species processes where there are more than one species of particles hopping and interacting on the line. There are also many models in which the pattern of interactions are more complex than simple exclusion, for example particles can annihilate each other or coalesce to each other or particle pairs can be created out of nothing. Some of these new models can be solved by the matrix product method, these are the ones which lead to elegant algebras,[24–26] others may lead to very complex algebras which cannot be solved in closed form.

11. Conclusion of Part II

There is a very reach literature on the subject of matrix product states in nonequilibrium matrix product states. Once we define a reaction-diffusion process, the Hamiltonian which resembles a quantum spin chain Hamiltonian is easily obtained. The next step is to form the matrix product algebra which is straightforward. However the solution of the MPS algebra is not an easy task. While the MPS formalism applies to both quantum many body systems and classical nonequilibrium systems, the difficulty of using it effectively shows itself in two different places, in each of these subjects. In quantum many body systems, we start from symmetry considerations and introduce a set of matrices from the outset, however the real challenge is to define these matrices such that they lead to a physically interesting Hamiltonian. On the other hand in non-equilibrium systems, we start from the Hamiltonian and then derive the algebra of the matrices. In this problem the challenging part is the solution of the MPS algebra. The ASEP is the simplest problem in this subject where the MPS algebra turns out to be rather simple. If you start from an arbitrary process, you can easily run into a highly formidable MPS algebra for which no easy solution exists. Therefore here the challenge is to design physical processes which lead to solvable MPS algebras, or the converse define simple MPS algebras which lead to interesting physical processes. Following this approach, many new interesting models have been introduced and solved. Obviously we cannot review these new models, and the interested reader is referred to the literature or the recent review articles for these developments.

12. Acknowledgements

We would like to thank the organizers of the summer school in Oxford-Kobe institute in Japan, specially Professor Mikio Nakahara for the very stimulating atmosphere of this gathering, which caused many refinements in the original draft of these notes. We specially thank Dr Akira Saitoh for his very valuable helps during our visit to Kinki University and Oxford-Kobe institute. Moreover V. K. would like to thank the hospitality of Abdus Salam International Center for Theoretical Physics (ICTP), Trieste, Italy, where these lecture notes were prepared.

References

1. S. Sachdev, Quantum Phase Transitions (Cambridge University Press, Cambridge, 1999).
2. M. Fannes, B. Nachtergaele and R. F. Werner, Commun. Math. Phys. 144, 443 (1992).
3. A. Klumper, A. Schadschneider and J. Sittartz, J. Phys. A (1991) L293; Z. Phys. B, 87 (1992) 281; Europhys. Lett., 24 (1993) 293.
4. M. M. Wolf, G. Ortiz, F. Verstraete and I. Cirac, Phys. Rev. Lett. 97, 110403 (2006).
5. M. A. Ahrens, A. Schadschneider, and J. Zittartz, Europhys. Lett. 59 6, 889 (2002);
6. E. Bartel, A. Schadschneider and J. Zittartz, Eur. Phys. Jour. B, 31, 2, 209-216 (2003).
7. M. A. Ahrens, A. Schadschneider, and J. Zittartz, Europhys. Lett. 59 6, 889 (2002).
8. H. Niggemann, and J. Zittarz, J. Phys. A31, 9819 (1998), [8] A. K. Kolezhuk and H. J. Mikeska, Phys. Rev. Lett. 80, 2709 (1998); Int. J. Mod. Phys. B, Vol.12, 2325-2348 (1998).
9. M. Asoudeh, V. Karimipour, and A. Sadrolashrafi, Phys. Rev. A 76, 012320(2007).
10. M. Asoudeh, V. Karimipour and A. Sadrolashrafi, Phys. Rev. B, 75, 224427 (2007).
11. M. Asoudeh, V. Karimipour, and A. Sadrolashrafi, Phys. Rev. B 76, 064433 (2007).
12. M. A. Ahrens, A. Schadschneider and J. Zittartz, Phys. Rev. B, 71, 174432 (2005).

13. F. Verstraete, J. I. Cirac, J. I. Latorre, E. Rico and M. M. Wolf, Phys. Rev. Lett. 94 (2005) 140601.
14. V. Murg, F. Verstraete, J. I. Cirac, Phys. Rev. A 75, 033605 (2007).
15. F. Verstraete, M. A. Martin-Delgado and J. I. Cirac, Phys. Rev. Lett. 92, 087201 (2004).
16. V. Karimipour, L. Memarzadeh, Phys. Rev. B 77, 094416 (2008)
17. S. Alipour, S. Baghbanzadeh, and V. Karimipour, Europhysics Letters (EPL), 84 (2008) 67006.
18. S. Alipour, V. Karimipour, and L. Memarzadeh, Phys. Rev. A 75, 052322 (2007); European Physical Journal B, 62, pp. 159-169, 2008.
19. C. K. Majumdar, J. Phys. C 3. 911(1969), J. Math. Phys. 10 (1969)1388.
20. C. K. Majumdar and D. P. Ghosh, J. Math. Phys. 10 (1969)1399.
21. I. Affleck, T. Kennedy, E.H. Lieb, H. Tasaki, Commun. Math. Phys. 115, 477 (1988); I. Affleck, E.H. Lieb, T. Kennedy, H. Tasaki, Phys. Rev. Lett. 59, 799 (1987).
22. N. G. van Kampen, Stochastic processes in physics and chemistry (North- Holland, Amsterdam, 1981).
23. B. Derrida, M.R. Evans, V. Hakim, V. Pasquier, J. Phys. A26, 1493-1517 (1993).
24. R. A. Blythe, M. R. Evans, J. Phys. A Math. Theor. 40 (2007) R333-R441.
25. V. Karimipour, Phys. Rev. E 59, 205, (1999).
26. M. Khorrami and V. Karimipour, J. Stat. Phys. 100, 999 (2000) 29.

ON A POSSIBLE DEFINITION OF ENTANGLEMENT IN ANTISYMMETRIC STATES

TSUBASA ICHIKAWA

Research Center for Quantum Computing,
Interdisciplinary Graduate School of Science and Engineering, Kinki University,
3-4-1 Kowakae, Higashi-Osaka, Osaka 577-8502, Japan
E-mail: tsubasa@alice.math.kindai.ac.jp

TOSHIHIKO SASAKI

Graduate School of Science, The University of Tokyo,
7-3-1 Hongo, Bunkyo-ku, Tokyo 113-8654, Japan

IZUMI TSUTSUI

Theory Center, Institute of Particle and Nuclear Studies,
High Energy Accelerator Research Organization (KEK),
1-1 Oho, Tsukuba, Ibaraki 305-0801, Japan

Keywords: Identical particles; entanglement.

According to Schrödinger,[1] entanglement is *the* characteristic trait of quantum mechanics, the one that enforces its entire departure from classical lines of thought. The entanglement exhibits non-locality of composite systems by violation of Bell inequality,[2] which gives an upper bound to a combination of correlation functions of local observables. Recently, as rapid growth of quantum information science, more attention has been paid to the entanglement. Theory of the entanglement has been developed by making use of the tensor product structure of the state space: For instance, for a bipartite system of distinguishable particles, the state space is given as $\mathcal{H}_A \otimes \mathcal{H}_B$. Then, the local observables are in the form of $O_A \otimes O_B$. The entangled states are characterized as those not written in the form of a direct product $|\psi\rangle_A \otimes |\phi\rangle_B$. All the ingredients of the theory can be specified by using the tensor product structure.

On the other hand, the tensor product structure is no longer available when indistinguishability is taken into account, since the state space of the indistinguishable particles is the (anti)symmetrized subspce of the tensor product space. Thus the conventional theory for the entanglement of the distinguishable particles is not applicable to that of the indistinguishable particles, and a new theory is highly desirable.

There are two primary approaches to this problem. The first one, in Refs,[3,4] judges the entanglement of the indistinguishable particles by using the Slater decomposition, which is an analogue to the Schmidt decomposition of a state vector of the distinguishable particles. The second approach, in Ref,[5] is based on the observation by EPR[6] that the entangled state of the distinguishable systems has no element of reality. Only bi-separability and full separability are discussed.

In this summer school, we showed a possible framework to characterize entanglement of the fermionic pure states. Our framework is based on correlation between subsystems, which is clearly experimentally testable. More precisely, we showed the followings:

- Construction of the local operators on the fermionic systems by using a direct sum decomposition of one-particle Hilbert space.
- Criterion of the entanglement. The state is called entangled if the correlation function associated with the local operators cannot be decoupled.
- Relation to the other approaches. In spite of the difference of the points of departure, our criterion becomes identical to that of Ref,[5] in case of the bi-separability and the full separability. Thus the our approach gives a suitable generalization of Ref.[5]

The discussion shown in this summer school results in Ref.[7]

Acknowledgements

This work is supported in part by 'Open Research Center' Project for Private Universities, and the Grant-in-Aid for Scientific Research (C), No. 20540391-H21 and Global COE Program "the Physical Sciences Frontier", all of MEXT, Japan.

References

1. E. Schrödinger, *Proc. Cambridge Phil. Soc.* **31** (1935) 555-563.
2. J. S. Bell, *Physics* **1** (1964) 195.

3. J. Schliemann, J. Ignacio Cirac, M. Kuś, M. Lewenstein and D. Loss, *Phys. Rev.* **A64** (2001) 022303.
4. K. Eckert, J. Schliemann, D. Bruß and M. Lewenstein, *Ann. Phys.* **299** (2002) 88.
5. G.-C. Ghirardi, L. Marinatto and T. Weber, *J. Stat. Phys.* **108** (2002) 49.
6. A. Einstein, B. Podolsky and N. Rosen, *Phys. Rev.* **47** (1935) 777.
7. T. Ichikawa, T. Sasaki and I. Tsutsui, *J. Math. Phys.* **51** (2010) 062202.

ENTANGLEMENT MEASURES FOR INTERMEDIATE SEPARABILITY

TOSHIHIKO SASAKI

Graduate School of Science, The University of Tokyo,
7-3-1 Hongo, Bunkyo-ku, Tokyo 113-8654, Japan
E-mail: tsasaki@post.kek.jp

TSUBASA ICHIKAWA

Research Center for Quantum Computing,
Interdisciplinary Graduate School of Science and Engineering, Kinki University,
3-4-1 Kowakae, Higashi-Osaka, Osaka 577-8502, Japan

IZUMI TSUTSUI

Theory Center, Institute of Particle and Nuclear Studies,
High Energy Accelerator Research Organization (KEK),
1-1 Oho, Tsukuba, Ibaraki 305-0801, Japan

We present a set of entanglement measures R_m for n-qubit quantum states. Since the measure R_m vanishes if and only if a state is separable into m subsystems, this measure R_m indicates whether the state can be decomposed into m subsystems. This R_m for $m = n$ is equivalent to Meyer–Wallach measure while R_m for $m = 2$ is effectively equivalent to the one recently introduced by Love *et al.* In order to show the utility of our measures for characterising globally entangled states, the measures R_m are evaluated explicitly for the GHZ, W and Dicke states.

Keywords: Entanglement measure; entanglement.

1. Introduction

Quantum entanglement is by now regarded as the most significant property of states in quantum mechanics which is distinguished from classical, common sensical pictures of our physical world. In the conceptual respect, it was the EPR paper[1] which pointed out first the intriguing nonlocal aspect of correlations implied by entangled states, which was later took up by Bell[2] and others[3] for actual verification in laboratory, leading to a series of experiments[4–6] which confirmed the distinctive correlations admitted by quantum states. More recently, it is reported that the speed of the 'influence'exerted by quantum entanglement is measured to be faster than that of light by at least four orders of magnitude.[8]

On account of this physical significance, entanglement will undoubtedly furnish a vital resource to our future technology, in particular in the realization of quantum computation and cryptography. Application in these fields requires us both to measure and manipulate the states of entangled n-qubit systems at a high level of accuracy. Over the last two decades, there have been a variety of proposals to characterize entangled states, such as the use of *canonical forms*, *entanglement witnesses* and *entanglement measures*.[10,11] These are in fact quite powerful for quantifying entanglement for a few small n cases, or essentially bipartite or tripartite systems, but the characterization becomes almost intractable for large n because of the possibly exponentially many structures of the entanglement the states can form, even restricting to pure states. It seems obvious, therefore, that for characterization of entanglement for generic n-qubit systems, we need to seek some means which are specifically designed for the purposes we want to achieve.

In this respect, we know that the entanglement measure proposed by Mayer and Wallach[12] is simple and yet handy for examining if the state is fully separable or not. Recently, Love *et al.*[13] proposed a measure which is also simple and acts as a means to examine if the state is globally entangled or not. In this article, we present a set of new entanglement measures which, in a sense, connect the two. More specifically, our measures can be used to examine if the state is separable into m subsystems for arbitrary $2 \leq m \leq n$, where the case $m = n$ yields exactly the Mayer–Wallach measure whereas the case $m = 2$ corresponds, essentially, to the measure proposed in.[13] Moreover, they can also be used to characterize globally entangled states in general, and this is demonstrated by the GHZ and W states in n-qubit systems, where we observe rather contrasting behaviors under the measures introduced here. Our measures are also evaluated for Wk states, which are introduced as extended W states, to show that the measures with different m reveal distinct properties of entanglement of the states with fixed n and k.

2. Entanglement Measures as Yardsticks of Separability in Arbitrary Subsystems

Consider an n-qubit system whose quantum states are described by the Hilbert space $\mathbb{C}^{2n}$. Let $\mathcal{P} = \{s_i\}_{i=1}^{m}$ be a partition of the set $\mathcal{N} = \{1, \ldots, n\}$, *i.e.*,

$$\bigcup_{i=1}^{m} s_i = \mathcal{N} \text{ and } s_i \cap s_j = \emptyset \text{ for } i \neq j. \tag{1}$$

Each subset s_i determines a corresponding subsystem of the total system $\mathbb{C}^{2n}$ by identifying the elements in s_i to the labels of the constituent subsystems (qubits) in the total system, and for this reason we also use s_i to refer to the subsystem specified by the subset. We denote by $\bar{s}_i$ the subset (and hence the corresponding subsystem) complementary to s_i in $\mathcal{N}$ with $s_i \cup \bar{s}_i = \mathcal{N}$.

Now, given a pure state $|\psi\rangle$, let ρ_{s_i} be the reduced density matrix in the subsystem s_i obtained by taking the trace of the density matrix $\rho = |\psi\rangle\langle\psi|$ over the complementary space $\bar{s}_i$. Letting also $|s_i|$ be the number of elements in the subset s_i, we then see that the quantity

$$\eta_{s_i}(\psi) = N(|s_i|)\left\{1 - \mathrm{tr}(\rho_{s_i})^2\right\} \qquad \text{where} \quad N(|s_i|) = \frac{2^{|s_i|}}{2^{|s_i|} - 1} \tag{2}$$

vanishes $\eta_{s_i}(\psi) = 0$ iff the state $|\psi\rangle$ is separable with respect to s_i and $\bar{s}_i$. Note that the normalization factor $N(|s_i|)$ in (2) ensures that $\eta_{s_i}(\psi) = 1$ when the reduced state is maximally mixed $\rho_{s_i} = \frac{1}{2^{|s_i|}} I$. Observe also from the fact $\mathrm{tr}(\rho_{s_i})^2 = \mathrm{tr}(\rho_{\bar{s}_i})^2$ that

$$(1 - 2^{-|s_i|})\eta_{s_i}(\psi) = (1 - 2^{-|\bar{s}_i|})\eta_{\bar{s}_i}(\psi), \tag{3}$$

showing that both $\eta_{s_i}(\psi)$ and $\eta_{\bar{s}_i}(\psi)$ carry essentially the same content of information about the state. From the quantities η_{s_i} in (2) considered for all the subsets s_i in $\mathcal{P}$, we evaluate the 'average' value for the partition $\mathcal{P}$ by the arithmetic mean,

$$\xi_{\mathcal{P}}(\psi) = \frac{1}{m} \sum_{i=1}^{m} \eta_{s_i}(\psi). \tag{4}$$

Clearly, we have $\xi_{\mathcal{P}}(\psi) = 0$ iff the state $|\psi\rangle$ is separable according exactly to the partition $\mathcal{P}$ of the total set $\mathcal{N}$.

Out of all possible partitions $\mathcal{P}$ of $\mathcal{N}$, we may choose those $\mathcal{P}$ consisting of m subsets for $2 \leq m \leq n$, and evaluate the geometric mean of the quantities $\xi_{\mathcal{P}}(\psi)$. Namely, if $d(\mathcal{P})$ is the number of subsets of the partition $\mathcal{P}$, we consider

$$R_m(\psi) := \left(\prod_{d(\mathcal{P})=m} \xi_{\mathcal{P}}(\psi) \right)^{1/S(n,m)}, \tag{5}$$

where

$$S(n,m) = \left\{ {n \atop m} \right\} = \sum_{k=1}^{m} \frac{(-1)^{m-k} k^{n-1}}{(k-1)!\,(m-k)!} \tag{6}$$

is the Stirling number in the second kind, which represents the number of all possible partitions of the integer n into m subsets, or the number of partitions $\mathcal{P}$ with $d(\mathcal{P}) = m$. The quantities $R_m(\psi)$ defined above are useful since

$$R_m(\psi) = 0 \quad \Longleftrightarrow \quad |\psi\rangle \text{ is separable (at least) in } m \text{ subystems in } \mathbb{C}^{2n}. \tag{7}$$

Now the point is that, since $R_m(\psi)$ are formed from $\eta_{s_i}(\psi)$ which are all entanglement monotone, each of them, $R_m(\psi)$, $m = 2, \ldots, n$, qualifies as an entanglement measure. In particular, for $m = n$ where the subsets s_i, $i = 1, \ldots, n$, correspond to all the constituent subsystems, we have

$$R_n(\psi) = \frac{1}{n} \sum_{i=1}^{n} 2 \left\{ 1 - \mathrm{tr}(\rho_{s_i})^2 \right\}, \tag{8}$$

which is precisely the Meyer-Wallach measure. At the other end $m = 2$, we have the partitions $S = \{s_1 = s, s_2 = \bar{s}\}$. Choosing the subset s so that $|s| \le |\bar{s}|$, and noting $S(n,2) = 2^{n-1} - 1$, we find

$$R_2(\psi) = \left(\prod_{1 \le |s| \le |\bar{s}|}{}' \; c(s)\, \eta_s(\psi) \right)^{1/(2^{n-1}-1)}, \tag{9}$$

where the prime indicates that both subsets s and $\bar{s}$ are included when $|s| = |\bar{s}|$, and the coefficients $c(s)$ in (9) are given by

$$c(s) = 1 - \frac{1}{2} \cdot \frac{2^{|\bar{s}|} - 2^{|s|}}{2^n - 2^{|s|}}. \tag{10}$$

The measure $R_2(\psi)$ is equivalent to the measure proposed recently by Love *et al.*, apart from the factor $c(s)$ which varies between the maximum value $c(s) = 1$ for $|s| = |\bar{s}|$ and the minimum value $c(s) = 1/2 + 1/(1 - 2^{1-n}) > 1/2$ for $|s| = 1$.

3. Illustrating R_m for some Classes of Globally Entangled States

We now evaluate the amount of entanglement possessed by the two familiar globally entangled states, the GHZ and the W states, by means of the new measures $R_m(\psi)$ introduced above. These are particular states which

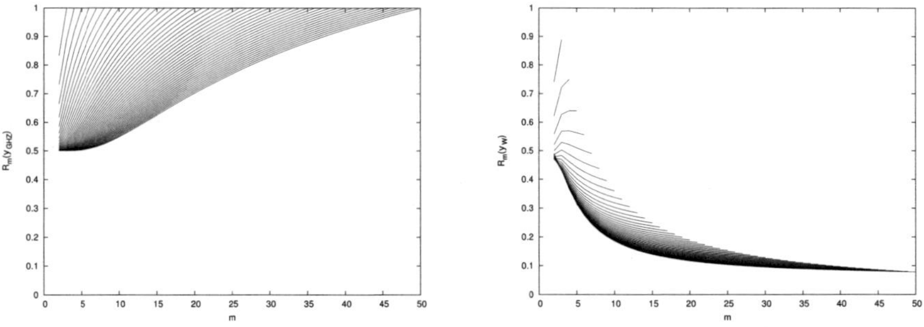

Fig. 1. The entanglement measures R_m evaluated for the GHZ state $\psi_{\rm GHZ}$ (left) and the W state $\psi_{\rm W}$(right) as functions of m for $3 \le n \le 50$. Each curve represents R_m for various $3 \le m \le n$ with a fixed value of n which can be read off by the right end value of m of the curve.

are invariant under all permutations of constituent subsystems, and this exchange symmetry allows us to facilitate the computation considerably.

To see this, we first note that for those symmetric states the quantity η_{s_i} in (2) depends only on the number of the elements $|s_i|$ of the subset s_i, not on the choice of the elements in s_i. To find the value of the measure $R_m(\psi)$, we need to consider all possible partitions $\mathcal{P}$ with $d(\mathcal{P}) = m$ to get the quantity $\xi_{\mathcal{P}}(\psi)$ in (4), but again the exchange symmetry implies that $\xi_{\mathcal{P}}(\psi)$ depends only on the way the partition $\mathcal{P}$ is formed in terms of the set of numbers $|s_i|$ of the elements in the subsets s_i comprising $\mathcal{P}$. To be more explicit, let us choose the numbering of the subsets s_i in the order $|s_1| \le |s_2| \le \cdots \le |s_m|$ and introduce the notation,

$$|\mathcal{P}| := \{|s_1|, |s_2|, \ldots, |s_m|\}. \tag{11}$$

Note that $|\mathcal{P}|$ furnishes an ordered partition of the integer n into $d(\mathcal{P}) = m$ nonvanishing integers by $n = |s_1| + |s_2| + \cdots + |s_m|$. For n and m with $2 \le m \le n$, let $P(n,m)$ be the set of all distinct ordered partitions of the integer n into m nonvanishing integers. Given some $|\mathcal{P}| \in P(n,m)$, we denote by $h(|\mathcal{P}|)$ the total number of partitions $\mathcal{P}$ sharing the same ordered partition $|\mathcal{P}|$. The measure $R_m(\psi)$ in (7) can then be calculated by the product of $\xi_{|\mathcal{P}|}(\psi) := \xi_{\mathcal{P}}(\psi)$ for all different $|\mathcal{P}|$ in $P(n,m)$, *i.e.*,

$$R_m(\psi) = \left(\prod_{|\mathcal{P}| \in P(n,m)} \left[\xi_{|\mathcal{P}|}(\psi)\right]^{h(|\mathcal{P}|)} \right)^{1/S(n,m)}. \tag{12}$$

Now we consider the n-qubit GHZ state,

$$\psi_{\rm GHZ} = \frac{1}{\sqrt{2}} \left(|11 \cdots 1\rangle + |00 \cdots 0\rangle \right). \tag{13}$$

The GHZ state is quite special since it has $\mathrm{tr}(\rho_{s_i})^2 = 1/2$ for all subsystems s_i, and from this we obtain

$$\eta_{s_i}(\psi_{\rm GHZ}) = \frac{1}{2} N(|s_i|) = \frac{2^{|s_i|-1}}{2^{|s_i|}-1}, \tag{14}$$

confirming that the quantity η_{s_i} in (2) depends on s_i only through the number $|s_i|$ of its elements. To illustrate our procedure to obtain the measures, let us consider, *e.g.*, the case $n = 4$, $m = 2$, for which the set $P(4,2)$ consists of the two elements, $|\mathcal{P}| = \{1,3\}$ and $\{2,2\}$. The numbers of partitions with the same $|\mathcal{P}|$ are, respectively, $h(\{1,3\}) = 4!/(1!3!) = 4$ and $h(\{2,2\}) = 4!/(2!2!2!) = 3$, confirming $S(4,2) = \sum_{|\mathcal{P}| \in P(4,2)} h(|\mathcal{P}|) = 7$. We then find

$$\xi_{\{1,3\}}(\psi_{\rm GHZ}) = \frac{11}{14}, \qquad \xi_{\{2,2\}}(\psi_{\rm GHZ}) = \frac{2}{3}, \tag{15}$$

and hence

$$R_2(\psi_{\rm GHZ}) = \left\{ \left(\frac{11}{14}\right)^4 \cdot \left(\frac{2}{3}\right)^3 \right\}^{\frac{1}{7}} \approx 0.732. \tag{16}$$

This procedure can be applied for any n and m, and the results up to $n = 50$ are shown in Fig. 1.

Next, we consider the W state,

$$\psi_{\rm W} = \frac{1}{\sqrt{n}} (|100 \cdots 0\rangle + |010 \cdots 0\rangle + \cdots + |000 \cdots 1\rangle), \tag{17}$$

for which we find

$$\eta_{s_i}(\psi_{\rm W}) = N(|s_i|) \, \frac{2|s_i|(n - |s_i|)}{n^2}. \tag{18}$$

For comparison, if we choose the case $n = 4$, $m = 2$, we find

$$\xi_{\{1,3\}}(\psi_{\rm W}) = \frac{33}{56}, \qquad \xi_{\{2,2\}}(\psi_{\rm W}) = \frac{2}{3}, \tag{19}$$

and hence

$$R_2(\psi_{\rm W}) = \left\{ \left(\frac{33}{56}\right)^4 \cdot \left(\frac{2}{3}\right)^3 \right\}^{\frac{1}{7}} \approx 0.621, \tag{20}$$

which is less than the value (16) of the GHZ state. As in the GHZ case, the results up to $n = 50$ are shown in Fig. 1.

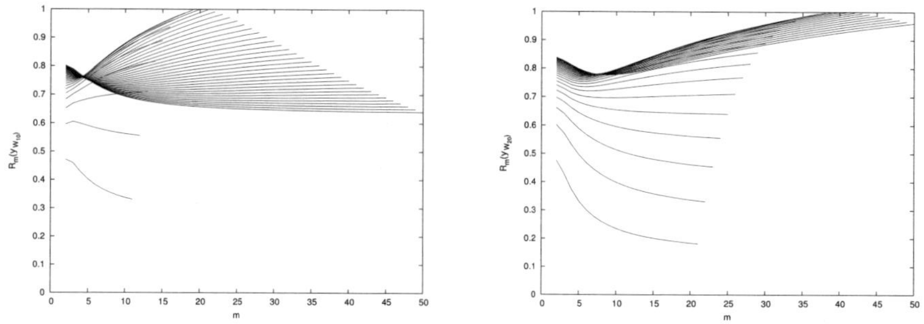

Fig. 2. The entanglement measures R_m evaluated for the W10 state (left) and the W20 state (right) as functions of m for $k+1 \leq n \leq 50$. Each curve represents R_m for various $k+1 \leq m \leq n$ with a fixed value of n which can be read off by the right end value of m of the curve.

It is clear from Fig. 1 that the GHZ and the W states exhibit rather contrasting behaviours for the entanglement measures R_m. Namely, for the GHZ state, R_m is a monotonically increasing function of m confined within $1/2 < R_m \leq 1$ approaching the value $R_n = 1$ at the right end $m = n$. In cotrast, for the W state, R_m is basically a decreasing function of m confined in $0 < R_m < 1/2$, except for the small $n < 9$ for which R_m can rise above the value $1/2$. These can also be seen directly from the formulae (14) and (18).

We may further consider a set of states which are totally symmetric with more than one $|1\rangle$ states in the subsystems. To be more explicit, we introduce the Wk states, or Dicke states, for $1 \leq k \leq n-1$ by

$$\psi_{\mathrm{W}k} := \binom{n}{k}^{-1/2} \left(|\underbrace{11\cdots1}_{k}\underbrace{00\cdots0}_{n-k}\rangle + \text{permutations} \right), \tag{21}$$

where 'permutations' means that all possible distinct terms possessing k '1's and $(n-k)$ '0's obtained by permutations of the first term are included, which altogether amount to $\binom{n}{k} = \frac{n!}{k!\,(n-k)!}$ terms. In particular, the $k = 1$ case gives the W state $\psi_{\mathrm{W}1} = \psi_{\mathrm{W}}$. For $k > 1$, the modified W states are slightly more involved than the W state but they are still manageable for the computation of the measures thanks to the symmetry.

To evaluate the measures, we first implement an appropriate unitary transformations to $\psi_{\mathrm{W}k}$ so that, for a given subsystem s_i, the state in s_i is represented by the left $|s_i|$ qubits in the n-qubit state. To proceed, it is more convenient to specify each of the terms in $\psi_{\mathrm{W}k}$ by the number of '1's, which is k for $\psi_{\mathrm{W}k}$, and an integer σ for $1 \leq \sigma \leq \binom{n}{k}$ labeling the

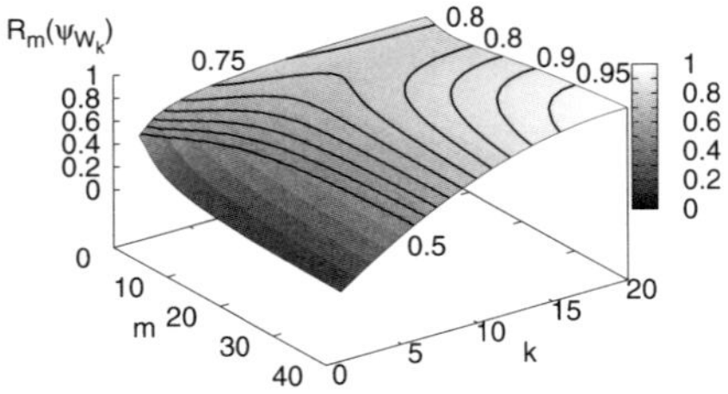

Fig. 3. The entanglement measures R_m evaluated for the Wk states as functions of m and k under the fixed value of $n = 40$.

distinct terms appearing in the permutations. Clearly, the same notation can be used for both of the subsystems s_i and $\bar{s}_i$ as well, and we may write an arbitrary term in (21) as the product of states in the two subsystems as $|k, \sigma\rangle = |r, \tau\rangle_{s_i} |k - r, \tau'\rangle_{\bar{s}_i}$, where $|r, \tau\rangle_{s_i}$ is a state of the subsystem s_i with r '1's and the label τ runs for $1 \le \tau \le \binom{|s_i|}{\tau}$, and similarly $|k - r, \tau'\rangle_{\bar{s}_i}$ is a state of the subsystem $\bar{s}_i$ with the label τ' running over $1 \le \tau' \le \binom{n-|s_i|}{k-r}$. This allows us to rewrite the Wk state (21) in the form,

$$\psi_{\mathrm{W}k} = \binom{n}{k}^{-1/2} \sum_{r=\max(|s_i|-(n-k),0)}^{\min(|s_i|,k)} \sum_{\tau,\tau'} |r, \tau\rangle_{s_i} |k - r, \tau'\rangle_{\bar{s}_i} . \tag{22}$$

The reduced density matrix is then found to be

$$\rho_{s_i} = \binom{n}{k}^{-1} \sum_{r=\max(|s_i|-(n-k),0)}^{\min(|s_i|,k)} \binom{n - |s_i|}{k - r} \sum_{\tau,\tau'} |r, \tau\rangle_{s_i} {}_{s_i}\langle r, \tau'| . \tag{23}$$

It is now straightforward to evaluate η_{s_i} to find

$$\eta_{s_i}(\psi_{\mathrm{W}k}) = N(|s_i|) \left\{ 1 - \binom{n}{k}^{-2} \sum_{r=\max(|s_i|-(n-k),0)}^{\min(|s_i|,k)} \binom{|s_i|}{r}^2 \binom{n - |s_i|}{k - r}^2 \right\} . \tag{24}$$

Based on the result (24), one can obtain the values of the measures R_m for the Wk states, and the outcomes are shown in Fig. 2 for the two cases $k = 10$ and $k = 20$. Interestingly, both of the $W10$ and $W20$ states show distinct behaviors which are also different from those of the GHZ states and W states mentioned above. Note that the measure R_m achieves the upper limit 1 when $n = m = 2k$. This can also be confirmed from (24) since for symmetric states $n = m$ implies $R_m(\psi) = \eta_{s_i}(\psi)$ with $|s_i| = 1$, which is 1 for $n = 2k$.

It is also interesting to look at the behaviors of the R_m measures for the Wk states as functions of k with some fixed n. This is done in Fig. 3, where we plot the values of R_m for Wk in the range $2 \leq m \leq 40$, $1 \leq k \leq 20$ when the total qubit is $n = 40$. We observe there that if we fix m, R_m as functions of k increases monotonically, indicating that, under any of the measures R_m, the Wk states are seen 'more entangled' for larger $k \leq n/2$. Moreover, one notices that the change in the values of the measure is in general more prominent for R_m with higher m. This implies that the Wk states change their entanglement property for larger k especially in the global aspect.

4. Physical Interpretations of the Measures R_m

Our measure R_m is one of the abstruct measures, which is compared with entanlement meausres defined by specific operation. Operationally specified entanglement measures have clear operational meanings by definition, although abstruct entanglement measures do not have clear physical meanings. Instead, some abstruct measures have scalability for the number of constituents. Since our measure R_m has this property, we can evaluate the state which consists of 50 constituents. In addition, we can relate the physical meaning to R_m.

The quantity η_{s_i}, or more generally the mean value $\xi_{\mathcal{P}}$ in (4) evaluated for the given partition $\mathcal{P}$, is related to the quantum Fisher information for the parameter estimation of the low-noise locally depolarizing channels whose actions for quantum states are specified by the partition $\mathcal{P}$.[14] Since the inverse of the quantum Fisher information gives the lower bound of the variance of estimators, we can provide the operational meaning for $\xi_{\mathcal{P}}$ as a measure of precision in the estimation of the strength of the low-noise locally depolarizing channels associated with $\mathcal{P}$. This relation between $\xi_{\mathcal{P}}$ and the quantum Fisher information implies that R_m may be interpreted as the quantum Fisher information for an assembly of the low-noise depolarizing channels under the condition that only the number of the local channels is known.

5. Conclusion

We have presented a set of entanglement measures R_m, $m = 2, \ldots, n$, for n-qubit states in which both the MW measure[12] and the measure proposed by Love *et al.*[13] arise as the two extreme cases $m = n$ and $m = 2$. Our measures R_m are scalable and can be used as indicators for separability of the n-qubit states into m subsystems. In addition, the value of R_m is operationally interpreted as the "averaged" Fisher information under low-noise depolarizing channels. We have seen by comparing the behaviors of the GHZ states, the W states and the Dicke states that the measures R_m are also useful to distinguish globally entangled states.

Acknowledgments

This work has been supported in part by the Grant-in-Aid for Scientific Research (C), No. 20540391-H20, and Global COE Program "the Physical Sciences Frontier", MEXT, Japan.

References

1. A. Einstein, B. Podolsky and N. Rosen, *Phys. Rev.* **47** 777 (1935).
2. J. S. Bell, *Physics* **1** 195 (1964).
3. J. F. Clauser, M. A. Horne, A. Shimony and R. A. Holt, *Phys. Rev. Lett.* **23** 880 (1969).
4. A. Aspect, J. Dalibard and G. Roger, *Phys. Rev. Lett.* **49** 1804 (1982).
5. G. Weihs, T. Jennewein, C. Simon, H. Weinfurter and A. Zeilinger, *Phys. Rev. Lett.* **81** 5039 (1998).
6. M. Rowe, D. Kielpinski, V. Meyer, C. Sackett, W. Itano, C. Monroe and D. Wineland, *Nature* **409** 791 (2001) .
7. G. C. Ghirardi, A. Rimini and T. Weber, *Lett. Nuovo Cimento* **27** 293 (1980).
8. D. Salart, A. Baas, C. Branciard, N. Gisin and H. Zbinden, *Nature* **454** 861 (2008).
9. M. C. Nielsen and I. L. Chuang, *Quantum Computaion and Quantum Information*, Cambridge university press, Cambridge, 2000.
10. R. Horodecki, P. Horodecki, M. Horodecki and K. Horodecki, Rev. Mod. Phys. **36** 865 (2009).
11. M. B. Plenio and S. Virmani, *Quant. Inf. Comput.* **7** 1 (2007).
12. D. A. Meyer and N. R. Wallach, *J. Math. Phys.* **43** 4273 (2002).
13. P. J. Love et al. *Quant. Inf. Processing* **6** 187 (2007).
14. S. Boixo and A. Monras, *Phys. Rev. Lett.* **100** 100503 (2008).

UNRUH EFFECT ON QUANTUM TELEPORTATION AND ENTANGLEMENT: IMPLICATIONS ON BLACK HOLE INFORMATION

K. SHIOKAWA

National Center for Theoretical Sciences,
National Cheng-Kung University, Tainan 701, Taiwan
and
Advanced Science Institute,
Institute for Physical and Chemical Research (RIKEN),
Wako-shi 351, Japan
E-mail: kshiok@riken.jp

Non-Markovian dynamics of quantum entanglement and teleportation process with continuous variables carried by accelerated observers under the influence of the environment is studied. In the full time nonperturbative evolution of quantum entanglement , the environment is found to play a dual role: It creates multipartite entanglement among observers, while suppressing them via decoherence. Motivated by the black hole information problem, we consider quantum teleportation between static and accelerated observers. Acceleration of the observer suppresses fidelity of teleportation. Information can escape outside of the horizon in the form of bipartite and tripartite entanglement during the teleportation process. Time dependence of information loss is calculated. In addition to the loss due to the interaction with the environment, the intrinsic loss in a measurement process was analysed.

Keywords: Entanglement; teleportation; Unruh effect; decoherence.

1. Introduction

The equivalence principle tells us that a particle under constant acceleration can be viewed as a particle under the influence of a static gravitational field. Explicit coordinate transformation shows that a particle moving in a hyperbolic trajectory due to constant acceleration in Minkowski spacetime is equivalent to a particle moving in a geodesic in Rindler spacetime. Under the coordinate tranformation, positive frequency modes in Minkowski spacetime are expressed as the mixture of positive and negative frequency Rindler modes. Thus the vacua for Minkowski modes and Rindler modes are not equivalent. As shown in,[1] an accelerated observer sees the Minkowski vacuum filled with Rindler particles. The spectrum of Rindler particles is thermal, similarly to the spectrum of the Hawking radiation from a black-hole. The Unruh effect modifies the quantum dynamics of accelerated systems in a nontrivial way causing apparent discrepancy between the interpretation of radiation emission and absorption between accelerated and static observers.[2]

Stimulated by recent progress in quantum information science, various attempts to understand relativistic quantum phenomena from the view of quantum information have been made.[3] Quantum teleportation with an accelerated observer was discussed in.[4] Their results, however, may not hold in the general situation.[5] The entanglement and fidelity in accelerated systems were discussed in.[6]

Semiclassical analysis predicts that the black hole emits thermal radiation similarly to the black body.[7] Coherence of the matter falling into the black hole appears to be lost.[8] Although entanglement in the Hawking pair is considered by many as crucial in order to rescue information from a black hole, it may not be sufficient for the complete retrieval of information.[9] Other nonlocal processes may also be required.[10] If the collapsing matter and incoming Hawking particle at the black hole singularity are maximally entangled, information may escape from a black hole[11] in the process analogous to quantum teleportation.[12] Although the original scenario in[11] is likely flawed by the interaction between the collapsing matter and the incoming Hawking flux,[13] qubit teleportation with the random unitary interaction between the original state and one of the entangled pair[14] indicates that nearly perfect information can be restored with proper encoding and quantum error correction. It is of interest to see how these scenarios based on quantum circuits are modified in curved spacetimes.

Motivated by these works, in this paper, we study the dynamical evolution of quantum entanglement and fidelity during quantum teleportation with accelerated motion of observers. The accelerated observer follows the hyperbolic trajectory which is a geodesic in the Rindler spacetime. Since the Rindler metric is equivalent to the Schwartzchild metric near horizon, our work can be viewed as simulating the information flow around the black hole horizon using the accelerated observer.

We start from developing a general framework to study the entanglement among an arbitrary number of observers generalizing the formalism in[15] to the relativistic setting in Sec. 2.[16] Each observer is carrying a quantum state moving in a four-dimensional spacetime interacting with quantum fields. Tracing out quantum fields, we obtain the reduced density matrix for $\mathcal{N}$ observers' quantum state. Each observer's state is assumed to be a continuous variable, which couples linearly with a scalar quantum field. In this work, we consider the vacuum state of the field as environment. We are primarily interested in the quantum state of observers and trace out all the field modes in our calculation. Then we study the entanglement dynamics among observers' states when each observer is allowed to move independently in a prescribed trajectory. When one of the observers is subjected to constant acceleration, his worldline in a Minkowski coordinate is a hyperbolic curve and asymptotically approaches the event horizon. The Unruh effect causes the observer to see the vacuum as a thermally excited state whose effective temperature depends on the acceleration. Entanglement among his and other observers' states will be modified accordingly.

Perturbative approximations such as Born or Born-Markov approximations are widely used for the study of decay processes and estimation of various quantities of open systems. Naive perturbation expansions, however, do not guarantee the positivity of the reduced density matrix. Entanglement properties are measured by the negativity of the partially transposed density matrix and sensitive to the positivity of the density matrix. For this reason, in this paper, we try to use an exact expression without making perturbative approximations. Nonperturbative analysis is one of the key achievements in solving the apparent paradox in the radiation-reaction process viewed in inertial and accelerated frames.[17]

Fluctuations of the canonical phase variables obey uncertainty relations. Uncertainties are modified in the presence of the environment. In the system out of equilibrium, they are time-dependent. In order to set up the quantum teleportation protocol[12] using our continuous variables, we introduce another observer who carries a state to be teleported. In Sec. 3, we obtain uncertainty relations for multi-modes from a correlation matrix of the system. Also from the correlation matrix after partial transpose, we calculate the bipartite entanglement measures for two and three modes from which we calculate genuine tripartite entanglement. We give a detailed study of the multi-mode uncertainties and follow their dynamical evolution. We are particularly interested in the case when one observer is under constant acceleration and study how his motion affects the uncertainties. For the same reason as entanglement mentioned above uncertainty relations are sensitive to the positivity of the density matrix. Thus although uncertainty relations should always be satisfied by any physical states in reality, they are no longer guaranteed under perturbative analysis. Next we study bipartite entanglement among these observers. As shown in,[18] uncertainty relations in the partially transposed density matrix give the necessary and sufficient condition for separability. Entanglement measures, negativity and the log-negativity, will also be calculated from the uncertainties.

Next we study the relation between bipartite and tripartite entanglement. For three modes, it is possible to find the genuine tripartite entanglement E_{ABC} from the monogamy relation.[19,20] In the conventional quantum optical setup for the teleportation,[21] the vacuum is meant for a ground state of one mode and treated locally. Thus there is no tripartite entanglement in the system. In our case, vacuum is a ground state of the quantum field which spreads nonlocally in spacetime. Therefore its influence on the observers can develop nonlocal correlation among their states. In order to see this, we set the initial state in a product state between a coherent and a two-mode squeezed state and see how the system develops tripartite entanglement dynamically in time.

In Sec. 4, we perform quantum teleportation using an entangled pair one of which is carried by a receiver under constant acceleration. We include the effect of quantum field continuously interacting with the system and calculate time-dependent teleportation fidelity as a measure of success of teleportation. We start from studying the generalized final state to which the sender's and one of the observers' state is projected after the measurement. While the experimental setting assumes the maximally entangled final state due to homodyne measurements, we generalize it to an arbitrary two-mode squeezed state and study its effect on the fidelity. We send the measurement result to the receiver followed by the appropriate unitary transformation on his state in order to recover the original state with generic initial and final squeezed states and evaluate how much information can be recovered. We consider both in the presence and absence of the interaction between the system and field. In all cases, we obtain the exact dynamical expressions for the fidelity. Finally we consider the maximally entangled initial and final states including the environment. We will study the full time evolution and the asymptotic long time, weak coupling limit of the fidelity. We discuss the implications of our results on the black hole information problem in Sec. 5. We will use natural units: $\hbar = c = k_B = 1$.

2. General Formulation

We consider the system composed of $\mathcal{N}$ observers moving in prescribed trajectories. Our Hamiltonian is a relativistic generalization of a $\mathcal{N}$-Brownian oscillator model studied in[15,22] and given by

$$H_S = \sum_{j=1}^{\mathcal{N}} \frac{P_j^2}{2M_j} + V_0(R_1, ..., R_{\mathcal{N}}) \,, \tag{1}$$

where the bare potential V_0 is a sum of the physical potential

$$V(R_1, ..., R_{\mathcal{N}}) = \sum_{j=1}^{\mathcal{N}} \frac{M_j \Omega_j^2}{2} R_j^2 \tag{2}$$

and the counter term ΔV

$$\Delta V(R_1, ..., R_{\mathcal{N}}) = \sum_{j=1}^{\mathcal{N}} \sum_{l=1}^{\mathcal{N}} \frac{\Delta V_{jl}}{2} R_j R_l. \tag{3}$$

We will use the form

$$\Delta V_{jl} = \frac{2\lambda_j \lambda_l \Lambda}{\pi} \tag{4}$$

with the frequency cutoff Λ for the scalar field in a three-dimensional space for our calculation.

Each oscillator variable R_j couples linearly with the scalar field ϕ_j at x_j as

$$H_I = \sum_{j=1}^{\mathcal{N}} \lambda_j \; R_j(\tau_j) \phi_j(x_j(\tau_j), t(\tau_j)). \tag{5}$$

Our convention includes the case in which different variables R_i and R_j $(i \neq j)$ couple different fields $\phi_i \neq \phi_j$ or the same field $\phi_i = \phi_j$. We assume that R_j is an internal coordinate of a j-th observer which depends on a proper time τ_j of the observer. Our total Hamiltonian is thus Lorentz invariant.

The field Hamiltonian for each ϕ_j in a D-dimensional space is

$$H_F = \frac{1}{2} \int d^D x \left[\Pi_{\phi_j}^2 + (\partial_x \phi_j)^2 + m_j^2 \phi_j^2 \right]. \tag{6}$$

Each field ϕ_j allows a mode decomposition:

$$\phi_j(x,t) = \int \frac{d^D k}{(2\pi)^{D/2} \sqrt{2\omega_k}} \left\{ b_k e^{-i\omega_k t + ikx} + b_k^\dagger e^{i\omega_k t - ikx} \right\} \tag{7}$$

with $\omega_k \equiv \sqrt{k^2 + m^2}$.

The Heisenberg equations that R_j satisfy are

$$M_j\ddot{R}_j(\tau) + M_j\Omega_j^2 R_j + \sum_{l=1}^{\mathcal{N}} \Delta V_{jl} R_l + 2\sum_{l=1}^{\mathcal{N}} \int_0^{\tau_j} ds \alpha_{Ijl}(\tau_j, s) R_l(s) = 0, \quad (8)$$

where

$$\alpha_{Ijl}(\tau_j, \tau_l') = -\lambda_j\lambda_l \sum_k \cos\left[k(x_j(\tau_j) - x_l(\tau_l'))\right] \sin\left[\omega_k(t(\tau_j) - t(\tau_l'))\right]/2\omega_k \quad (9)$$

is an imaginary part of the response function[24] defined as $\alpha_{jl}(\tau_j, \tau_l') \equiv \lambda_j\lambda_l \sum_k \cos\left[k(x_j(\tau_j) - x_l(\tau_l'))\right] e^{-i\omega_k(t(\tau_j)-t(\tau_l'))}/2\omega_k$, where $\sum_k \equiv \int d^D k/(2\pi)^D$. We set the initial condition at $t = 0$ hypersurface when all observers are at rest with their proper times set to $\tau_j = 0$ for all j. Eq. (8) contains nonlocal kernels $\alpha_{Ijl}(\tau_j, \tau_l')$ and the exact time evolution of R_j at τ depends on its and other observer's past history.

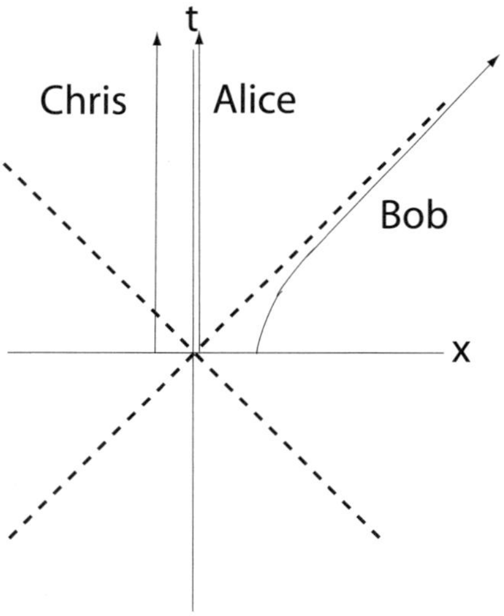

Fig. 1. The world line of three observers.

Let us consider the case $\mathcal{N} = 3$ in which the first observer Bob is accelerated with a constant acceleration a in an accelerated trajectory so that its coordinate is given by $t(\tau) = \sinh(a\tau)/a$ and $x_1(\tau) = \cosh(a\tau)/a$. The second observer Alice and the third observer Chris's locations are fixed at $x_2 = 0$ and $x_3 = -L < 0$, respectively. We assume that Alice and Chris are close to each other ($L << 1$) and share a common environment

($\phi_2 = \phi_3$), while the accelerated observer Bob is located in a distance such that R_1 can be treated as coupling independently to an environment. The response functions (9) are then given by

$$\alpha_{11}(\tau,\tau') = \lambda_1^2 \sum_k \cos\left[k(x_1(\tau) - x_1(\tau'))\right] e^{-i\omega_k(t(\tau)-t(\tau'))}/2\omega_k,$$

$$\alpha_{jj}(\tau,\tau') = \lambda_j^2 \sum_k e^{-i\omega_k(\tau-\tau')}/2\omega_k \text{ (for } j=2,3),$$

$$\alpha_{23}(\tau,\tau') = \alpha_{32}(\tau,\tau') = \lambda_2\lambda_3 \sum_k \cos(Lk) e^{-i\omega_k(\tau-\tau')}/2\omega_k, \qquad (10)$$

and $\alpha_{12} = \alpha_{13} = \alpha_{21} = \alpha_{31} = 0$. Note that the coordinate time t and the proper time are the same for Alice and Chris but not for Bob. The most interesting term above is $\alpha_{11}(\tau)$ where each mode in Minkowski spacetime is expressed in Rindler coordinates. The conversion of modes in different spacetimes is given by the Bogoliubov transformation between annihilation/creation operators.[23] Writing the momentum space integral in (7) in terms of Rindler modes reproduces a standard derivation of the Unruh effect based on the Bogoliubov transformation.[23] For a massless field $\phi_{1,2}$ in a three dimensional space, the kernels in (10) can be brought to the following form:

$$\alpha_{11}(\tau) = \frac{\lambda_1^2}{2\pi^2}\int_0^\Lambda dkk\left[\cos(k\tau)\coth(\frac{\pi k}{a}) - i\sin(k\tau)\right],$$

$$\alpha_{jj}(\tau) = \frac{\lambda_j^2}{2\pi^2}\int_0^\Lambda dkk\left[\cos(k\tau) - i\sin(k\tau)\right], \text{ (for } j=2,3),$$

$$\alpha_{23}(\tau) = \alpha_{32}(\tau) = \frac{\lambda_2\lambda_3}{\pi^2 L}\int_0^\Lambda dk \sin(Lk)\left[\cos(k\tau) - i\sin(k\tau)\right]. \qquad (11)$$

The kernel α_{23} is responsible for the interaction between Alice and Chris's state that becomes negligible for large separation $L >> 1/\Lambda$, which is a characteristic decay of a massless field in three dimension and justifies neglecting the interaction between the accelerated observer and the rest of observers. Comparing the first kernel α_{11} with that of the influence functional,[24] we see that the first observer sees the vacuum as a thermal bath with the Unruh temperature $T_U = a/2\pi$.[25,26] Writing the influence kernel directly in the Rindler time has advantages in studying quantum dynamical aspects of observer's quantum states under the influence of the Unruh effect, which we will describe below in details.

In this paper, we consider the small separation limit $L \to 0$ for simplicity. Bipartite entanglement dynamics for the finite L case was studied in[15] and our result can be easily extended to this case. For large Λ, the Heisenberg equations of motion for three coordinate variables can be written in the local form:

$$M_1\ddot{R}_1(t) + M_1\Omega_1^2 R_1(t) + \gamma_1\dot{R}_1(t) = 0, \tag{12}$$

$$M_2\ddot{R}_2(t) + M_2\Omega_2^2 R_2(t) + \gamma_2\dot{R}_2(t) + \gamma\dot{R}_3(t) = 0, \tag{13}$$

$$M_3\ddot{R}_3(t) + M_3\Omega_3^2 R_3(t) + \gamma_3\dot{R}_3(t) + \gamma\dot{R}_2(t) = 0,$$

where we write $\gamma_j \equiv \lambda_j^2/\pi$ (for $j = 1, 2, 3$) and $\gamma \equiv \lambda_2\lambda_3/\pi$. The divergence from nonlocal terms are canceled by counter terms. For our case, $\Delta V_{12} = \Delta V_{21} = \Delta V_{13} = \Delta V_{31} = 0$. Hereafter we will set $M_1 = M_2 = M_3 = 1$.

We write the solution for the first equation in (12) with initial conditions $R_1(0) = 0$ and $\dot{R}_1(0) = 1$ as $g_1(\tau)$:

$$g_1(\tau) = \frac{\sin(\Omega_{1r}\tau)}{\Omega_{1r}} e^{-\gamma_1\tau}, \tag{14}$$

where $\Omega_{1r}^2 \equiv \Omega_1^2 - \gamma_1^2$. A pair of solutions of (13) R_2 and R_3 with initial conditions $R_2(0) = R_3(0) = 0$ and $\dot{R}_2(0) = 1$, $\dot{R}_3(0) = 0$ will be written as $h_3(\tau)$ and $h_5(\tau)$. For $\Omega_2 = \Omega_3$ and $\lambda_2 = \lambda_3$, the solutions are given by $h_3(t) \equiv (g_3(t) + g_0(t))/2$ and $h_5(t) \equiv (g_3(t) - g_0(t))/2$, where

$$g_3(t) = \frac{\sin(\Omega_{2r}t)}{\Omega_{2r}} e^{-\gamma t} \text{ and } g_0(t) = \frac{\sin(\Omega_2 t)}{\Omega_2} \tag{15}$$

are the solutions corresponding to two normal modes of a coupled oscillator and $\Omega_{2r}^2 \equiv \Omega_2^2 - \gamma^2$.

General solutions with arbitrary initial conditions R_{j0} and P_{j0} of the Heisenberg equations (12) and (13) for $j = 1, 2, 3$ are

$$R_j(\tau) = \sum_{k=1}^{3} C_{R_jR_k} R_{k0} + \sum_{k=1}^{3} C_{R_jP_k} P_{k0} + \lambda_j \int_0^\tau ds g_{2l_j-1}(\tau - s)\phi_j(s),$$

$$P_j(\tau) = \sum_{k=1}^{3} C_{P_jR_k} R_{k0} + \sum_{k=1}^{3} C_{P_jP_k} P_{k0} + \lambda_j \int_0^\tau ds g_{2l_j}(\tau - s)\phi_j(s), \tag{16}$$

where $g_{2j} \equiv \dot{g}_{2j-1}$ and $l_1 = 1, l_{2,3} = 2$.

The expectation value of phase space variables can be expressed in the matrix form:

$$\begin{pmatrix} \langle R_1 \rangle \\ \langle P_1 \rangle \\ \langle R_2 \rangle \\ \langle P_2 \rangle \\ \langle R_3 \rangle \\ \langle P_3 \rangle \end{pmatrix} = \mathcal{C} \begin{pmatrix} R_{10} \\ P_{10} \\ R_{20} \\ P_{20} \\ R_{30} \\ P_{30} \end{pmatrix} = \begin{pmatrix} C_{R_1R_1} & C_{R_1P_1} & 0 & 0 & 0 & 0 \\ C_{P_1R_1} & C_{P_1P_1} & 0 & 0 & 0 & 0 \\ 0 & 0 & C_{R_2R_2} & C_{R_2P_2} & C_{R_2R_3} & C_{R_2P_3} \\ 0 & 0 & C_{P_2R_2} & C_{P_2P_2} & C_{P_2R_3} & C_{P_2P_3} \\ 0 & 0 & C_{R_3R_2} & C_{R_3P_2} & C_{R_3R_3} & C_{R_3P_3} \\ 0 & 0 & C_{P_3R_2} & C_{P_3P_2} & C_{P_3R_3} & C_{P_3P_3} \end{pmatrix} \begin{pmatrix} R_{10} \\ P_{10} \\ R_{20} \\ P_{20} \\ R_{30} \\ P_{30} \end{pmatrix}$$

The time evolution matrix $\mathcal{C}$ for our solutions is given by

$$\mathcal{C} \equiv \begin{pmatrix} f_1 & g_1 & 0 & 0 & 0 & 0 \\ f_2 & g_2 & 0 & 0 & 0 & 0 \\ 0 & 0 & f_3 & h_3 & f_5 & h_5 \\ 0 & 0 & f_4 & h_4 & f_6 & h_6 \\ 0 & 0 & f_5 & h_5 & f_3 & h_3 \\ 0 & 0 & f_6 & h_6 & f_4 & h_4 \end{pmatrix}, \tag{17}$$

where $f_{2j} \equiv \dot{f}_{2j-1}$, and $f_{2j-1} \equiv g_{2j} - 2\dot{g}_{2j}(0)g_{2l_j-1}$ for $j = 1$ and $f_{2j-1} \equiv h_{2j} - 2\dot{h}_{2j}(0)g_{2l_j-1}$ for $j = 2, 3$.

It is convenient for our purpose to use a characteristic function for the Wigner representation.[27] For the $\mathcal{N}$-particle system:

$$\chi_W(\mathcal{Y}, \tau) = \mathrm{Tr}\left[\rho(0)e^{i\sum_{j=1}^{2\mathcal{N}} Y_j X_j(\tau)}\right], \tag{18}$$

where we defined $X_{2j-1} \equiv \sqrt{\Omega_j}R_j$, $X_{2j} \equiv P_j/\sqrt{\Omega_j}$ for $j = 1, 2, ..., \mathcal{N}$ and $\mathcal{Y} \equiv (Y_1 ... Y_{2\mathcal{N}})$. The matrix $\mathcal{C}$ is scaled accordingly. Now we trace out the field ϕ in order to obtain the reduced dynamics of the system. In our case $\mathcal{N} = 3$ with a factorized initial condition: $\rho(0) = \rho_S(0) \otimes \prod_{j=1}^{2} \rho_{\phi_j}(0)$, χ_W is also factorized into two parts as $\chi_W(\mathcal{Y}, \tau) = \chi_W^S(\mathcal{Y}, \tau) \prod_{j=1}^{2} \chi_W^{\phi_j}(\mathcal{Y}, \tau)$. The system part $\chi_W^S(\mathcal{Y}, \tau) = \mathrm{Tr}_S\left[\rho_S(0)e^{i\sum_{j=1}^{2\mathcal{N}} Y_j X_{Cj}(\tau)}\right]$, where $X_{Cj}(\tau)$ are solutions of the Heisenberg equations with $\phi_j = 0$, and the environment parts

$$\chi_W^{\phi_1}(\mathcal{Y}, \tau) = \mathrm{Tr}_{\phi_1}\left[\rho_{\phi_1}(0) \exp\left[i\lambda_1 \sum_{l=1}^{2} \left\{ Y_l \int_0^\tau ds g_l(\tau - s) \right\} \phi_1(s) \right]\right],$$
$$\chi_W^{\phi_2}(\mathcal{Y}, \tau) = \mathrm{Tr}_{\phi_2}\left[\rho_{\phi_2}(0) \exp\left[i \left\{ \sum_{l=3}^{6} \lambda_{[l]} Y_l \int_0^\tau ds g_{(l)}(\tau - s) \right\} \phi_2(s) \right]\right], \tag{19}$$

where $[l]$ is equal to $l/2$ (for even l) and $(l+1)/2$ (for odd l) and $(l) = l$ for $l = 3, 4$ and $(l) = l - 2$ for $l = 5, 6$. In response to the state of the system, each field mode is shifted by system-bath interaction. After the time scale of each field mode, the back action of this process changes the system's state depending on the shift of the field mode. The collective effect of these processes leads to the non-Markovian evolution of the reduced system. Our results are valid for arbitrary temperatures. In the presence of ultra-short pulses whose duration is comparable to the cutoff frequency, nonlocal dissipation kernels in the Heisenberg equations will be modified and need to be treated explicity.[28]

The field characteristic function in (19) can be evaluated exactly. We assume the environment is initially in a Minkowski vacuum state. Its density matrix is given as $\rho_\phi(0) = \mid 0_M\rangle\langle 0_M \mid$. The environment characteristic function can be written as

$$\chi_W^{\phi_1}(\mathcal{Y},t)\chi_W^{\phi_2}(\mathcal{Y},t) = \exp\left[-\frac{1}{2}\mathcal{Y}^T\mathbf{\Sigma}\mathcal{Y}\right] \tag{20}$$

$$= \exp\left[-\frac{1}{2}(Y_1...Y_6)^T\begin{pmatrix} \Sigma_{11} & \Sigma_{12} & 0 & 0 & 0 & 0 \\ \Sigma_{21} & \Sigma_{22} & 0 & 0 & 0 & 0 \\ 0 & 0 & \Sigma_{33} & \Sigma_{34} & \Sigma_{35} & \Sigma_{36} \\ 0 & 0 & \Sigma_{43} & \Sigma_{44} & \Sigma_{45} & \Sigma_{46} \\ 0 & 0 & \Sigma_{53} & \Sigma_{54} & \Sigma_{55} & \Sigma_{56} \\ 0 & 0 & \Sigma_{63} & \Sigma_{64} & \Sigma_{65} & \Sigma_{66} \end{pmatrix}\begin{pmatrix} Y_1 \\ ... \\ Y_6 \end{pmatrix}\right],$$

where

$$\Sigma_{jl}(\tau) = \begin{cases} \frac{\lambda_1^2}{4\pi}\int_{-\infty}^{\infty}\frac{dk}{\omega_k}\int_0^\tau ds\int_0^\tau ds' g_j(\tau-s)g_l(\tau-s')e^{ik(x_1(s)-x_1(s'))}e^{-i\omega_k(t_1(s)-t_1(s'))} \\ \qquad \text{for } j,l=1,2 \\ \\ \frac{\lambda_{[j]}\lambda_{[l]}}{2\pi}\int_0^\infty\frac{d\omega}{\omega}\int_0^\tau ds\int_0^\tau ds' g_{(j)}(\tau-s)\cos\omega(s-s')g_{(l)}(\tau-s') \\ \qquad \text{for } j,l=3,...,6 \end{cases}$$

are time-dependent, nonequilibrium fluctuations of the system variables induced from the environment. Nonvanishing off-diagonal correlations Σ_{jl} for $j \neq l$ generated from the interaction with the environment are signatures of induced correlation and entanglement among observer's quantum states. For $j,l = 1,2$,

$$\Sigma_{jl}(\tau) = \frac{\lambda_1^2}{2\pi}\int_0^\infty d\omega\omega\coth(\frac{\omega\pi}{a})\int_0^\tau ds\int_0^\tau ds' g_j(\tau-s)g_l(\tau-s')\cos\omega(s-s').$$

If the initial states of system variables are all Gaussian states with vanishing mean positions and momenta, $\langle\mathcal{X}(0)\rangle = 0$, the system characteristic function also takes the Gaussian form:

$$\chi_W^S(\mathcal{Y},t) = \exp\left[-\frac{1}{2}\mathcal{Y}^T(\Delta\mathcal{X})_C^2(\tau)\mathcal{Y}\right] \tag{21}$$

$$\equiv \exp\left[-\frac{1}{2}(Y_1...Y_6)^T\begin{pmatrix} \langle\{X_{1C},X_{1C}\}\rangle & ... & \langle\{X_{C1},X_{C6}\}\rangle \\ & ... & \\ \langle\{X_{C6},X_{C1}\}\rangle & ... & \langle\{X_{C6},X_{C6}\}\rangle \end{pmatrix}\begin{pmatrix} Y_1 \\ ... \\ Y_6 \end{pmatrix}\right],$$

where $\{A,B\} \equiv (AB+BA)/2$ is an anticommutator and $\mathcal{X}_C = (X_{C1}...X_{C6})$ satisfy the equations of motion (12) and (13) for damped harmonic oscillators. $(\Delta\mathcal{X})_C^2(\tau)$ are essentially the initial fluctuations of the system variables shifted by the damped oscillatory motion of a coupled harmonic oscillator. Combining with the characteristic function for the field, we obtain

$$\chi_W(\mathcal{Y},t) = \exp\left[-\frac{1}{2}\mathcal{Y}^T\langle\{\mathcal{X},\mathcal{X}^T\}\rangle\mathcal{Y}\right] \tag{22}$$

$$= \exp\left[-\frac{1}{2}(Y_1...Y_6)^T\begin{pmatrix}\langle\{X_1,X_1\}\rangle & ... & \langle\{X_1,X_6\}\rangle \\ & ... & \\ \langle\{X_6,X_1\}\rangle & ... & \langle\{X_6,X_6\}\rangle\end{pmatrix}\begin{pmatrix}Y_1\\ ...\\ Y_6\end{pmatrix}\right]$$

$$= \exp\left[-\frac{1}{2}(Y_1...Y_6)^T\begin{pmatrix}\langle\{X_{1C},X_{1C}\}\rangle+\Sigma_{11} & ... & \langle\{X_{C1},X_{C6}\}\rangle+\Sigma_{16} \\ & ... & \\ \langle\{X_{C6},X_{C1}\}\rangle+\Sigma_{61} & ... & \langle\{X_{C6},X_{C6}\}\rangle+\Sigma_{66}\end{pmatrix}\begin{pmatrix}Y_1\\ ...\\ Y_6\end{pmatrix}\right].$$

The total correlation given by the sum of system uncertainty and induced fluctuations from environment $\langle\{\mathcal{X},\mathcal{X}^T\}\rangle = \langle\{\mathcal{X}_C,\mathcal{X}_C^T\}\rangle + \mathbf{\Sigma}$ should be understood as the fluctuation-dissipation relation generalized to systems arbitrarily far from equilibrium. If the system eventually reaches equilibrium, the usual fluctuation-dissipation theorem can be obtained in the long time limit. In our model, however, the interaction between Alice and Chris's state prevents the system from reaching equilibrium and the fluctuation-dissipation theorem will be violated similarly to two harmonic oscillators coupled to the same field studied in.[15]

3. Entanglement Dynamics of Accelerated Oscillators

A separability criterion for a bipartite two-level-system can be naturally extended to continuous Gaussian variables.[18] The necessary and sufficient condition for separability of the density matrix is to have only non-negative eigenvalues after the partial transpose of one of its subsystem. For Gaussian variables in the Wigner distribution, the partial transpose of a density matrix in one of the oscillator component is equivalent to a mirror reflection of that component. The necessary and sufficient condition for the general $(1+\mathcal{N})$ bipartite Gaussian modes to be separable is that the partially mirror reflected state is still a physical quantum state that satisfies the uncertainty principle.[29] For instance, for three oscillators, the bipartite entanglement between the first and the rest of oscillators can be measured by first taking a partial mirror reflection on the first variable in the phase space as $(R_1,P_1,R_2,P_2,R_3,P_3) \to (R_1,-P_1,R_2,P_2,R_3,P_3)$, then seeing if the resulting state satisfies or violates the uncertainty relation. Note that the total mirror reflection does not change the entanglement properties thus the reflection on the second and third variables yields the same result. In terms of $\mathcal{X}$, this can be expressed as a matrix multiplication by the matrix

$\eta \equiv \mathrm{diag}(1, -1, 1, 1, 1, 1)$ as $\mathcal{X} \to \eta\mathcal{X}$. It follows that the partial mirror reflection transforms the covariance matrix as

$$(\Delta\mathcal{X})^2 \to \eta(\Delta\mathcal{X})^2\eta^T. \tag{23}$$

From Williamson's theorem,[30] there exists a symplectic transformation that diagonalizes any positive-definite $2\mathcal{N} \times 2\mathcal{N}$ symmetric matrix into the following form:

$$(\Delta\mathcal{X}_D)^2 = \begin{pmatrix} \zeta_1 & 0 & \ldots & 0 \\ 0 & \zeta_1 & \ldots & 0 \\ \ldots & \ldots & \ldots & \ldots \\ 0 & \ldots & \zeta_{\mathcal{N}} & 0 \\ 0 & \ldots & 0 & \zeta_{\mathcal{N}} \end{pmatrix}. \tag{24}$$

Although such a symplectic transformation does not preserve the eigenvalue spectrum in general, the diagonal components ζ_l for $l = 1...\mathcal{N}$ can be calculated as follows. Writing a commutation relation in a $2\mathcal{N} \times 2\mathcal{N}$ matrix form as $[X_i, X_j] = i\Gamma_{ij}$ with

$$\Gamma = \begin{pmatrix} 0 & 1 & \ldots & 0 \\ -1 & 0 & \ldots & 0 \\ \ldots & \ldots & \ldots & \ldots \\ 0 & \ldots & 0 & 1 \\ 0 & \ldots & -1 & 0 \end{pmatrix},$$

we construct a real symmetric matrix $\Delta\mathcal{X}\Gamma(\Delta\mathcal{X})^2\Gamma^T\Delta\mathcal{X}$. This matrix has an eigenvalue spectrum ζ_l^2 $(l = 1...\mathcal{N})$.[31] The uncertainty relation can be generalized to a symplectic invariant form $(\Delta\mathcal{X})^2 + i\Gamma/2 \geq 0$. By changing to the diagonalized form $(\Delta\mathcal{X}_D)^2$ above, the uncertainty relation is equivalent to saying that $\zeta_l \geq 1/2$ for all l.

3.1. *Bipartite entanglement*

For a bipartite continuous variable system, there is a criteria for a necessary and sufficient condition for separability.[18,32,33] We write the 2×2 covariance matrix here as

$$(\Delta\mathcal{X})^2 \equiv \langle\{\mathcal{X}, \mathcal{X}^T\}\rangle \equiv \begin{pmatrix} D_1 & A \\ A^T & D_2 \end{pmatrix}, \tag{25}$$

then eigenvalues ζ_l^2 can be written explicitly in terms of the following symplectic invariants $\Delta_{1,2}$ constructed from the determinants of covariances

$|A|$, $|D_{1,2}|$, $|(\Delta\mathcal{X})^2|$ as

$$\Delta_1 = |D_1| + |D_2| + 2|A|,$$
$$\Delta_2 = |(\Delta\mathcal{X})^2|.$$

Then

$$\zeta_\pm^2 = \tfrac{1}{2}\left[\Delta_1 \pm \sqrt{\Delta_1^2 - 4\Delta_2}\right]. \tag{26}$$

Under the partial transpose (23) without the third varible, $A \to -A$ and these eigenvalues will be changed to

$$\lambda_\pm^2 = \tfrac{1}{2}\left[\Delta_1 \pm \sqrt{\tilde{\Delta}_1^2 - 4\Delta_2}\right], \tag{27}$$

where $\tilde{\Delta}_1 \equiv |D_1| + |D_2| - 2|A|$. The separability conditions are

$$\lambda_\pm^2 \geq \tfrac{1}{4}. \tag{28}$$

Note that the inequalities for ζ_+ and λ_+ follow automatically from those for ζ_- and λ_-. Thus λ_- carries the essential information on the separability of quantum states.

Let us consider a two-mode squeezed state with a squeezing parameter r as an initial state.[34] Its correlation matrix is

$$(\Delta\mathcal{X})_C^2(0) \equiv \langle\{\mathcal{X}_C(0), \mathcal{X}_C^T(0)\}\rangle = \frac{1}{2}\begin{pmatrix} \cosh(2r)\mathbf{1} & -\sinh(2r)\sigma_3 \\ -\sinh(2r)\sigma_3 & \cosh(2r)\mathbf{1} \end{pmatrix}.$$

In the Wigner representation, the same state can be expressed as

$$W(R_1, R_2, P_1, P_2) = \frac{4}{\pi^2} e^{-e^{2r}\left[\Omega(R_1-R_2)^2+(P_1+P_2)^2/\Omega\right]-e^{-2r}\left[\Omega(R_1+R_2)^2+(P_1-P_2)^2/\Omega\right]}.$$

This state can be obtained by acting a squeezing operator $e^{ir(R_1P_2-P_1R_2)}$ on the vacuum. For a large squeezing $r \to \infty$, $W(R_1, R_2, P_1, P_2) \sim \delta(R_1 - R_2)\delta(P_1+P_2)$, thus it becomes the EPR (Einstein-Podolsky-Rosen) state.[35]

In Fig. 2, the temporal behavior of ζ_- (uncertainty) and λ_- (uncertainty after the partial transpose) is plotted. The initial state is a pure two mode squeezed state with $r = 0.5$. The state exhibits environment-induced disentanglement at $\Omega t \sim 0.4$ with no acceleration and at $\Omega t \sim 0.31$ with acceleration. The role of acceleration is similar to that of temperature that increases the disentanglement rate.

Now we study entanglement measures; the negativity and the log negativity. The negativity $\mathcal{N}$[36,37] for our system can be defined as

$$\mathcal{N} = \frac{||\rho_r^T|| - 1}{2}, \tag{29}$$

where ρ_r^T is the reduced density matrix after the partial transpose. $\mathcal{N}$ is equal to the sum of all negative eigenvalues of ρ_r^T. Thus it measures how much ρ_r^T fails to be positive. The Peres criteria[32] tell us that this can be used as a measure of entanglement. It is an entanglement monotone, which does not increase under local operations and classical communications. The logarithmic negativity $E_{\mathcal{N}}$ defined as

$$E_{\mathcal{N}} = \ln ||\rho_r^T|| \tag{30}$$

is also an entanglement monotone. The diagonalization of $(\Delta\mathcal{X})^2$ changes the original state into the thermal state. Then the partially transposed density matrix ρ_r^T after the same transformation also has the thermal form. It can be written in terms of the symplectic invariants $\lambda_\pm$ as

$$\rho_r^T = \prod_{\pm}\left[\left(\frac{2}{2\lambda_\pm + 1}\right)\sum_{n=0}^{\infty}\left(\frac{2\lambda_\pm - 1}{2\lambda_\pm + 1}\right)^n |n_\pm\rangle\langle n_\pm|\right]. \tag{31}$$

For separable states, $\lambda_\pm \geq 1/2$. Then $||\rho_r^T|| = 1$ and $\mathcal{N} = E_{\mathcal{N}} = 0$. For entangled states, $\lambda_- < 1/2$ but $\lambda_+ \geq 1/2$. $\mathcal{N}$ and $E_{\mathcal{N}}$ are monotonic functions of λ_- as

$$\begin{aligned} \mathcal{N} &= \max\left[0, \frac{1-2\lambda_-}{4\lambda_-}\right], \\ E_{\mathcal{N}} &= \max\left[0, -\ln(2\lambda_-)\right]. \end{aligned} \tag{32}$$

In Fig. 3, the negativity $\mathcal{N}$ and the logarithmic negativity $E_{\mathcal{N}}$ are shown as a function of the proper time. The initial state is a two mode squeezed state. They both vanish at the same time since they both give the necessary and sufficient condition of bipartite entanglement. The larger acceleration yields the larger disentanglement rate. The long time behavior of two-mode uncertainties is damped oscillatory and the exact disentangling behavior shows a series of decays and revivals. The detailed study of bipartite entanglement at low temperatures is available in[15] and we will not discuss further here.

3.2. *Tripartite entanglement dynamics*

It is convenient to characterize an n-mode uncertainty relation as a sum of symplectic invariants[29] as

$$\Sigma_n \equiv \sum_{j=0}^{n} 4^{j-n}(-1)^{n+j}\Delta_j^n, \tag{33}$$

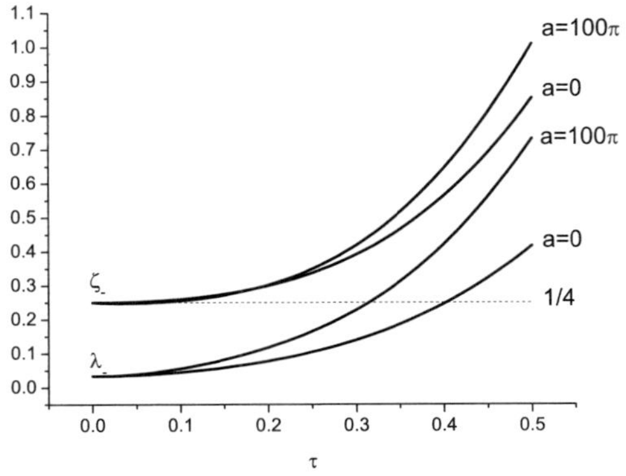

Fig. 2. The temporal evolution of uncertainties (ζ_- in Eq. (26) and λ_- in Eq. (27)) before and after the partial transpose in the presence and absence of Bob's acceleration is plotted. The initial condition is a two mode squeezed state with $r = 0.5$. Other parameters are $\gamma = 0.05$ and $\Lambda = 50$. Parameter values in each figure is chosen such that the characteristic behavior is most visible.

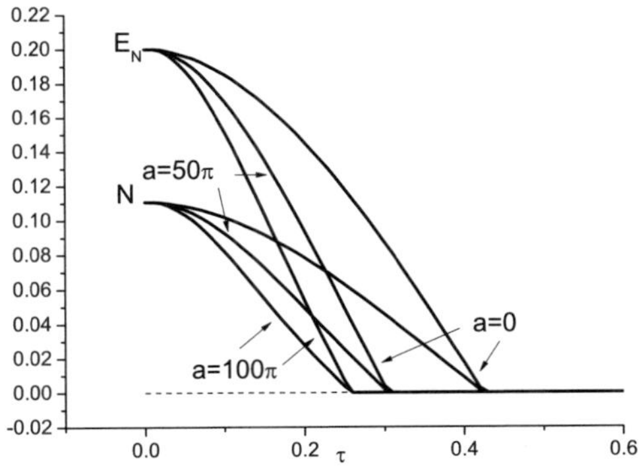

Fig. 3. The temporal evolution of the negativity (29) and the lognegativity (30) for different acceleration before and after the partial transpose is plotted. The initial condition is a two mode squeezed state with $r = 0.1$. Other parameters are $\gamma = 0.1$, $\Lambda = 50$.

where $\Delta_j^n (j = 1, ..., n)$ are principal minors[38] of the matrix $\Gamma(\Delta\mathcal{X})^2$ of order $2j$. We defined $\Delta_0^n \equiv 1$.

For the matrix in the Williamson normal form (24), $\Delta_j^n = \sum_{\{k_1,...,k_j\}} \zeta_{k_1}^2 ... \zeta_{k_j}^2 (j = 1, ..., n)$, where $\{k_1, ..., k_j\}$ is an unordered set of different integers between 1 and n. Then we have

$$\Sigma_n = \frac{(-1)^n}{4^n}(1 - \sum_{k_1=1}^{n} 4\zeta_{k_1}^2 + ... + (-4)^n \sum_{k_1=...=k_n=1}^{n} \zeta_{k_1}^2 ... \zeta_{k_n}^2) \tag{34}$$
$$= (\zeta_1^2 - \frac{1}{4})...(\zeta_n^2 - \frac{1}{4}).$$

Thus the uncertainty relation implies $\Sigma_n \geq 0$.

For the tripartite case $n = 3$, we write

$$(\Delta\mathcal{X})^2 = \begin{pmatrix} D_1 & A_{12} & A_{13} \\ A_{12}^T & D_2 & A_{23} \\ A_{13}^T & A_{23}^T & D_3 \end{pmatrix}. \tag{35}$$

Then

$$\begin{aligned} \Delta_1^3 &= |D_1| + |D_2| + |D_3| + 2|A_{12}| + 2|A_{13}| + 2|A_{23}|, \\ \Delta_2^3 &= |\hat{D}_1| + |\hat{D}_2| + |\hat{D}_3| + 2|\hat{A}_{12}| + 2|\hat{A}_{13}| + 2|\hat{A}_{23}|, \\ \Delta_3^3 &= |(\Delta\mathcal{X})^2|, \end{aligned} \tag{36}$$

where ˆstands for the codeterminant. In Fig. 4, Σ_3 is plotted as a function of time. $\Sigma_3 \geq 0$ throughout the entire evolution indicates that the uncertainty relation is always satisfied. The effect of acceleration is again similar to the effect of temperature which increases the uncertainty.

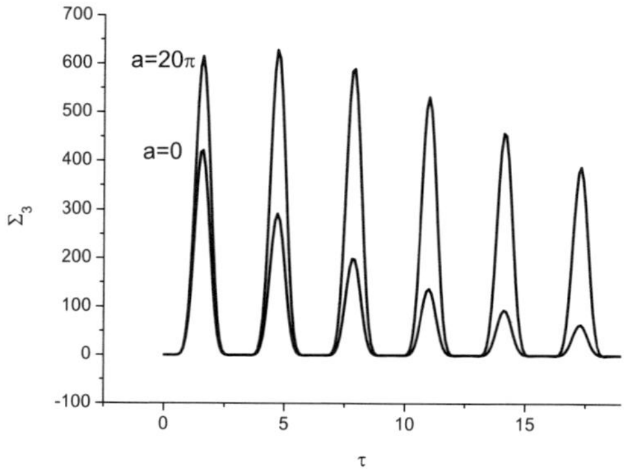

Fig. 4. The temporal evolution of the three-body uncertainty Σ_3 is shown. The initial condition is a two mode squeezed state with $r = 0.2$ and a ground state of Chris. Other parameters are $\gamma = 0.005$, $\Lambda = 50$. $\Sigma_3 \geq 0$ indicates that the uncertainty relation is always satisfied.

Under the partial transpose on the first variable, $|A_{12}|$,$|A_{13}|$,$|\hat{A}_{12}|$,$|\hat{A}_{13}|$ change signs and the uncertainty after the partial transpose $\tilde{\Sigma}_3$ becomes

$$\tilde{\Sigma}_3 \equiv -\frac{1}{64} + \frac{\tilde{\Delta}_1^3}{16} - \frac{\tilde{\Delta}_2^3}{4} + \tilde{\Delta}_3^3, \tag{37}$$

where

$$\begin{aligned}
\tilde{\Delta}_1^3 &= |D_1| + |D_2| + |D_3| - 2|A_{12}| - 2|A_{13}| + 2|A_{23}|, \\
\tilde{\Delta}_2^3 &= |\hat{D}_1| + |\hat{D}_2| + |\hat{D}_3| - 2|\hat{A}_{12}| - 2|\hat{A}_{13}| + 2|\hat{A}_{23}|, \\
\tilde{\Delta}_3^3 &= |(\Delta\mathcal{X})^2|.
\end{aligned}$$

From (37),

$$\tilde{\Sigma}_3 = (\lambda_1^2 - \frac{1}{4})...(\lambda_3^2 - \frac{1}{4}). \tag{38}$$

For separable states, each term on the right hand side is nonnegative, thus $\tilde{\Sigma}_3 \geq 0$ follows. For our interest of $(1+2)$ mode bipartite system, or more generally, $(1+n)$ bipartite system, $\tilde{\Sigma}_{(1+n)} \geq 0$ is shown to give a necessary and sufficient condition for separability.[29]

Three-mode entanglement for continuous variables can be categorized into the following sets[39] based on the values of the bipartite entanglement measure E_i with $i = A(BC), (AB)C, (AC)B$ ($E_{A(BC)}$ is bipartite entanglement between Alice and Bob-Chris together): (1) fully entangled states (none of E_i is vanishing.), (2) one-mode biseparable states (only one of E_i is vanishing.), (3) two-mode biseparable states (only two of E_i are vanishing.), (4) three-mode biseparable states (all of E_i are vanishing but these states cannot be written as a mixture of tripartite product states), (5) fully separable states (mixture of tripartite product states).

For bipartite entanglement for three qubits, the square of the concurrence C is known to satisfy the following monogamy inequality[19]

$$C^2_{A(BC)} \geq C^2_{AB} + C^2_{AC}, \tag{39}$$

where C_{AB} is the concurrence between A and B, etc. The inequality also holds for Gaussian states if we take the square of the logarithmic negativity.[20] Genuine tripartite entanglement for continuous variables E_{ABC} can be defined by minimizing the difference between the left-hand and the right-hand side for all possible combinations of three subsystems, namely,

$$E_{ABC} = \min_{ijk} \left\{ E_{i(jk)} - E_{ij} - E_{ik} \right\}. \tag{40}$$

Note that the monogamy inequality guarantees $E_{ABC} \geq 0$. In Fig. 5, the temporal evolution of bipartite entanglement E_{AB},E_{AC},E_{BC} are shown. The initial state is chosen to be the product state of a two-mode squeezed

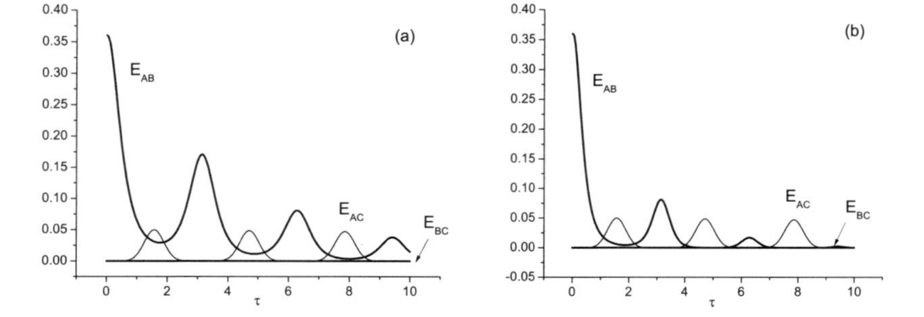

Fig. 5. The temporal evolution of bipartite entanglement E_{AB}, E_{AC}, and E_{BC} are plotted as a function of time. The initial condition is a product state of a two-mode squeezed state of Alice and Bob with $r = 0.3$ and a ground state of Chris. $a = 0$ in Fig. 6(a) and $a = 50\pi$ in Fig. 6(b). Other parameters are $\gamma = 0.001$, $\Lambda = 50$.

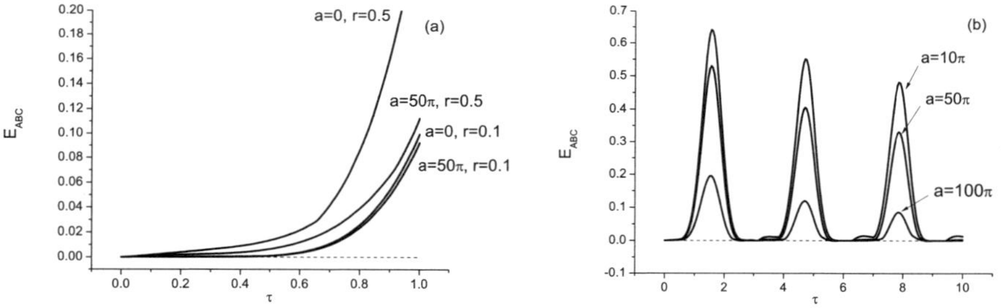

Fig. 6. The temporal evolution of the genuine tripartite entanglement E_{ABC} is plotted as a function of time. The initial condition and unshown parameters are the same as in Fig. 6. The short time behavior is shown in (a). While squeezing of the initial state enhances E_{ABC}, acceleration suppresses it. The long time behavior in (b) is damped oscillatory.

state of Alice and Bob and a coherent state of Chris. Thus initially $E_{AC} = E_{BC} = 0$. As time evolves, decoherence due to the interaction with the environment[40] causes E_{AB} to damp out. The interaction with the environment induces entanglement between Alice and Chris's state, which evolves periodically in time and some of entanglement in E_{AB} tranfers to E_{AC}. There is no entanglement between Bob and Chris's state during the entire period since there is neither initial nor interaction-induced entanglement between them. Note a dual role of the environment. It is both the source of multipartite entanglement between observers and of decoherence that decreases entanglement. The former causes initial increase and the latter causes damping of entanglement in time. The similar behavior is already seen in two static Brownian oscillators.[15] In Fig. 6, the temporal evolution of E_{ABC} is shown. The initial state is the same as in Fig. 5, a product state of a two-mode squeezed state of Alice and Bob and a ground state of Chris. Thus initially there is no genuine tripartite entanglement. The nonlocal vacuum generates entanglement among three variables. The same vacuum also induces decoherence which eventually washes away all entanglement. The Unruh effect suppresses both bipartite and tripartite entanglement.

4. Quantum Teleportation Dynamics with an Accelerated Observer

Quantum teleportation with continuous variables based on[21] has been demonstrated experimentally.[43] Initially Alice and Bob share a two-mode squeezed state. The mode carried by Alice is mixed with Chris's state prepared in a coherent state through the 50 % beam-splitter. Their state is measured by the homodyne detector and the result is sent to Bob through the classical channel. Bob's state is shifted according to the measurement result in order to reproduce the same state as Chris's.

In the scheme of quantum teleportation for qubits or continuous variables, after Chris's state is mixed with Alice's, their states are measured and the measurement result is send to Bob in a distance. In the continuous variable quantum teleportation, the homodyne measurement on Alice and Chris's state projects their state to the maximally entangled EPR state. The imprecision involved in the measurement can be included, for instance, by making Gaussian smearing of the EPR state. Here we take into account the finite uncertainty in the projected state by considering a two-mode squeezed final state with a squeezing parameter r_2. We denote the measurement result as a complex number $\beta = \beta_R + i\beta_I$, where β_R and β_I are both real numbers. Thus after the measurement, the original quantum state of Alice and Chris is projected to

$$|\beta\rangle_{AC} = \cosh^{-1} r_2 \sum_{n=0}^{\infty} \tanh^n r_2 \hat{D}(\beta)|n\rangle_A |n\rangle_C, \tag{41}$$

where the translation operator $\hat{D}(\beta)$ is given by

$$\hat{D}(\beta) = e^{\sqrt{2}i\beta_I \hat{x}_C - \sqrt{2}i\beta_R \hat{p}_C}. \tag{42}$$

In the limit $r_2 \to \infty$, $|\beta\rangle_{AC}$ becomes the maximally-entangled EPR state.

After the measurement of Alice and Chris's state, the density matrix for Bob becomes

$$\hat{\rho}_B = N_B {}_{AC}\langle\beta|\alpha\rangle_C \hat{\rho}_{AB} \, {}_C\langle\alpha|\beta\rangle_{AC}, \tag{43}$$

where $N_B = \cosh(r_1 - r_2)\cosh(r_1 + r_2)/\pi$ is a normalization constant. With this normalization, $P(\beta) = Tr_B \hat{\rho}_B$ is equal to the probability to obtain the measurement result β. Bob's state is shifted depending on the measurement result β so that the final state of Bob is $\hat{\rho}_{out} = \hat{D}(\beta)\hat{\rho}_B\hat{D}^\dagger(\beta)$. The teleportation fidelity F is an overlap between Bob's final state $\hat{\rho}_{out}$ and Chris's initial state $|\alpha_C\rangle$ and given by

$$F = \frac{{}_B\langle\alpha|\hat{\rho}_{out}|\alpha\rangle_B}{P(\beta)}. \tag{44}$$

Suppose we do not make post-selection following the measurement, the fidelity becomes, instead of (44),

$$F = \frac{{}_B\langle\alpha|\hat{\rho}_B|\alpha\rangle_B}{P(\beta)}. \tag{45}$$

In this case, the averaged fidelity over all measurement outcome β is

$$F_{av} = \int d^2\beta P(\beta)F(\beta) = \int d^2\beta_B\langle\alpha|\hat{\rho}_B|\alpha\rangle_B. \tag{46}$$

Similarly to (41) we write the initial two-mode squeezed state (29) in the Fock space representation as

$$|AB\rangle = \cosh^{-1} r_1 \sum_{n=0}^{\infty} \tanh^n r_1 |n\rangle_A |n\rangle_B. \tag{47}$$

In the absence of the environment, from (41), we obtain $F_{av} = (\cosh r_1)^{-2} \exp\left[-|\alpha|^2/\cosh^2 r_1\right]$ in the $r_2 \to \infty$ limit. Then $F_{av} \to 0$ as initial squeezing r_1 gets larger. This tells us that quantum teleportation cannot be successful without post-selection.

On the other hand, with post selection, the average fidelity can be shown to be

$$F_{av} = \frac{1 + \tanh r_1}{2}. \tag{48}$$

For a coherent state r_1, $F_{av} = 1/2$ and for any positive squeezing, $F_{av} > 1/2$. Since $F_{av} = 1/2$ is a maximal value achieved by the classical method.[44] We see that the initial entanglement in Alice and Bob's state is essential for quantum teleportation.

Now we consider the effect of the environment on the performance of teleportation. In our situation, the environment acts continuously on the system so that the final density matrix of Bob depends on how we send the measurement result of Alice and Chris to Bob. In order to eliminate this ambiguity, here we consider the effect of the environment up to the point when Alice and Chris's state is measured. We introduce the complex variables Δ_i $(i = 1, ..., 6)$ to simplify the expression as follows:

$$\begin{aligned}
\Delta_1 &= (\Delta\mathcal{X})^2_{11} - (\Delta\mathcal{X})^2_{22} + 2i(\Delta\mathcal{X})^2_{12},\\
\Delta_2 &= (\Delta\mathcal{X})^2_{11} + (\Delta\mathcal{X})^2_{22},\\
\Delta_3 &= (\Delta\mathcal{X})^2_{33} - (\Delta\mathcal{X})^2_{44} + 2i(\Delta\mathcal{X})^2_{34},\\
\Delta_4 &= (\Delta\mathcal{X})^2_{33} + (\Delta\mathcal{X})^2_{44},\\
\Delta_5 &= (\Delta\mathcal{X})^2_{13} - (\Delta\mathcal{X})^2_{24} + 2i((\Delta\mathcal{X})^2_{14} + (\Delta\mathcal{X})^2_{23}),\\
\Delta_6 &= (\Delta\mathcal{X})^2_{13} + (\Delta\mathcal{X})^2_{24} + 2i((\Delta\mathcal{X})^2_{14} - (\Delta\mathcal{X})^2_{23}).
\end{aligned} \tag{49}$$

With a measurement outcome β, ${}_B\langle\alpha|\hat{\rho}_{out}|\alpha\rangle_B$ and $P(\beta)$ can be calculated straightforwardly. Both can be written in the Gaussian form:

$${}_B\langle\alpha|\hat{\rho}_{out}|\alpha\rangle_B = N_1 \exp\left[-\Gamma_1(\alpha-\beta)^2 - \bar{\Gamma}_1(\bar{\alpha}-\bar{\beta})^2 - 2\Gamma_2|\alpha-\beta|^2\right]$$

and

$$P(\beta) = N_{P1} \exp\left[-\Gamma_3(\alpha-\beta)^2 - \bar{\Gamma}_3(\bar{\alpha}-\bar{\beta})^2 - 2\Gamma_4|\alpha-\beta|^2\right]. \tag{50}$$

The normalization factors and coefficients in exponents are both dependent on the correlation matrix and initial condition. As a result, they have the complicated time dependence as follows. We will write $c_i = \cosh(2r_i)$ and

$s_i = \sinh(2r_i)$ for $i = 1, 2$ below:

$$
\begin{aligned}
N_1 &= \frac{8N_B}{(d_2 d_3)^{1/2}}, \\
\Gamma_1 &= \frac{\bar{p} f_1^2 + p\bar{f}_2^2 + 2q f_1 \bar{f}_2}{d_3 d_2^2 d_1^2} - \frac{s_2^2 \Delta_1}{d_2}, \\
\Gamma_2 &= \frac{\bar{p} f_1 f_2 + p\bar{f}_1 \bar{f}_2 + q(|f_1|^2 + |f_2|^2)}{d_3 d_2^2 d_1^2} - \frac{s_2^2 c_4}{d_2}, \\
c_3 &= c_2 + \Delta_2, \\
c_4 &= c_2 + \Delta_2 - \frac{d_1}{c_2 - 1}, \\
b_1 &= \bar{\Delta}_1 \Delta_5^2 + \Delta_1 \Delta_6^2 + 2c_3 \Delta_5 \Delta_6, \\
b_2 &= \bar{\Delta}_1 \Delta_5 \bar{\Delta}_6 + \Delta_1 \bar{\Delta}_5 \Delta_6 + c_3(|\Delta_5|^2 + |\Delta_6|^2), \\
d_1 &= (c_2 + \Delta_2)^2 - |\Delta_1|^2, \\
d_2 &= s_2^4 + d_1(c_2 + 1)^2 - 2s_2^2(c_2 + 1)(c_2 + \Delta_2), \\
d_3 &= q^2 - |p|^2, \\
d_4 &= (d_1(\Delta_4 + 1) - b_2)^2 - (d_1 \Delta_3 - b_1)^2, \\
p &= \Delta_3 - d_1^{-1}[\bar{\Delta}_1 \Delta_5^2 + \Delta_1 \Delta_6^2 + 2c_3 \Delta_5 \Delta_6], \\
&\quad - d_2^{-1} d_1^{-2} s_2^4[(|\Delta_1|^2 + c_3(c_3 + c_4))(\Delta_5^2 \bar{\Delta}_1 + \Delta_6^2 \Delta_1) \\
&\quad + 2\Delta_5 \Delta_6(|\Delta_1|^2(2c_3 + c_4) + c_3^2 c_4)], \\
q &= \Delta_4 + 1 - d_1^{-1}[\bar{\Delta}_1 \Delta_5 \bar{\Delta}_6 + \Delta_1 \bar{\Delta}_5 \Delta_6 + c_3(|\Delta_5|^2 + |\Delta_6|^2)] \\
&\quad + d_2^{-1} d_1^{-2} s_2^4[(|\Delta_1|^2 + c_3(c_3 + c_4))(\Delta_5 \bar{\Delta}_1 \bar{\Delta}_6 + \bar{\Delta}_5 \Delta_1 \Delta_6) \\
&\quad + (|\Delta_1|^2(2c_3 + c_4) + c_3^2 c_4)(|\Delta_5|^2 + |\Delta_6|^2)], \\
f_1 &= d_2 d_1 - s_2^3[\Delta_5(|\Delta_1|^2 + c_3 c_4) + \Delta_1 \Delta_6(c_3 + c_4)], \\
f_2 &= -s_2^3[\Delta_5 \bar{\Delta}_1(c_3 + c_4) + \Delta_6(|\Delta_1|^2 + c_3 c_4)], \\
\Gamma_3 &= d_1(d_1 \Delta_3 - b_1)/d_4, \\
\Gamma_4 &= d_1(d_1(\Delta_4 + 1) - b_2)/d_4, \\
N_{P1} &= \frac{4c_3 N_B}{d_4^{1/2}}.
\end{aligned} \tag{51}
$$

The averaged fidelity over all measurement outcome β is

$$
\begin{aligned}
F_{av} &= \int d^2\beta P(\beta) F(\beta) = \int d^2\beta_B \langle \alpha | \hat{\rho}_{out} | \alpha \rangle_B \\
&= \frac{4\pi N_B}{(d_2 d_3 d_5)^{1/2}},
\end{aligned} \tag{52}
$$

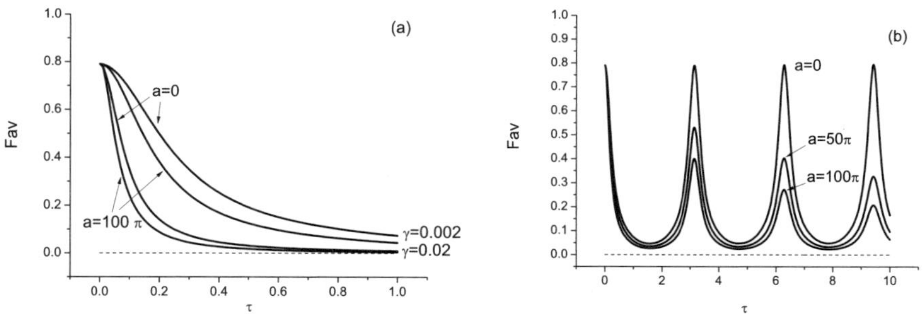

Fig. 7. The temporal evolution of the fidelity for the two mode squeezed final state with the squeezing parameter $r_2 = 1.0$ is shown. The initial state is a two mode squeezed state with the squeezing parameter $r_1 = 1.0$. $\gamma = 0.002$, and $\Lambda = 300$. The short time behavior in (a) shows that strong system-bath coupling and acceleration both suppress the fidelity. The long time behavior in (b) is damped oscillatory.

where

$$d_5 = \Gamma_2^2 - |\Gamma_1|^2. \tag{53}$$

In Fig. 7, the time evolution of the averaged fidelity is plotted. For the final state to be in a squeezed state with a finite squeezing parameter r_2, the initial fidelity is not one. The short time decay rate is increased with stronger system-field coupling and with larger acceleration of Bob as seen in Fig. 7(a). The long time evolution exhibits damped oscillations. The larger the acceleration, the larger the damping rate results.

The expressions in (51) become much simpler if there is no interaction with the environment. In this case, from (44),

$$F = \exp\left[-2|\alpha - \beta|^2 \frac{c_1 c_2 - s_1 s_2 + 1}{(c_1 + 1)(c_2 + 1)}\right] \tag{54}$$

and

$$F_{av} = \frac{2\cosh(r_1 - r_2)\cosh(r_1 + r_2)}{(c_1 + 1)(c_2 + 1) - s_1 s_2} \tag{55}$$

follow. When the initial and final states become maximally entangled, $r_1, r_2 \to \infty$, then $F_{av} \to 1$. Thus there is no loss of fidelity in this limit. In the absence of the environment, we achieve the unit fidelity for the quantum teleportation with initial and final EPR states, which is the case originally studied in.[45] We now consider the effect of the environment in this special case. Bob's state after measurement can be written similarly as in (43) as

$$\hat{\rho}_B = \frac{1}{\pi} \sum_{n,m=0}^{\infty} {}_A\langle n|_C\langle n|\hat{D}^\dagger(\beta)\hat{\rho}_{ABC}\hat{D}(\beta)|m\rangle_A|m\rangle_C. \tag{56}$$

Here we consider the influence of environment on all three observer's states and wrote their state in the density matrix form as $\hat{\rho}_{ABC}$. We find the averaged fidelity

$$F_{av} = \frac{1}{\left[(\tilde{\Sigma}_{11}+1)(\tilde{\Sigma}_{22}+1) - \tilde{\Sigma}_{12}^2\right]^{1/2}},$$

where $\tilde{\Sigma}_{ij}$ are components of a 2×2 matrix made up of the components of the 6×6 matrix $\mathbf{\Sigma}$ in (21). We first write $\mathbf{\Sigma}$ as

$$\begin{pmatrix} \Sigma_B & 0 & 0 \\ 0 & \Sigma_A & \Sigma_{AC} \\ 0 & \Sigma_{AC}^T & \Sigma_C \end{pmatrix} = \begin{pmatrix} \Sigma_{11} & \Sigma_{12} & 0 & 0 & 0 & 0 \\ \Sigma_{21} & \Sigma_{22} & 0 & 0 & 0 & 0 \\ 0 & 0 & \Sigma_{33} & \Sigma_{34} & \Sigma_{35} & \Sigma_{36} \\ 0 & 0 & \Sigma_{43} & \Sigma_{44} & \Sigma_{45} & \Sigma_{46} \\ 0 & 0 & \Sigma_{53} & \Sigma_{54} & \Sigma_{55} & \Sigma_{56} \\ 0 & 0 & \Sigma_{63} & \Sigma_{64} & \Sigma_{65} & \Sigma_{66} \end{pmatrix}. \tag{57}$$

Then

$$\tilde{\Sigma} = \sigma_3 \Sigma_B \sigma_3 + \Sigma_A - \Sigma_{AC}\sigma_3 - \sigma_3 \Sigma_{AC}^T + \sigma_3 \Sigma_C \sigma_3. \tag{58}$$

In Fig. 8, the time evolution of the average fidelity for the maximally entangled initial and final state is plotted. The decay is faster for larger acceleration. After the damped oscillatory behavior, all curves appear to reach nonvanishing asymptotic values. We will now show this is indeed the case.

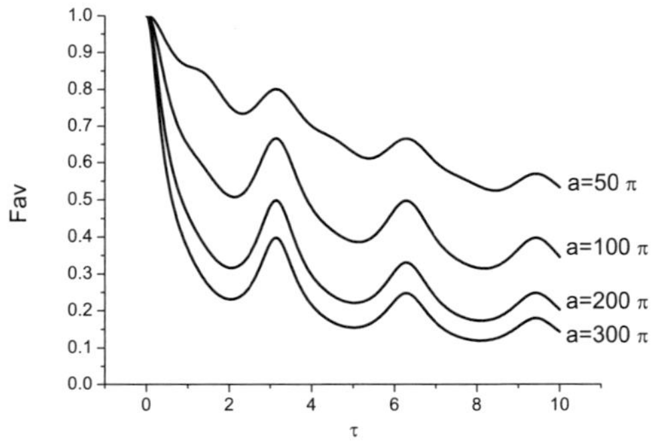

Fig. 8. The temporal evolution of the fidelity for the maximally entangled initial and final state is shown. $\gamma = 0.001$, $\Lambda = 50$.

In the long time limit, the state of Bob in the absence of Alice and Chris will thermalize at the Unruh temperature T_U similarly to the single particle quantum brownian motion at finite temperatures.[46] On the other hand, the state of Alice and Chris in the absence of Bob will be in the steady oscillatory state (see the discussion in[15]). Now let us first assume the interaction of Bob with the environment is negligible. In this case, $\Sigma_B = 0$. In the weak coupling limit,[15]

$$\Sigma_A = \Sigma_{AC} = \Sigma_C = \frac{1}{4}\begin{pmatrix} 1 & 0 \\ 0 & 1 \end{pmatrix}. \tag{59}$$

Then

$$\tilde{\Sigma} = \begin{pmatrix} 0 & 0 \\ 0 & 1 \end{pmatrix} \tag{60}$$

follows. This will give the averaged fidelity

$$F_{av} \to \frac{1}{\sqrt{2}} \sim 0.70. \tag{61}$$

Thus although there is still information loss, a part of quantum information manages to escape making quantum teleportation successful.[43] This case can be compared to,[13] where the authors pointed out that the unitary quantum gates between the incoming multi-level quantum state with one of the EPR pair prior to the measurement ruin the unitarity of the transfer matrix, and to,[14] where the author included random unitary gates in quantum circuits estimated the fidelity to be 0.85.

Now we take into account the interaction between Bob and the environment. This will add the thermal factor in induced fluctuations of Bob's state in the long time limit as

$$\Sigma_B = \frac{\coth(\Omega/2T_U)}{4}\begin{pmatrix} 1 & 0 \\ 0 & 1 \end{pmatrix}. \tag{62}$$

Then

$$\tilde{\Sigma} = \frac{\coth(\Omega/2T_U)}{4}\begin{pmatrix} 1 & 0 \\ 0 & 1 \end{pmatrix} + \begin{pmatrix} 0 & 0 \\ 0 & 1 \end{pmatrix}. \tag{63}$$

The averaged fidelity in this case will decrease to

$$F_{av} = \frac{2}{(\coth(\Omega/2T_U)+4)^{1/2}(\coth(\Omega/2T_U)+8)^{1/2}} \le \frac{2}{\sqrt{45}} \sim 0.30 < 0.5, \tag{64}$$

where the equality holds for zero acceleration ($T_U = 0$). We see that quantum teleportation fails in this case. In this sense, quantum information will be totally lost.

5. Implications

Many interacting quantum systems can be simulated with a proper design of quantum circuits,[41] where the interaction among subsystems are mimicked by quantum gates. With the suitably chosen measurement and post-selection scheme, we can simulate many interesting systems under nontrivial situations. In,[11] similarity between information escaping from a black hole and a quantum teleportation protocol was pointed out. In their analysis, the generic interaction between collapsing matter and a Hawking particle which can ruin unitarity was neglected.[13] With quantum error correction and proper encoding of quantum states, one may still be able to recover full information.[14] These authors speculate the possible quantum process in and around a black hole using analogies with quantum circuits.

From the equivalence principle, the acceleration due to gravitational force can be locally transformed away by the appropriate coordinate transformation. In the absence of other forces, the observer then follows a geodesic in a curved spacetime. The Unruh effect on a constantly accelerated observer can be identified as the effect on the static observer in Rindler spacetime. Since near the Rindler horizon, the Rindler spacetime looks similar to the Schwartzchild black hole spacetime, this is equivalent to studying the Hawking effect on the quantum dynamics of a static observer near a black hole event horizon.

With our general formulation in Sec. 2, we can study the quantum dynamics of many observers moving in an arbitrary fashion in arbitrary spacetime dimension. As a special case, we focused our study on the Unruh effect on various quantum teleportation protocols. The equivalence principle further allows our model to simulate nontrivial curved spacetime effects including black hole problems.

In the present work, we view our scheme as simulating a possibly occurring process around a black hole and are only interested until the time of the measurement. Bob can stop accelerating eventually and receive the message from Alice. This is indeed the situation analogous to a decaying black hole.[42] Although our system does not include gravitational interaction between observers, as we saw in Sec. 3, the background field induces the effective interaction and entanglement between them.

Our nonperturbative dynamical analysis on quantum entanglement and teleportation with accelerated systems is a priori independent from the black hole physics. Based on final state projection, nevertheless, there are some implications on the black hole information problem that can be summarized as follows. There are two fundamental sources of nonunitary evolutions and information loss studied in our model. (1) The intrinsic loss due to the measurement imposed on the evolution of the total system. In a black hole spacetime, the space-like singularity may provide an effective region on which the state of infalling matter and the incoming Hawking particle are projected. (2) The induced loss due to interaction of the quantum system with the surrounding quantum field. This is present in any spacetime and responsible for the quantum/classical correspondence of quantum systems of large degrees of freedom.[40] The loss is from the interaction among infalling matter, Hawking particles, and gravitational field of a black hole. Thus in addition to the correlation in a Hawking pair, there is induced three-body entanglement between collapsing matter and a Hawking pair, which varies dynamically due to gravitational interaction. In the practical setting of the laboratory, there are many other phenomenological loss of information such as photon loss during the optical transmission, leakage, etc. We did not consider these effects in this work.

At the present stage without complete knowledge of quantum gravity, we do not know the final state of the black hole. Assuming a maximally entangled final state, we may retrieve quantum information from a black hole just as in teleportation provided that (I) the induced loss due to vacuum is negligible. Since the environment is always present, this type of loss is unavoidable; (II) The postselection in the teleportation itself requires some information to leak outside the black hole. This may involve non-trivial spacetime topology with worm holes or baby universe. Imposing the condition on the final state in addition to the initial state can be viewed as an extension of the time-symmetric formulation of quantum mechanics to spacetime physics.[47,48] In the meantime, the postselection has now become a widely used technique in the laboratory. We hope that our attempts using accelerated observers provide a practical setting to understand the black hole information problem.

6. Acknowledgments

The author thanks H. Hofmann, A. Hosoya, and S. Lloyd, for interesting discussions.

References

1. W.G. Unruh, Phys. Rev. D**14**,870 (1976). Luis C.B. Crispino, Atsushi Higuchi, and George E.A. Matsas, Rev. Mod. Phys. **80**, 787-838 (2008) for a recent review.
2. W. G. Unruh and R. M. Wald, Phys. Rev. D**29**, 1047 (1984).

3. A. Peres and D. Terno, Rev. Mod. Phys. **76**, 93 (2004).
4. P. M. Alsing and G. Milburn, Phys. Rev. Lett. **91**, 180404 (2003).
5. R. Schützhold and W. G. Unruh, e-print quantph/0506028.
6. I. Fuentes-Schuller and R. B. Mann, Phys. Rev. Lett. **95**, 120404 (2005); F. Benatti and R. Floreanini, J. Phys. A**39**, 2689 (2006); S. Masser and P. Spindel, Phys. Rev. D**74**, 085031 (2006); G. Adesso, I. Fuentes-Schuller, M. Ericsson, Phys. Rev. A**76**, 062112 (2007); S. Y. Lin, C. H. Chou, and B. L. Hu, Phys. Rev. D**78**, 125025 (2008); Andre G. S. Landulfo and George E. A. Matsas, Physical Review A**80**, 032315 (2009).
7. S. W. Hawking, Nature (London) **248**, 30 (1974).
8. S. W. Hawking, Phys. Rev. D**14**, 2460 (1976).
9. D. Page, in *Proceedings of the 5th Canadian Conference on General Relativity and Astrophysics, University of Waterloo, 13-15 May, 1993*, edited by R. B. Mann an R. G. McLenaghan (World Scientific, Singapore, 1994) and references therein.
10. U. H. Danielsson and M. Schiffer, Phys. Rev. D**48**, 4779 (1993).
11. G. Horowitz and J. Maldacena, JHEP **02**, 008 (2004).
12. C. H. Bennett, G. Brassard, C. Crepeau, R. Jozsa, A. Peres, and W. K. Wootters, Phys. Rev. Lett. **70**, 1895 (1993).
13. D. Gottesman and J. Preskill, JHEP **03**, 026 (2004).
14. S. Lloyd, Phys. Rev. Lett. **96**, 061302 (2006).
15. K. Shiokawa, Phys. Rev. A**79**, 012308 (2009).
16. K. Shiokawa, gr-qc/0910.1715.
17. D. J. Raine, D. W. Sciama, and P. G. Grove, Proc. R. Soc. Lond, A**435**, 205 (1991).
18. R. Simon, Phys. Rev. Lett. **84**, 2726 (2000).
19. V. Coffman, J. Kundu, W. K. Wootters Phys. Rev. A **61**, 052306 (2000).
20. G. Adesso, A. Serafini, and F. Illuminati, Phys. Rev. A**73**, 032345 (2006).
21. S. L. Braunstein and H. J. Kimble, Phys. Rev. Lett. **80**, 869 (1998).
22. K. Shiokawa, Jour. Math. Chem. **45**, 175 (2009).
23. N. D. Birrel and P. C. W. Davis, *Quantum Fields in Curved Space*, (Cambridge University Press, Cambridge, 1982).
24. R. P. Feynman and F. L. Vernon, Ann. Phys. **24**, 118 (1963).
25. J. Anglin, Phys. Rev. D **47**, 4525 (1993).
26. A. Raval, B. L. Hu, and J. Anglin, Phys. Rev. D **53**, 7003 (1996).
27. E. Wigner, Phys. Rev. **40**, 749 (1932).
28. K. Shiokawa and B. L. Hu, Quantum Information Processing, **6**, 55 (2007).
29. A, Serafini, Phys. Rev. Lett. **96**, 110402 (2006).
30. J. Williamson, Amer. J. Math. 58, 141 (1936).
31. R. Simon, N. Mukunda, and B. Dutta, Phys. Rev. A **49**, 1567 (1994).
32. A. Peres, Phys. Rev. Lett. **77**, 1413 (1996).
33. P. Horodecki, Phys. Lett. A **232**, 333 (1997).
34. L. Mandel and E. Wolf, *Optical Coherence and Quantum Optics* (Cambridge University Press, Cambridge, 1995); D. Walls and G. J. Milburn, *Quantum Optics* (Springer, Berlin, 1995).
35. A. Einstein, B. Podolsky, and N. Rosen, Phys. Rev. **47**, 777 (1935).

36. G. Vidal and R. F. Werner, Phys. Rev. A**65**, 032314 (2002).
37. G. Adesso, A. Serafini, and F. Illuminati, Phys. Rev. A**70**, 022318 (2004).
38. B. Jeffreys, Methods of Mathematical Physics (Cambridge University Press 1999).
39. G. Giedke, B. Kraus, M. Lewenstein, and J. I. Cirac, Phys. Rev. A**64**, 052303-1 (2001).
40. *Decoherence and the Appearance of the Classical World in Quantum Theory*, eds. D. Giulini, *et al.*, (Springer, Berlin, 1996).
41. M. A. Nielsen and I. L. Chuang, *Quantum Computation and Quantum Information*, (Cambridge University Press, Cambridge, 2000).
42. R. D. Carlitz and R. S. Wiley, Phys. Rev. D**36**, 2336 (1987).
43. A. Furusawa, J. L. Sorensen, S. L. Braunstein, C. A. Fuchs, H. J. Kimble, E. S. Polzik, Science **282**, 706 (1998); H. F. Hofmann, T. Ide, T. Kobayashi, and A. Furusawa, Phys. Rev. A**62**, 062304 (2000).
44. S. L. Braunstein, C. A. Fuchs, and H. J. Kimble, Jour. Mod. Opt. **47**, 267 (2000).
45. L. Vaidman, Phys. Rev. A**49**, 1473 (1994).
46. A. O. Caldeira and A. J. Leggett,Physica A**121**, 587 (1983).
47. Y. Aharonov, P. G. Bergmann, and J. L. Lebowitz, Phys. Rev. **134**, B1410 (1964).
48. M. Gell-mann and J. B. Hartle, in *The Physical Origins of Time Asymmetry*, edited by J. Halliwell, J. Perez-Mercader, and W. H. Zurek (Cambridge University Press, Cambridge, 1994).

SYSTEMATIC CONSTRUCTION OF GENERALIZED BELL INEQUALITIES*

SHOGO TANIMURA

Department of Applied Mathematics and Physics, Graduate School of Informatics, Kyoto University, Kyoto 606-8501, Japan
E-mail: tanimura@i.kyoto-u.ac.jp

Although the violation of the Bell inequality is usually attributed to the nonlocality of quantum state, we explain that it is attributed to the noncommutativity of quantum observables. From this point of view we provide a method for systematic construction of generalized Bell inequalities and derive a new inequality that belongs to a different type from the traditional Bell inequality. The new inequality provides a severer and fairer test of quantum mechanics.

Keywords: Hidden-variable theory; Bell inequality.

1. Introduction

In the history of quantum mechanics, doubts about the validity or completeness of quantum mechanics have been raised repeatedly. In the early time of the history, Einstein, Podolsky and Rosen[2] pointed out that a correlation between far-separated particles seems to violate the uncertainty relation and they thought that they found a defect of quantum mechanics. Bohr[3] rebutted their point by referring his complementarity principle. However, nowadays EPR's finding is called entanglement and regarded as a starting point of quantum information science. On the other hand, de Broglie[1] proposed the pilot-wave interpretation of wave functions, which was later overcome by the Copenhagen interpretation. Bohm[4] and other people[9] developed de Broglie's idea to build the hidden-variable theory as an alternative to quantum mechanics. Thus, study of hidden-variable theory means an attempt for making a theory that has stronger validity than conventional quantum mechanics.

*This work was first presented at Summer School on Decoherence, Entanglement and Entropy, which was held at Kobe in August 2009. The original title of the presentation was "A new version of Bell's inequality". The full paper[31] is also available.

Bell[5] formulated his famous inequality in the context of the hidden-variable theory. His inequality enables to make an experimental test to compare the prediction of quantum mechanics with predictions of other alternative theories. Since his proposal, many experiments have been performed[11,14] and they support the validity of quantum mechanics with increasing accuracy.[27] Many people[12,22,24] also have proposed generalizations of the Bell inequality. Thus, today there seems no room for doubt on the validity of quantum mechanics.

However, in this paper, we attempt to propose a new test for checking the validity of quantum mechanics from a different viewpoint. At least, we provide a new explanation on the Bell inequality. This explanation gives an insight for understanding the implication of quantum mechanics and suggests a way for producing various types of paradoxical inequalities that are not equivalent to the original Bell inequality.

The hidden-variable theory is built on the following assumptions: (1) The state of a physical system is completely specified not by wave function or state vector ψ, but by a variable λ or a set of variables $\lambda = (\lambda_1, \lambda_2, \cdots, \lambda_n)$. (2) Any physical observable A is a function of λ. Once the value of the hidden variable λ are specified, the value $A(\lambda) \in \mathbb{R}$ of the observable is uniquely determined. (3) A precise value of the variable λ in each experiment is not known. The variable λ obeys a probability distribution $P(\lambda)$, which is nonnegative, $P \geq 0$, and normalized, $\int P d\lambda = 1$. (4) Additionally, in the local hidden-variable theory, the value or the distribution of the hidden variable cannot be affected by superluminous signals.

In the scheme of the hidden-variable theory, the expectation value of the observable A is calculated as

$$\langle A \rangle = \int A(\lambda) P(\lambda)\, d\lambda \tag{1}$$

with the probability distribution $P(\lambda)$ of the hidden variable λ. Thus, a question arises: *can we distinguish the hidden-variable theory from quantum mechanics by a physical experiment?* At the first glance, there are huge degrees of freedom in the choice of probability distribution of the hidden variables. Hence, with the hidden-variable theory, it seems always possible to reproduce the same prediction as the quantum-mechanical prediction. There seems no experimental test that can judge any discrepancy between quantum mechanics and the hidden-variable theory.

Bell proved an inequality that bounds the expectation value of a certain observable in the scheme of the local hidden-variable theory. Clauser, Horne, Shimony and Holt[8,10] reformulated Bell's inequality in a form more suitable for experimental tests. They introduced four observables A_1, A_2, B_1, B_2, which take ± 1 as their values, and defined a quantity $S := A_1B_1 + A_1B_2 + A_2B_1 - A_2B_2$ by combining them. They proved that the expectation value of S must be in the range

$$-2 \leq \langle S \rangle \leq 2 \qquad : \text{hidden-variable theory,} \tag{2}$$

if the local hidden-variable theory is correct. We call (2) the BCHSH inequality abbreviating the names of the authors, Bell, Clauser, Horne, Shimony and Holt. On the other hand, quantum mechanics predicts that

$$-2\sqrt{2} \leq \langle S \rangle \leq 2\sqrt{2} \qquad : \text{quantum mechanics.} \tag{3}$$

If experiments yield the value of $\langle S \rangle$ in the range $-2\sqrt{2} \leq \langle S \rangle < -2$ or in $2 < \langle S \rangle \leq 2\sqrt{2}$, the BCHSH inequality is violated and we can conclude that the hidden-variable theory is wrong and quantum mechanics is correct. After the proposal of the BCHSH inequality, many experiments have been performed[11,14,15,27] and they revealed violations of the BCHSH inequality. Thus, there is no doubt of the failure of the hidden-variable theory.

However, the meaning of the quantity $S = A_1B_1 + A_1B_2 + A_2B_1 - A_2B_2$ seems obscure. Many generalizations of the BCHSH inequality have been proposed by other researchers[12,22,24] and a systematic method for generalization has also been given by Avis, Moriyama and Owari,[29] who used the methods of operations research. However, it is still desirable to construct Bell-like inequalities with the understanding of the physical principle that enables the construction. In addition, it is noted that the BCHSH inequality is not a fair test for the hidden-variable theory in a sense explained below.

The types of tests of the hidden-variable theory and quantum mechanics are classified as the following. In general, for a physical quantity T, each theory predicts that the expectation value of T falls in some range as

$$a \leq \langle T \rangle \leq b \qquad : \text{hidden-variable theory,} \tag{4}$$

$$c \leq \langle T \rangle \leq d \qquad : \text{quantum mechanics.} \tag{5}$$

Then, by measuring the experimental value, we can judge the validities of the two theories. We classify the tests into four types:

$$\begin{aligned} &\text{type 1: } c < a < b < d \\ &\text{type 2: } c < a < d < b, \quad \text{or} \quad a < c < b < d \\ &\text{type 3: } a < c < d < b \\ &\text{type 4: } c < d < a < b, \quad \text{or} \quad a < b < c < d. \end{aligned} \tag{6}$$

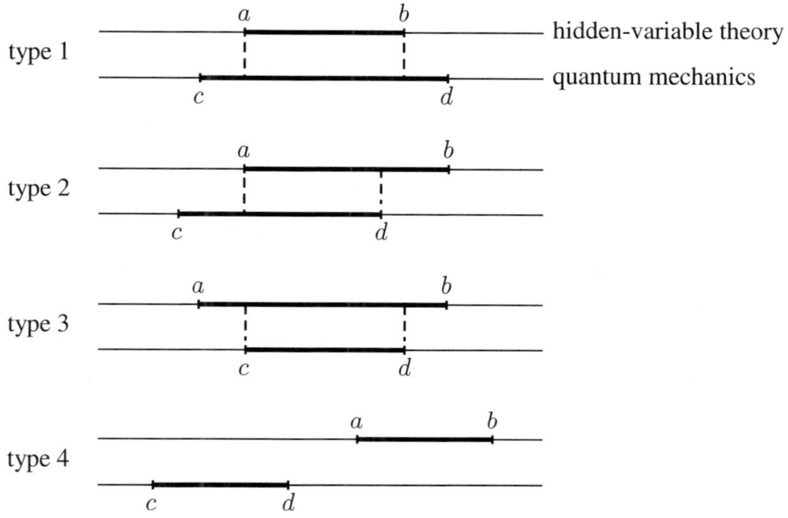

Fig. 1. Classification of tests of the hidden-variable theory and quantum mechanics. Ranges of the predictions of the two theories are indicated by thick bars.

According to this scheme, the BCHSH inequality belongs to the type 1, where the range of prediction of the hidden-variable theory is included in the range of quantum mechanics. Thus, in any experiment, it cannot happen that only the hidden-variable theory is correct and quantum mechanics is wrong. However, if we have a test of the type 2 with $c < a < d < b$ and if we obtain the experimental value in $d < \langle T \rangle \leq b$, we should conclude that the hidden-variable theory is correct and quantum mechanics is wrong. On the other hand, the Kochen-Specker theorem[7] and the Greenberger-Horne-Shimony-Zeilinger test[18,20,21] belong to the type 4, where the ranges of the hidden-variable theory and quantum mechanics are completely disjoint. Of course, we do not expect that any experiments invalidate quantum mechanics in the real world. However, it is desirable for strengthening the validity of quantum mechanics to have a test that can reveal even the failure of quantum mechanics and success of the hidden-variable theory. By passing such a severe test like type 2 or type 4, quantum mechanics will become more persuasive and reliable.

In this paper, we explain several mathematical reasons for the violation of the BCHSH inequality. We also introduce a method for making systematic generalizations of the BCHSH inequality; this method is the main result of this paper. As a product of the main result, we invent a quantity

$$\begin{aligned} T = A_1B_3 + A_1B_6 + A_2B_3 - A_2B_6 \\ +A_3B_2 + A_3B_5 + A_1B_2 - A_1B_5 \\ +A_2B_1 + A_2B_4 + A_3B_1 - A_3B_4, \end{aligned} \tag{7}$$

where the observables A_i $(i = 1, 2, 3)$ and B_j $(j = 1, \cdots, 6)$ take their values in $\{1, -1\}$ and A_i commutes with B_j. We will show that the two theories predict the range of the expectation value as

$$\begin{aligned} -6 \le \langle T \rangle \le 6 \qquad &: \text{hidden-variable theory,} \\ -6\sqrt{2} \le \langle T \rangle \le 2\sqrt{2} \quad &: \text{quantum mechanics.} \end{aligned} \tag{8}$$

Hence, this set of inequalities belongs to the type 2, which offers a severer and fairer test for comparing quantum mechanics and the hidden-variable theory than the conventional BCHSH inequality.

2. Bell-Clauser-Horne-Shimony-Holt Inequality

2.1. *Setting up*

The scheme of a Bell-type experiment is briefly sketched in this section for a pedagogical purpose. Of course, this scheme is already well known and hence the reader may skip this part.

Assume that two spin-$\frac{1}{2}$ particles are emitted simultaneously from a common source. The particle pairs are generated repeatedly and they are prepared to be in the same state. The two particles fly in opposite directions, left and right. Two detectors are placed far from the particle source. Each detector receives one particle and measures a component of the spin of the particle. If the spin received by the left detector is in the up state, the value of an observable A is defined to be $A = +1$. If the spin received

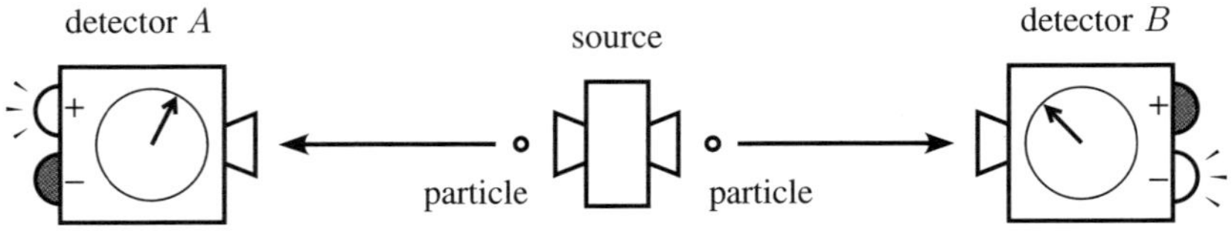

Fig. 2. A setup of Bell-type experiment is depicted à la Mermin.[13,18,19] Two particles are emitted simultaneously from a common source. One detector has a dial and two lamps. The angle of the dial is variable. If the spin of the received particle is parallel to the dial, the plus sign lights. If the spin is antiparallel to the dial, the minus sign lights.

by the left detector is in the down state, the value of A is defined to be $A = -1$. Similarly, the spin measured by the right detector defines an observable B, whose value is ± 1 according to up or down state of the detected particle. Assume that one can rearrange the detector to choose the spin component to be measured. The left detector has an adjustable angle variable α and defines the observable $A(\alpha)$ of the spin component in the angle α. Similarly, the right one has an angle variable β and defines the spin observable $B(\beta)$. For selected angles $\alpha_1, \alpha_2, \beta_1, \beta_2$, the four observables $A_1 = A(\alpha_1)$, $A_2 = A(\alpha_2)$, $B_1 = B(\beta_1)$ and $B_2 = B(\beta_2)$ are defined. Each observable takes $+1$ or -1 as its value according to up or down state of the spin of the detected particle.

The quantity to be calculated is the average of

$$S = A_1 B_1 + A_1 B_2 + A_2 B_1 - A_2 B_2. \qquad (9)$$

The outputs A_1 and B_1 from a particle pair define the value of the product $A_1 B_1$. This procedure is repeated over particle pairs and the average $\langle A_1 B_1 \rangle$ is determined. In another run, measurements of A_1 and B_2 are repeated, and the average $\langle A_1 B_2 \rangle$ is determined. In a similar way, averages of other terms are also measured and added up for defining

$$\langle S \rangle = \langle A_1 B_1 \rangle + \langle A_1 B_2 \rangle + \langle A_2 B_1 \rangle - \langle A_2 B_2 \rangle. \qquad (10)$$

This is the BCHSH quantity.

2.2. *Hidden-variable theory*

Here we present a quick derivation of the BCHSH inequality (2). In the context of the hidden-variable theory, it is assumed that the state of the pair of particles is uniquely specified by a hidden variable λ. The observable A is assumed to be a function $A(\alpha, \lambda)$ of the parameter α that specifies the arrangement of the detector and the hidden variable λ. Once the value of λ is fixed, the measured values of A_1, A_2, B_1, B_2 are fixed to be $+1$ or -1. Actually, it is assumed that the hidden variable is not fixed but it obeys a probability distribution $P(\lambda)$, which is non-negative and normalized. The average of the observable A is defined with the probability $P(\lambda)$ as

$$\langle A_\alpha \rangle = \int A(\alpha, \lambda) P(\lambda)\, d\lambda. \qquad (11)$$

This is the scheme of the hidden-variable theory.

Using the above scheme we can deduce the upper and lower bounds of the quantity

$$S = A_1(B_1 + B_2) + A_2(B_1 - B_2).$$

As stated abobe, the values of $A_i(\lambda) = A(\alpha_i, \lambda)$ and $B_i(\lambda) = B(\beta_i, \lambda)$ are in $\{+1, -1\}$. When $B_1 + B_2 = \pm 1 \pm 1 = \pm 2$, the other term is $B_1 - B_2 = 0$. When $B_1 + B_2 = \pm 1 \mp 1 = 0$, the other term is $B_1 - B_2 = \pm 2$. Thus, one of $(B_1 + B_2)$ or $(B_1 - B_2)$ is always 0 and the other is ± 2. The coefficients A_1, A_2 are also $+1$ or -1. Hence, the possible values of the quantity $S = A_1(B_1 + B_2) + A_2(B_1 - B_2)$ are ± 2. Since the probability is assumed to be nonnegative and normalized, the average of S is between the minimum and the maximum of S,

$$-2 \leq \langle S \rangle = \int (A_1 B_1 + A_1 B_2 + A_2 B_1 - A_2 B_2)\, P(\lambda)\, d\lambda \leq 2. \qquad (12)$$

Thus, the BCHSH inequality (2) is proved.

2.3. *Quantum mechanics*

Let us turn to quantum mechanics. In the context of quantum mechanics, the state of a pair of spin-$\frac{1}{2}$ particles are described by a vector in the two-qubit Hilbert space $\mathbb{C}^2 \otimes \mathbb{C}^2$ and the observables of the pair are described by operators on $\mathbb{C}^2 \otimes \mathbb{C}^2$. The spin observable measured at the left detector or the right detector is defined to be

$$A(\alpha) = (\sigma_z \cos\alpha + \sigma_x \sin\alpha) \otimes I, \qquad (13)$$
$$B(\beta) = I \otimes (\sigma_z \cos\beta + \sigma_x \sin\beta), \qquad (14)$$

respectively. Here I is the two-dimensional identity matrix and σ_i are the Pauli matrices. The parameters α and β are adjustable for specifying the spin component to be measured. Substituting functions of a common angle variable θ into them, four observables

$$A_1 = A(0) = \sigma_z \otimes I, \qquad (15)$$
$$A_2 = A(2\theta) = (\sigma_z \cos 2\theta + \sigma_x \sin 2\theta) \otimes I, \qquad (16)$$
$$B_1 = B(\theta) = I \otimes (\sigma_z \cos\theta + \sigma_x \sin\theta), \qquad (17)$$
$$B_2 = B(-\theta) = I \otimes (\sigma_z \cos\theta - \sigma_x \sin\theta) \qquad (18)$$

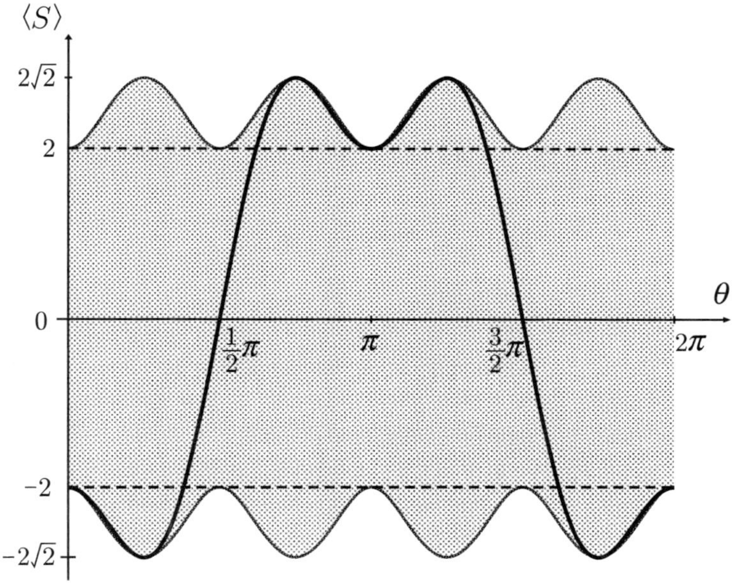

Fig. 3. A test of the ordinary Bell-Clauser-Horne-Shimony-Holt inequality. The hidden-variable theory predicts that $-2 \leq \langle S \rangle \leq 2$. Quantum mechanics predicts that $\langle S \rangle$ is in the gray area B_S, which is defined in (23). The thick curve is the expectation value (26) associated with the singlet state.

are defined. From the matrix representation for S

$$\begin{aligned} S_\theta &= A_1B_1 + A_1B_2 + A_2B_1 - A_2B_2 \\ &= 2\cos\theta \begin{pmatrix} 1 & 0 & 0 & 0 \\ 0 & -1 & 0 & 0 \\ 0 & 0 & -1 & 0 \\ 0 & 0 & 0 & 1 \end{pmatrix} \\ &\qquad + 2\cos 2\theta \sin\theta \begin{pmatrix} 0 & 1 & 0 & 0 \\ 1 & 0 & 0 & 0 \\ 0 & 0 & 0 & -1 \\ 0 & 0 & -1 & 0 \end{pmatrix} + 2\sin 2\theta \sin\theta \begin{pmatrix} 0 & 0 & 0 & 1 \\ 0 & 0 & 1 & 0 \\ 0 & 1 & 0 & 0 \\ 1 & 0 & 0 & 0 \end{pmatrix}, \end{aligned} \tag{19}$$

the eigenvalues of S_θ are calculated to be

$$\begin{aligned} &\{s_1(\theta), s_2(\theta), s_3(\theta), s_4(\theta)\} \\ &= \left\{2\cos 2\theta,\ -2\cos 2\theta,\ 2\sqrt{1+\sin^2 2\theta},\ -2\sqrt{1+\sin^2 2\theta}\right\}. \end{aligned} \tag{20}$$

For a general state including a mixed state, the expectation value of S is given by a convex combination of the eigenvalues, $\langle S_\theta \rangle = \sum_{i=1}^4 w_i s_i(\theta)$. Then, the sets of expectation values are put in order as

$$\begin{aligned} Q_S(\theta) &:= \Big\{ \sum_{i=1}^4 w_i s_i(\theta) \,\Big|\, w_i \in \mathbb{R},\, 0 \le w_i \le 1,\, \sum_{i=1}^4 w_i = 1 \Big\} \\ &= \Big[-2\sqrt{1+\sin^2 2\theta},\, 2\sqrt{1+\sin^2 2\theta} \Big] \subset \mathbb{R}, \end{aligned} \tag{21}$$

$$Q_S := \bigcup_{0 \le \theta \le 2\pi} Q_S(\theta) = \big[-2\sqrt{2},\, 2\sqrt{2} \big] \subset \mathbb{R}, \tag{22}$$

$$B_S := \big\{ (\theta, s) \,\big|\, 0 \le \theta \le 2\pi,\, s \in Q_S(\theta) \big\} \subset \mathbb{R}^2. \tag{23}$$

The set $Q_S(\theta)$ is all of possible expectation values of S_θ for a fixed angle θ, and the set Q_S is all the possible expectation values of S_θ for an arbitrary angle θ. The band B_S formed by values allowed by quantum mechanics is shown in Fig. 3 as the shaded domain. It is clear that the set Q_S gives the bound $-2\sqrt{2} \le \langle S \rangle \le 2\sqrt{2}$ in the context of quantum mechanics. Thus the inequality (3) is proved.

On the other hand, in the previous subsection, it was proved that the range of the expectation value allowed by the hidden-variable theory is

$$H_S := [-2,\, 2] \subset \mathbb{R}. \tag{24}$$

Then, it holds that $H_S \subset Q_S(\theta)$ for any θ. Thus, the quantity S provides only the type 1 test.

In particular, if the particle pair is the spin singlet state

$$\begin{aligned} \psi &= \psi_1 + \psi_2 = \frac{1}{\sqrt{2}} \big(|\uparrow\downarrow\rangle - |\downarrow\uparrow\rangle \big) \\ &= \frac{1}{\sqrt{2}} \begin{pmatrix} 1 \\ 0 \end{pmatrix} \otimes \begin{pmatrix} 0 \\ 1 \end{pmatrix} - \frac{1}{\sqrt{2}} \begin{pmatrix} 0 \\ 1 \end{pmatrix} \otimes \begin{pmatrix} 1 \\ 0 \end{pmatrix} = \frac{1}{\sqrt{2}} \begin{pmatrix} 0 \\ 1 \\ -1 \\ 0 \end{pmatrix}, \end{aligned} \tag{25}$$

the expectation value becomes

$$\langle \psi | S_\theta | \psi \rangle = -2\cos\theta - 2\sin 2\theta \sin\theta. \tag{26}$$

At $\theta = \pm\frac{\pi}{4}$, it becomes $\langle \psi | S_\theta | \psi \rangle = -2\sqrt{2}$. At $\theta = \pm\frac{3}{4}\pi$, it becomes $\langle \psi | S_\theta | \psi \rangle = 2\sqrt{2}$. Thus, the maximum violation of the BCHSH inequality is attained at these angles.

2.4. *Role of locality*

Here, we explain the implication of locality. In the context of the hidden-variable theory, locality or nonlocality is formulated as follows. Locality requests that the probability distribution is independent of the directions of the detectors, which are placed far from each other and from the source of the particle pair. Namely, in the local theory, the average is given by the formula

$$\langle A_\alpha B_\beta \rangle = \int A(\alpha, \lambda) B(\beta, \lambda) P(\lambda)\, d\lambda \quad : \text{local hidden variable.} \tag{27}$$

On the other hand, the nonlocal hidden-variable theory allows that the probability distribution depends on the directions of the distantly separated detectors. Hence, the above formula is replaced by

$$\langle A_\alpha B_\beta \rangle = \int A(\alpha, \lambda) B(\beta, \lambda) P(\alpha, \beta, \lambda)\, d\lambda \quad : \text{nonlocal hidden variable.} \tag{28}$$

The calculation (12) is based on the local hidden-variable theory, not on the nonlocal theory. It is to be emphasized that the BCHSH inequality is proved in the context of the local hidden-variable theory.

In the context of quantum mechanics, we adopt the conventional interpretation, which tells that the locality implies the commutativity of observables separated by a spacelike distance. The operator A act on the Hilbert space of the particle flying to the left, while the operator B act on the Hilbert space of the particle flying to the right. Thus, they are acting on different spaces, and hence, they commute.

3. Why Is the BCHSH Inequality Violated?

3.1. *Interference effect*

In this section, we give several explanations for the violation of the BCHSH inequality. Here, we consider in the context of quantum mechanics.

One of the explanations for the violation is given in terms of the interference effect. If the system is in a mixed state described by the density matrix

$$\begin{aligned}\rho &= |\psi_1\rangle\langle\psi_1| + |\psi_2\rangle\langle\psi_2| \\ &= \frac{1}{2}\begin{pmatrix} 1 & 0 \\ 0 & 0 \end{pmatrix} \otimes \begin{pmatrix} 0 & 0 \\ 0 & 1 \end{pmatrix} + \frac{1}{2}\begin{pmatrix} 0 & 0 \\ 0 & 1 \end{pmatrix} \otimes \begin{pmatrix} 1 & 0 \\ 0 & 0 \end{pmatrix} = \frac{1}{2}\begin{pmatrix} 0 & 0 & 0 & 0 \\ 0 & 1 & 0 & 0 \\ 0 & 0 & 1 & 0 \\ 0 & 0 & 0 & 0 \end{pmatrix},\end{aligned} \tag{29}$$

the expectation value of S_θ becomes

$$\mathrm{Tr}\,(S_\theta\,\rho) = \langle\psi_1|S_\theta|\psi_1\rangle + \langle\psi_2|S_\theta|\psi_2\rangle = -2\cos\theta, \tag{30}$$

and it stays in the range $-2 \leq \mathrm{Tr}\,(S_\theta\,\rho) \leq 2$ being consistent with the BCSHS inequality. Instead of the mixed state, if the pure state $\rho = |\psi\rangle\langle\psi|$ with (25) is substituted, the off-diagonal elements of S_θ in (19) contribute to the expectation value as

$$\begin{aligned}\langle\psi|S_\theta|\psi\rangle &= \langle\psi_1|S_\theta|\psi_1\rangle + \langle\psi_2|S_\theta|\psi_2\rangle + \langle\psi_1|S_\theta|\psi_2\rangle + \langle\psi_2|S_\theta|\psi_1\rangle \\ &= -2\cos\theta - 2\sin 2\theta\sin\theta.\end{aligned} \tag{31}$$

The cross terms $\langle\psi_1|S_\theta|\psi_2\rangle + \langle\psi_2|S_\theta|\psi_1\rangle$ represent the interference effect, which causes the violation of the BCHSH inequality. The interference of the two terms in the superposed state $\psi = \frac{1}{\sqrt{2}}(\,|\uparrow\downarrow\rangle - |\downarrow\uparrow\rangle\,)$ is also called entanglement effect. This kind of explanation for the violation of the BCHSH inequality can be found in recent textbooks.[23]

3.2. *Noncommutativity and a useful trick*

Another explanation for the violation of the BCHSH inequality is given in terms of the noncommutativity of quantum observables. A basic fact underlying the violation is that *the eigenvalues of a sum of noncommutative operators are not equal to the sum of the eigenvalues of the respective operators in general.* Let us observe this fact via a simple example. The eigenvalues of the two Pauli matrices

$$\sigma_x = \begin{pmatrix} 0 & 1 \\ 1 & 0 \end{pmatrix} \qquad \sigma_z = \begin{pmatrix} 1 & 0 \\ 0 & -1 \end{pmatrix} \tag{32}$$

are ± 1. Does the matrix $C := \sigma_x + \sigma_z$ have ± 2 or 0 as its eigenvalues? *No!* Through the calculation

$$\det(sI - C) = \det\begin{pmatrix} s-1 & -1 \\ -1 & s+1 \end{pmatrix} = s^2 - 2, \tag{33}$$

it is easily seen that eigenvalues of C are $\pm\sqrt{2}$. We may say symbolically that *in quantum mechanics 1+1 is not 2 but* $\sqrt{2}$.

This kind of phenomenon can happen since noncommuting observables do not have certain values simultaneously. This property of the noncommuting matrices is summarized as a useful trick that is written in the typical form as

$$\frac{1}{\sqrt{2}}(\sigma_z + \sigma_x) + \frac{1}{\sqrt{2}}(\sigma_z - \sigma_x) = \sqrt{2}\,\sigma_z. \tag{34}$$

Each term $\frac{1}{\sqrt{2}}(\sigma_z \pm \sigma_x)$ in the left-hand side has the spectrum $\{1, -1\}$. Hence, a naive application of the arithmetic rule tells that the value of the left-hand side should be in $\{2, 0, -2\}$. But, actually the right term $\sqrt{2}\,\sigma_z$ has the spectrum $\{\sqrt{2}, -\sqrt{2}\}$. This discrepancy is the basic mechanism that causes the violation of the BCHSH inequality.

The hidden-variable theory assumes that the observables A_α and B_β have certain values $A_\alpha(\lambda)$ and $B_\beta(\lambda)$ at any time even when the observables are not measured. It also assumes that the values obey the ordinary arithmetic rule. The reasoning based on these assumptions leads to the BCHSH inequality (12). However, in quantum mechanics, values cannot be assigned to the noncommuting observables simultaneously and the naive arithmetic rule is not applicable to their values. Hence, the BCHSH inequality is violated.

This type of reasoning reveals the trick for making the BCHSH quantity S. The starting point is

$$S = \sqrt{2}\,(\sigma_z \otimes \sigma_z + \sigma_x \otimes \sigma_x). \tag{35}$$

Note that the spectra of $\sigma_z \otimes \sigma_z$ and $\sigma_x \otimes \sigma_x$ are both $\{1, -1\}$. Since $\sigma_z \otimes \sigma_z$ and $\sigma_x \otimes \sigma_x$ commute, the naive arithmetic rule is applicable to them and the spectrum of S should be a subset of $\{2\sqrt{2},\ 0,\ -2\sqrt{2}\}$. Actually, the spectrum of S is $\{2\sqrt{2},\ -2\sqrt{2}\}$. Then, by applying the trick (34) for rewriting S, we obtain

$$\begin{aligned}
S &= \sqrt{2}\,(\sigma_z \otimes \sigma_z + \sigma_x \otimes \sigma_x) \\
&= \sqrt{2}\Big[\sigma_z \otimes \frac{1}{2}\{(\sigma_z + \sigma_x) + (\sigma_z - \sigma_x)\} + \sigma_x \otimes \frac{1}{2}\{(\sigma_z + \sigma_x) - (\sigma_z - \sigma_x)\}\Big] \\
&= \sigma_z \otimes \frac{1}{\sqrt{2}}(\sigma_z + \sigma_x) + \sigma_z \otimes \frac{1}{\sqrt{2}}(\sigma_z - \sigma_x) \\
&\quad + \sigma_x \otimes \frac{1}{\sqrt{2}}(\sigma_z + \sigma_x) - \sigma_x \otimes \frac{1}{\sqrt{2}}(\sigma_z - \sigma_x) \\
&= A_1 B_1 + A_1 B_2 + A_2 B_1 - A_2 B_2,
\end{aligned} \tag{36}$$

which is the quantity maximally violating the BCHSH inequality. By introducing the adjustable parameter θ, we obtain the quantity S_θ expressed in terms of the observables (15)–(18)

Actually, Bell[6,21] himself noticed that the values of noncommuting observables do not obey the naive additive law but he did not utilize this property to derive his inequality. Seevinck and Uffink[28] argued that the noncommutativity is related to the violation of the original BCHSH inequality, but they did not consider a method for generating variations of the BCHSH inequality.

4. New Test

4.1. *New quantity*

Here, we build a new quantity T that satisfies a new type of the Bell-like inequality. Our starting point is the quantity

$$T = 2\sqrt{2}\,(\sigma_x \otimes \sigma_x + \sigma_y \otimes \sigma_y + \sigma_z \otimes \sigma_z). \tag{37}$$

The three terms $\sigma_a \otimes \sigma_a$ $(a = x, y, z)$ are mutually commutative and their spectra are $\{1, -1\}$. Therefore, the spectrum of T should be a subset of $\{6\sqrt{2},\, 2\sqrt{2},\, -2\sqrt{2},\, -6\sqrt{2}\}$. The matrix representation of T is calculated as

$$\begin{aligned}
\frac{1}{2\sqrt{2}} T &= \begin{pmatrix} 0 & 1 \\ 1 & 0 \end{pmatrix} \otimes \begin{pmatrix} 0 & 1 \\ 1 & 0 \end{pmatrix} + \begin{pmatrix} 0 & -i \\ i & 0 \end{pmatrix} \otimes \begin{pmatrix} 0 & -i \\ i & 0 \end{pmatrix} + \begin{pmatrix} 1 & 0 \\ 0 & -1 \end{pmatrix} \otimes \begin{pmatrix} 1 & 0 \\ 0 & -1 \end{pmatrix} \\
&= \begin{pmatrix} 0 & 0 & 0 & 1 \\ 0 & 0 & 1 & 0 \\ 0 & 1 & 0 & 0 \\ 1 & 0 & 0 & 0 \end{pmatrix} + \begin{pmatrix} 0 & 0 & 0 & -1 \\ 0 & 0 & 1 & 0 \\ 0 & 1 & 0 & 0 \\ -1 & 0 & 0 & 0 \end{pmatrix} + \begin{pmatrix} 1 & 0 & 0 & 0 \\ 0 & -1 & 0 & 0 \\ 0 & 0 & -1 & 0 \\ 0 & 0 & 0 & 1 \end{pmatrix} \\
&= \begin{pmatrix} 1 & 0 & 0 & 0 \\ 0 & -1 & 2 & 0 \\ 0 & 2 & -1 & 0 \\ 0 & 0 & 0 & 1 \end{pmatrix}.
\end{aligned} \tag{38}$$

It is easily seen that the eigenvalues of T are $2\sqrt{2}$ (threefold degeneracy) and $-6\sqrt{2}$ (no degeneracy); the corresponding eigenvectors are

$$2\sqrt{2} : \; a \begin{pmatrix} 1 \\ 0 \\ 0 \\ 0 \end{pmatrix} + b \begin{pmatrix} 0 \\ 1 \\ 1 \\ 0 \end{pmatrix} + c \begin{pmatrix} 0 \\ 0 \\ 0 \\ 1 \end{pmatrix}, \qquad -6\sqrt{2} : \begin{pmatrix} 0 \\ 1 \\ -1 \\ 0 \end{pmatrix}. \tag{39}$$

In other words, the triplet spin states have the eigenvalue $T = 2\sqrt{2}$, while the singlet spin state has the eigenvalue $T = -6\sqrt{2}$. Hence, quantum mechanics predicts the bound

$$-6\sqrt{2} \;\leq\; \langle T \rangle \;\leq\; 2\sqrt{2} \qquad : \text{quantum mechanics} \tag{40}$$

of the expectation value for any state.

Next, by applying the trick (34) several times, we rewrite T as

$$
\begin{aligned}
T &= 2\sqrt{2}\,(\sigma_x \otimes \sigma_x + \sigma_y \otimes \sigma_y + \sigma_z \otimes \sigma_z) \\
&= \frac{2\sqrt{2}}{4}\Big[\sigma_x \otimes \{(\sigma_x + \sigma_y) + (\sigma_x - \sigma_y) + (\sigma_z + \sigma_x) - (\sigma_z - \sigma_x)\} \\
&\qquad + \sigma_y \otimes \{(\sigma_y + \sigma_z) + (\sigma_y - \sigma_z) + (\sigma_x + \sigma_y) - (\sigma_x - \sigma_y)\} \\
&\qquad + \sigma_z \otimes \{(\sigma_z + \sigma_x) + (\sigma_z - \sigma_x) + (\sigma_y + \sigma_z) - (\sigma_y - \sigma_z)\}\Big]. \qquad (41)
\end{aligned}
$$

By introducing

$$
\begin{aligned}
A_1 &= \sigma_x \otimes I, & A_2 &= \sigma_y \otimes I, & A_3 &= \sigma_z \otimes I, \\
B_1 &= \frac{1}{\sqrt{2}} I \otimes (\sigma_y + \sigma_z), & B_2 &= \frac{1}{\sqrt{2}} I \otimes (\sigma_z + \sigma_x), \\
B_3 &= \frac{1}{\sqrt{2}} I \otimes (\sigma_x + \sigma_y), & B_4 &= \frac{1}{\sqrt{2}} I \otimes (\sigma_y - \sigma_z), \\
B_5 &= \frac{1}{\sqrt{2}} I \otimes (\sigma_z - \sigma_x), & B_6 &= \frac{1}{\sqrt{2}} I \otimes (\sigma_x - \sigma_y), \qquad (42)
\end{aligned}
$$

we reach the expression

$$
\begin{aligned}
T &= A_1(B_3 + B_6 + B_2 - B_5) + A_2(B_1 + B_4 + B_3 - B_6) \\
&\quad + A_3(B_2 + B_5 + B_1 - B_4) \\
&= A_1(B_3 + B_6) + A_2(B_3 - B_6) \\
&\quad + A_3(B_2 + B_5) + A_1(B_2 - B_5) \\
&\quad + A_2(B_1 + B_4) + A_3(B_1 - B_4), \qquad (43)
\end{aligned}
$$

which was shown at (7) in the Introduction.

4.2. *Hidden-variable theory vs. quantum mechanics*

The hidden-variable theory is applicable to T in this form (43). In the context of the hidden-variable theory, A_α and B_β are functions of the hidden variable λ, and the values of $A_\alpha(\lambda)$ and $B_\beta(\lambda)$ are in $\{1, -1\}$. Then, it is easily seen from the expression (43) that the possible values of $T(\lambda)$ are in $\{6, 2, -2, -6\}$. Hence, the average $\langle T \rangle = \int T(\lambda)P(\lambda)d\lambda$ is in the range

$$
-6 \le \langle T \rangle \le 6 \qquad : \text{hidden-variable theory.} \qquad (44)
$$

The range (40) of the prediction of quantum mechanics and the range (44) of the hidden-variable theory have the overlap $[-6, 2\sqrt{2}]$ where both the two theories hold, and the region $[-6\sqrt{2}, -6) \cup (2\sqrt{2}, 6]$ where only one of the two theories holds. Thus, T is a quantity that realizes the type 2 test.

Moreover, we introduce a parameter θ and define T_θ as the polynomial (43) of

$$A_1 = (\sigma_z \cos 2\theta + \sigma_x \sin 2\theta) \otimes I, \tag{45}$$

$$A_2 = (\sigma_z \cos 2\theta + \sigma_y \sin 2\theta) \otimes I, \tag{46}$$

$$A_3 = \sigma_z \otimes I, \tag{47}$$

$$B_1 = I \otimes (\sigma_y \cos\theta + \sigma_z \sin\theta), \tag{48}$$

$$B_2 = I \otimes (\sigma_z \cos\theta + \sigma_x \sin\theta), \tag{49}$$

$$B_3 = I \otimes (\sigma_x \cos\theta + \sigma_y \sin\theta), \tag{50}$$

$$B_4 = I \otimes (\sigma_y \cos\theta - \sigma_z \sin\theta), \tag{51}$$

$$B_5 = I \otimes (\sigma_z \cos\theta - \sigma_x \sin\theta), \tag{52}$$

$$B_6 = I \otimes (\sigma_x \cos\theta - \sigma_y \sin\theta). \tag{53}$$

The parameter θ is interpreted as an angle that specifies the directions of the detectors. The matrix representation of T_θ is

$$T_\theta = 2(\cos\theta + \sin\theta) \times \begin{pmatrix} 1 & (1-i)\cos 2\theta & 0 & 0 \\ (1+i)\cos 2\theta & -1 & 2\sin 2\theta & 0 \\ 0 & 2\sin 2\theta & -1 & -(1-i)\cos 2\theta \\ 0 & 0 & -(1+i)\cos 2\theta & 1 \end{pmatrix}. \tag{54}$$

When $\theta = \frac{\pi}{4}$, T_θ is reduced to the original form (38). The eigenvalues of T_θ are

$$\begin{aligned} &\{t_1(\theta), t_2(\theta), t_3(\theta), t_4(\theta)\} \\ &= \Big\{ 2(\cos\theta + \sin\theta)\big(-\sin 2\theta \pm \sqrt{\cos^2 2\theta + 2 + 2\sin 2\theta}\big), \\ &\quad 2(\cos\theta + \sin\theta)\big(\sin 2\theta \pm \sqrt{\cos^2 2\theta + 2 - 2\sin 2\theta}\big) \Big\}. \end{aligned} \tag{55}$$

The sets of values allowed by quantum mechanics are arranged as

$$Q_T(\theta) := \Big\{ \sum_{i=1}^{4} w_i t_i(\theta) \,\Big|\, w_i \in \mathbb{R},\ 0 \le w_i \le 1,\ \sum_{i=1}^{4} w_i = 1 \Big\} \subset \mathbb{R}, \tag{56}$$

$$Q_T := \bigcup_{0 \le \theta \le 2\pi} Q_T(\theta) = [-6\sqrt{2},\ 6\sqrt{2}\,] \subset \mathbb{R}, \tag{57}$$

$$B_T := \{(\theta, t) \,|\, 0 \le \theta \le 2\pi,\ t \in Q_T(\theta)\} \subset \mathbb{R}^2. \tag{58}$$

And the range allowed by the hidden-variable theory is denoted as

$$H_T := [-6,\ 6\,] \subset \mathbb{R}. \tag{59}$$

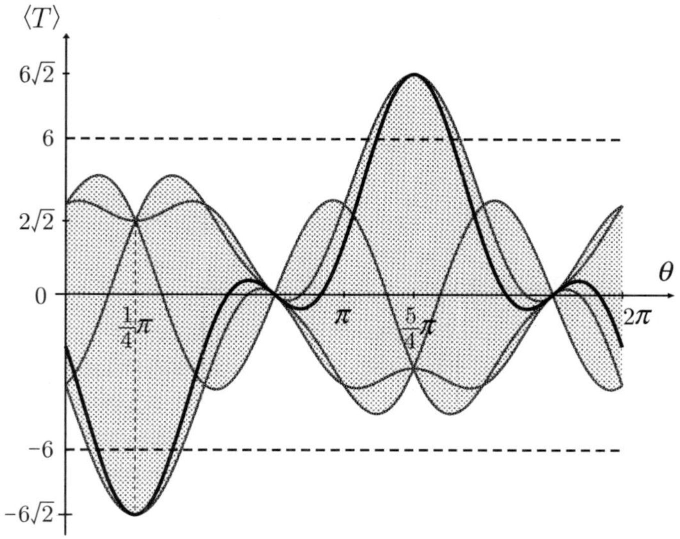

Fig. 4. A test of the new inequality. The hidden-variable theory predicts that $-6 \leq \langle T \rangle \leq 6$. Quantum mechanics predicts that $\langle T \rangle$ is in the gray area B_T, which is defined in (58). The thick curve is the expectation value (60) associated with the singlet state.

In Fig. 4, the band B_T is shown as the shaded region. At any fixed value of θ, the predictions of the two theories have some overlap, namely, we have $Q_T(\theta) \cap H_T \neq \varnothing$. It also happens that $Q_T(\theta) \subset H_T$ for some θ. However, it never happens that $Q_T(\theta) \supset H_T$ at any θ. In this sense, the tests of type 2 and type 3 are realized. It is also to be noted that $Q_T \supset H_T$, namely, the whole set of quantum predictions is wider than the predictions of the hidden-variable theory.

In particular, the singlet state ψ of (25) yields the expectation value

$$f(\theta) = \langle \psi | T_\theta | \psi \rangle = -2(\cos\theta + \sin\theta)(1 + 2\sin 2\theta). \tag{60}$$

The range $F := \{ f(\theta) \,|\, 0 \leq \theta \leq 2\pi \} = [-6\sqrt{2}, 6\sqrt{2}]$ covers H_T completely. This means that the violation of the bound of the hidden-variable theory are observed with the state ψ.

If we take another entangled state

$$\chi = \frac{1}{\sqrt{2}}(\,|\uparrow\uparrow\rangle + |\downarrow\downarrow\rangle\,) = \frac{1}{\sqrt{2}}\begin{pmatrix}1\\0\end{pmatrix} \otimes \begin{pmatrix}1\\0\end{pmatrix} + \frac{1}{\sqrt{2}}\begin{pmatrix}0\\1\end{pmatrix} \otimes \begin{pmatrix}0\\1\end{pmatrix} = \frac{1}{\sqrt{2}}\begin{pmatrix}1\\0\\0\\1\end{pmatrix} \tag{61}$$

instead of the singlet state ψ, it yields

$$g(\theta) = \langle \chi | T_\theta | \chi \rangle = 2(\cos\theta + \sin\theta). \tag{62}$$

The range $G := \{g(\theta) \,|\, 0 \leq \theta \leq 2\pi\} = [-2\sqrt{2}, 2\sqrt{2}]$ is included in H_T completely. This means that the violation of the hidden-variable theory will never be observed in the measurement of T with the state χ.

5. Concluding Remarks

Here, we give a summary of this study. In this paper, we pointed out that the violation of the BCHSH inequality can be understood as a result of the noncommutativity of quantum observables. For noncommuting operators, an eigenvalue of a sum of operators does not coincide with a sum of eigenvalues of the respective operators. Using this property, we invented a method of building systematically Bell-like observables and showed that the conventional BCHSH inequality is reconstructed by this method. This diminished the ad hoc nature of the BCHSH observable.

We classified possible tests of the hidden-variable theory and quantum mechanics. We pointed out that there was no chance in the conventional BCHSH test to reveal the invalidity of quantum mechanics with the validity of the hidden-variable theory. By applying our method, we constructed the new observable T and calculated the range of its average in the contexts of the hidden-variable theory and quantum mechanics, respectively. It was shown that there is a chance that the new test with T reveals the invalidity of quantum mechanics with the validity of the hidden-variable theory.

Of course, we do not aim to deny the validity of quantum mechanics, but we aim to support it by passing the new severer test. It is also our purpose to clarify the implication of violations of Bell-like inequalities of various types.

There are several remaining problems concerning the generalized inequality. The first one is the existence problem in a mathematical sense. The Bell-like inequality is a necessary condition for the existence of the probability distribution that satisfies (27). In the case of the conventional BCHSH inequality, Fine[16] proved that the set of inequalities

$$-2 \leq \langle A_1 B_1 \rangle + \langle A_1 B_2 \rangle + \langle A_2 B_1 \rangle - \langle A_2 B_2 \rangle \leq 2, \tag{63}$$

$$-2 \leq \langle A_1 B_2 \rangle + \langle A_2 B_1 \rangle + \langle A_2 B_2 \rangle - \langle A_1 B_1 \rangle \leq 2, \tag{64}$$

$$-2 \leq \langle A_2 B_1 \rangle + \langle A_2 B_2 \rangle + \langle A_1 B_1 \rangle - \langle A_1 B_2 \rangle \leq 2, \tag{65}$$

$$-2 \leq \langle A_2 B_2 \rangle + \langle A_1 B_1 \rangle + \langle A_1 B_2 \rangle - \langle A_2 B_1 \rangle \leq 2 \tag{66}$$

is a necessary and sufficient condition for the existence of the probability distribution of the hidden variable. Although there are studies on the tightness of some variations of the BCHSH inequalities,[17,25,30] finding the necessary and sufficient condition for the existence of the probability distribution of the hidden variable for our observable T is left as an open problem.

The second one is a practical problem. In principle, the test that we proposed can be implemented in an experiment using pairs of photons or spin-half particles. But our choice involves nine observables (45)–(53) with one variable angle. Hence, our scheme is still too cumbersome for practical use. It is more desirable to reduce the number of observables to make the experiments easier.

The third one is the extensibility of our scheme. The proposed observable T is defined in the two-qubit Hilbert space. It is possible to extend our scheme to multiqubit systems. It is also desirable to construct a Bell-like quantity with less number of observables.

Acknowledgements

We thank Prof. M. Nakahara for his encouragement on our study. We are indebted to Prof. Valerio Scarani, who introduced to us the work of A. Fine[16] and the work of D. Collins and N. Gisin.[24] We thank T. Iwai, Y. Y. Yamaguchi, I. Tsutsui, T. Ichikawa, M. Ozawa, A. Hosoya, M. Kitano, H. Kobayashi and S. Ogawa for their interest in our study and for their helpful comments. This work was supported by a Grant-in-Aid for Scientific Research of the Japan Society for the Promotion of Science (Grant No. 22540410).

References

1. 1927 Fifth Solvay International Conference on Electrons and Photons.
2. A. Einstein, B. Podolsky and N. Rosen, *Phys. Rev.* **47**, 777 (1935).
3. N. Bohr, *Phys. Rev.* **48**, 696 (1935).
4. D. Bohm, *Phys. Rev.* **85**, 166 (1952).
5. J. S. Bell, *Physics* **1**, 195 (1964).
6. J. S. Bell, *Rev. Mod. Phys.* **38**, 447 (1966).
7. S. Kochen and E. P. Specker, *J. Math. Mech.* **17**, 59 (1967).
8. J. F. Clauser, M. A. Horne, A. Shimony and R. A. Holt, *Phys. Rev. Lett.* **23**, 880 (1969).
9. For a review, see F. J. Belinfante, *A Survey of Hidden-Variables Theories* (Pergamon Press, 1973).
10. J. F. Clauser and M. A. Horne, *Phys. Rev. D* **10**, 526 (1974).

11. J. F. Clauser and A. Shimony, *Rep. Prog. Phys.* **41**, 1881 (1978).
12. M. Froissart, *Il Nuovo Cim. B* **64**, 241 (1981).
13. N. D. Mermin *Am. J. Phys.* **49**, 940 (1981).
14. A. Aspect, P. Grangier and G. Roger, *Phys. Rev. Lett.* **47**, 460 (1981).
15. A. Aspect, P. Grangier and G. Roger, *Phys. Rev. Lett.* **49**, 91 (1982).
16. A. Fine, *Phys. Rev. Lett.* **48**, 291 (1982). [Comments: A. Garg and N. D. Mermin, *Phys. Rev. Lett.* **49**, 242 (1982). A. Fine, *Phys. Rev. Lett.* **49**, 243 (1982).]
17. A. Garg and N. D. Mermin, *Found. Physics* **14**, 1 (1984).
18. D. Mermin, *Am. J. Phys.* **58**, 731 (1990).
19. N. D. Mermin, *Boojums All the Way Through* (Cambridge University Press, 1990).
20. D. M. Greenberger, M. A. Horne, A. Shimony and A. Zeilinger, *Am. J. Phys.* **58**, 1131 (1990).
21. D. Mermin, *Rev. Mod. Phys.* **65**, 803 (1993).
22. A. Peres, *Found. Phys.* **29**, 589 (1999).
23. For example, see p. 229 of A. Shimizu, *Foundations of Quantum Theory* (in Japanese, *Ryoshiron-no Kiso*), section 8.7 in the revised edition (Saiensu-sha, 2004).
24. D. Collins and N. Gisin, *J. Phys. A* **37**, 1775 (2004).
25. W. Laskowski, T. Paterek, M. Żukowski and Č. Brukner, *Phys. Rev. Lett.* **93**, 200401 (2004).
26. D. Deutsch, e-print arXiv: quant-ph/0401024v1.
27. H. Sakai et al., *Phys. Rev. Lett.* **97**, 150405 (2006).
28. M. Seevinck and J. Uffink, *Phys. Rev. A* **76**, 042105 (2007).
29. D. Avis, S. Moriyama and M. Owari, *IEICE Trans. Fundamentals* **E92-A**, 1254 (2009).
30. K. F. Pál and T. Vértesi, *Phys. Rev. A* **79**, 022120 (2009).
31. T. Isobe and S. Tanimura, e-print arXiv: 1005.4966, to appear in *Prog. Theor. Phys.* **124**, No. 2, August 2010.

ON 3-VARIABLE EXPONENTIAL POLYNOMIALS AND QUANTUM ALGORITHMS

YASUO OHNO*, YOSHITAKA SASAKI† and CHIKA YAMAZAKI‡

*Department of Mathematics, Kinki University,
Higashi-Osaka, Osaka 577-8502, Japan
E-mail: ohno@math.kindai.ac.jp

†Interdisciplinary Graduate School of Science and Engineering,
Kinki University, Higashi-Osaka, Osaka 577-8502, Japan
E-mail: sasaki@alice.math.kindai.ac.jp

‡Department of Mathematics, Kinki University,
Higashi-Osaka, Osaka 577-8502, Japan
E-mail: chika@math.kindai.ac.jp

We calculate the zeros of an exponential polynomial of three variables by a classical and quantum algorithms which are based on the method of van Dam and Shparlinski, they treated the case of two variables, and compare with the time complexity of those cases. Further we compare the case of van Dam and Shparlinski with our case by considering the ratio (classical/quantum) of the time complexity. Then we can observe the ratio decreases.

Keywords: Quantum computing; exponential congruence; discrete logarithm; character sum.

1. Introduction

For a prime number p, we put $q = p^\nu$, where ν is a certain positive integer. Then we denote the finite field by $\mathbb{F}_q$ which has q elements. Namely, $\mathbb{F}_q$ forms an additive group and $\mathbb{F}_q^\times := \mathbb{F}_q\backslash\{0\}$ forms a multiplicative group, where 0 is the zero element in $\mathbb{F}_q$. Any element of $\alpha \in \mathbb{F}_q^\times$ have a periodicity, that is there exists a smallest natural number s such that $\alpha^s = 1$. We call such s the "multiplicative order" of α. It is known that the multiplicative order is a divisor of $\#\mathbb{F}_q^\times = q - 1$. See Ref. 5 or Appendix in this article for further explanation of finite fields.

In mathematics, it is important to calculate common solutions of equations

$$F_j(x_1, \ldots, x_m) = 0 \quad (j = 1, \ldots, l). \tag{1}$$

Here F_j's are polynomials with $\mathbb{F}_q$ coefficients. We denote the number of common solutions of (1) with coordinates in $\mathbb{F}_{q^n}$ by N_n ($\mathbb{F}_{q^n}$ is the n dimensional algebraic extension of $\mathbb{F}_q$). The generating function of N_n is given by the zeta function

$$Z(T) := \exp\left(\sum_{n=1}^{\infty} \frac{N_n}{n} T^n\right)$$

and several significant properties of the above zeta function were showed.

In Ref. 3, van Dam and Shparlinski treated the following exponential polynomial

$$f(x, y) = a_1 g_1^x + a_2 g_2^y - b \tag{2}$$

and calculated the zeros of (2) by quantum algorithms. We can see that this problem implies a straightforward generalization of the discrete logarithm problem. They compared the time complexity of a classical algorithm with that of a quantum algorithm. Then the "cubic" speed-up was observed.

In this article, we treat the three variable case

$$f_b(x_1, x_2, x_3) := a_1 g_1^{x_1} + a_2 g_2^{x_2} + a_3 g_3^{x_3} - b \tag{3}$$

and calculate the solutions of $f_b(x_1, x_2, x_3) = 0$ by using quantum algorithms which are natural generalizations of the method of van Dam and Shparlinski. Here, a_i, $g_j \in \mathbb{F}_q^{\times}$ $(i, j = 1, 2, 3)$ and $b \in \mathbb{F}_q$. Further we also compare the time complexity of a classical algorithm with that of a quantum algorithm. Then exponentially "5/2 times" speed-up is observed.

In the next section, we introduce some notation and give the considerable lemma which supports whether there exist solutions of (3). In Section 3, we calculate the time complexity of a classical algorithm. Further in Section 4, we calculate the time complexity of a quantum algorithm.

2. The Number of Solutions of Equation

In this section, we give an important formula which expresses the density of solutions of

$$f_b(x_1, x_2, x_3) := a_1 g_1^{x_1} + a_2 g_2^{x_2} + a_3 g_3^{x_3} - b = 0 \tag{4}$$

as Lemma 2.1, below. To state it, we introduce some notation.

Let each s_i be the multiplicative order of g_i $(i = 1, 2, 3)$ in (4). We put

$$\begin{aligned} X_i &:= \{0, 1, \ldots, s_i - 1\} \cong \mathbb{Z}/s_i\mathbb{Z}, \quad (i = 1, 2, 3), \\ X_3(r) &:= \{0, 1, \ldots, r - 1\} \subseteq X_3 \quad (r = 1, 2, \ldots, s_3), \\ \boldsymbol{X}^3(r) &:= X_1 \times X_2 \times X_3(r) \end{aligned}$$

and

$$\boldsymbol{X}^3 := \boldsymbol{X}^3(s_3) = X_1 \times X_2 \times X_3.$$

Then we define

$$\begin{aligned} S_{f_b}(r) &:= \{(x_1, x_2, x_3) \in \boldsymbol{X}^3(r) \mid f_b(x_1, x_2, x_3) = 0\}, \\ N_{f_b}(r) &:= \#S_{f_b}(r) \end{aligned}$$

for $r = 1, \ldots, s_3$.

By using above notation, we can state the following result:

Lemma 2.1. *Let δ be a parameter satisfying $\delta = o(q)$. For $r > \delta^2 q^3 (s_1 s_2)^{-2}$, we have*

$$N_{f_b}(r) = \frac{s_1 s_2 r}{q} + O(\delta\sqrt{rq}), \tag{5}$$

except for at most q/δ^2 exceptional b's. Further O-constant can be taken 1.

Choosing $\delta = (\log q)^{1/2}$ in Lemma 2.1, we have

Corollary 2.1. *If $q^3(s_1 s_2)^{-2} \log q < r \leq s_3$, then we see that $S_{f_b}(r) \neq \phi$ holds except for at most $q/\log q$ exceptional b's.*

Remark 2.1. The above lemma and corollary make the point that the solutions of (4) exist only when

$$\frac{s_1 s_2}{q} \geq \left(\frac{q}{s_3 - 2} \log q\right)^{1/2} (> 1).$$

This inequality implies that the multiplicative orders s_1 and s_2 are somewhat large.

Remark 2.2. The exponent $1/2$ of $\delta = (\log q)^{1/2}$ is not necessary. In fact, $\delta = (\log q)^{\varepsilon}$ with any $\varepsilon > 0$ is sufficient.

Proof of Lemma 2.1. Let ψ be a non-trivial additive character over $\mathbb{F}_q$, in fact, any additive character over $\mathbb{F}_q$ can be given as a map $\mathbb{F}_q \to \mathbb{C}_1^*$, where $\mathbb{C}_1^* := \{z \in \mathbb{C} \mid |z| = 1\}$ (see Theorem 5.7 in Ref. 5). To evaluate $N_{f_b}(r)$, we use the following formula which plays as a counting function:

$$\frac{1}{q}\sum_{\mu\in\mathbb{F}_q}\psi(u\mu) = \begin{cases} 1 & \text{if } u = 0, \\ 0 & \text{otherwise.} \end{cases} \tag{6}$$

Then we have

$$\begin{aligned} N_{f_b}(r) &= \sum_{(x_1,x_2,x_3)\in\boldsymbol{X}^3(r)} \frac{1}{q}\sum_{\mu\in\mathbb{F}_q}\psi(\mu(f_b(x_1,x_2,x_3))) \\ &= \frac{s_1s_2r}{q} + \frac{1}{q}\sum_{\mu\in\mathbb{F}_q^*}\sum_{(x_1,x_2,x_3)\in\boldsymbol{X}^3(r)}\psi(\mu(f_b(x_1,x_2,x_3))) \\ &=: \frac{s_1s_2r}{q} + \Delta_b(r). \end{aligned} \tag{7}$$

If the contribution from the second term on the right-hand side of the above formula can be estimated by $o(s_1s_2r/q)$, the above formula tells us the existence of the solution of $f_b(x_1,x_2,x_3)$. To consider it, we evaluate the mean value of the second term on the right-hand side of (7) with respect to b. Namely, we evaluate

$$E(r) := \sum_{b\in\mathbb{F}_q}|\Delta_b(r)|^2.$$

From (6) and some properties of the additive character over $\mathbb{F}_q$, we obtain

$$\begin{aligned} E(r) =& \frac{1}{q^2}\sum_{\mu,\mu'\in\mathbb{F}_q^\times}\left(\prod_{j=1}^{2}\left(\sum_{x_j,x_j'\in X_j}\psi(a_j(\mu g_j^{x_j} - \mu' g_j^{x_j'}))\right)\right) \\ &\times\sum_{x_3,x_3'\in X_3(r)}\psi(a_3(\mu g_3^{x_3} - \mu' g_3^{x_3'}))\sum_{b\in\mathbb{F}_q}\psi(b(\mu'-\mu)) \\ =& \frac{1}{q}\sum_{\mu\in\mathbb{F}_q^\times}\left(\prod_{j=1}^{2}\left(\sum_{x_j,x_j'\in X_j}\psi(a_j\mu(g_j^{x_j} - g_j^{x_j'}))\right)\right) \\ &\times\sum_{x_3,x_3'\in X_3(r)}\psi(a_3\mu(g_3^{x_3} - g_3^{x_3'})) \\ =& \frac{1}{q}\sum_{\mu\in\mathbb{F}_q^\times}\left(\prod_{j=1}^{2}\left|\sum_{x_j\in X_j}\psi(a_j\mu g_j^{x_j})\right|^2\right)\left|\sum_{x_3\in X_3(r)}\psi(a_3\mu g_3^{x_3})\right|^2. \end{aligned}$$

It is known that

$$\left|\sum_{x_j \in X_j} \psi(a_j \mu g_j^{x_j})\right| \le \sqrt{q} \quad \text{for } j = 1, 2 \text{ and any } \mu \in \mathbb{F}_q^\times$$

(see Theorem 8.78 in Ref. 5). Hence we have

$$E(r) < q \sum_{\mu \in \mathbb{F}_q} \left| \sum_{x_3 \in X_3(r)} \psi(a_3 \mu f^{x_3}) \right|^2 = q^2 r.$$

Therefore, if we put $\delta = o(q)$, then we can see that there exist at most q/δ^2 exceptional b's such that

$$\left| \frac{1}{q} \sum_{\mu \in \mathbb{F}_q^*} \sum_{(x_1, x_2, x_3) \in X_3(r)} \psi(\mu(f_b(x_1, x_2, x_3))) \right| \ge \delta \sqrt{rq}. \tag{8}$$

Hence we obtain

$$N_{f_b}(r) = \frac{s_1 s_2 r}{q} + O(\delta \sqrt{qr})$$

for other b's. Now, the proof of Lemma 2.1 is completed. □

3. Calculation of the Deterministic Time for a Classical Algorithm

We follow the method of van Dam and Shparlinski in Ref. 3. Then we have

Theorem 3.1. *Except for at most $q/\log q$ exceptional b's, we can either find a solution $(x_1, x_2, x_3) \in \boldsymbol{X}^3$ of the equation (4) or decide that it does not have a solution in deterministic time $q^{3/2}(\log q)^{O(1)}$ as a classical computer.*

In the proof of the above theorem, we use the symbol $A \ll B$ which means $A = O(B)$.

Proof. Using a standard deterministic factorization algorithm, we factorize $q - 1$ and find the orders s_j $(j = 1, 2, 3)$ of g_j in time $q^{1/2}(\log q)^{O(1)}$. We may assume without loss of generality that $s_1 \ge s_2 \ge s_3$. For calculated orders s_1 and s_2, we put

$$r = \lceil q^3 (s_1 s_2)^{-2} \log q \rceil. \tag{9}$$

Then we see that the solution of (4) certainly exists when $r \le s_3$. However, when $r > s_3$, we do not know whether such solutions exits. Therefore we have to consider those two cases.

For each $(x_2, x_3) \in X_2 \times X_3(r)$, we calculate the deterministic time of the discrete logarithm x_1 such that $g_1^{x_1} = a_1^{-1}(b - a_2 g_2^{x_2} - a_3 g_3^{x_3})$. It is known that the deterministic time for this case is $s_1^{1/2}(\log q)^{O(1)}$ (see Section 5.3 in Ref. 2).

(1) The case $r \leq s_3$. We have

$$(s_2 r) s_1^{1/2} (\log q)^{O(1)} \ll q^{3/2} (\log q)^{O(1)},$$

since $s_1^{1/2} s_2 r < (s_1^2 s_2^2 r)^{1/2}$.

(2) The case $r > s_3$. Similarly, we see that the deterministic time is

$$(s_2 s_3) s_1^{1/2} (\log q)^{O(1)} \ll q^{3/2} (\log q)^{O(1)},$$

since $s_1^{1/2} s_2 s_3 < (s_1^2 s_2^2 s_3)^{1/2} < (s_1^2 s_2^2 r)^{1/2}$. □

4. Calculation of the Time Complexity for a Quantum Algorithm

In this section, we describe quantum algorithms which are based on the method of Ref. 3.

Theorem 4.1. *Except for at most $q/\log q$ exceptional b's, we can either find a solution $(x_1, x_2, x_3) \in \boldsymbol{X}^3$ of the equation (4) or decide that it does not have a solution in time $q^{3/5}(\log q)^{O(1)}$ as a quantum computer.*

Proof. Using Shor's algorithm,[6] we can obtain the multiplicative orders s_j's $(j = 1, 2, 3)$ in polynomial time in $\log q$. We may assume without loss of generality that $s_1 \geq s_2 \geq s_3$. As in the proof of Theorem 3.1, we put r as (9). Further, we consider a polynomial time quantum subroutine $\mathcal{S}(x_2, x_3)$ which either finds and returns $x_1 \in X_1$ with

$$g_1^{x_1} = a_1^{-1}(b - a_2 g_2^{x_2} - a_3 g_3^{x_3})$$

or reports that no such x_1 exists for a given $(x_2, x_3) \in X_2 \times X_3(r)$ by using Shor's discrete logarithm algorithm.

(1) The case $r \leq s_3$. Using Grover's search algorithm,[4] we search the subroutine $\mathcal{S}(x_2, x_3)$ for all $(x_2, x_3) \in X_2 \times X_3(r)$ in time

$$(x_2 r)^{1/2} (\log q)^{O(1)} \ll q^{3/5} (\log q)^{O(1)},$$

since $x_2 r \leq (s_1^2 s_2^2 r)^{2/5}$.

(2) The case $r > s_3$. Similarly, we search the $\mathcal{S}(x_2, x_3)$ for all $(x_2, x_3) \in X_2 \times X_3$ in time

$$(x_2x_3)^{1/2}(\log q)^{O(1)} \ll q^{3/5}(\log q)^{O(1)},$$

since $x_2x_3 \le (s_1^2s_2^2s_3)^{2/5} < (s_1^2s_2^2r)^{2/5}$. □

In Ref. 3, van Dam and Shparlinski mentioned when the multiplicative orders are large, there is a more efficient quantum algorithm. Similarly, we can also consider a more efficient quantum algorithm.

Theorem 4.2. *If we assume*

$$(s_1s_2)^2s_3 > q^3 \log q,$$

then we can either find a solution $(x_1, x_2, x_3) \in \boldsymbol{X}^3$ *of the equation* (4) *or decide that it does not have a solution in time* $q^2(s_1^2s_2^2s_3)^{-1/10}(\log q)^{O(1)}$ *as a quantum computer, except for at most* $q/\log q$ *exceptional* b*'s.*

Remark 4.1. The upper bound of the running time of the algorithm of Theorem 4.2 is

$$O(q^{1/5}(\log q)^{O(1)}).$$

Proof of Theorem 4.2. We may assume without loss of generality that $s_1 \ge s_2 \ge s_3$. We put

$$r = \lfloor q^3(s_1s_2)^{-2} \log q \rfloor \tag{10}$$

Then from the assumption of the theorem we see that $r \le s_3$. Hence there are some solutions of (4) in $\boldsymbol{X}^3(r)$ and we denote the number of the solutions of (4) by M. Note that $M \asymp (s_1s_2r)/q$.

As in the case of Ref. 3, we use the version of Grover's algorithm as described in Ref. 1 that finds one out of m matching items in a set of size t by using only $O(\sqrt{t/m})$ queries. We search the subroutine $\mathcal{S}(x_2, x_3)$ for all $(x_2, x_3) \in X_2 \times X_3(r)$. Then the time complexity is

$$\left(\frac{s_2r}{M}\right)^{1/2}(\log q)^{O(1)} \le q^{1/2}(s_1^2s_2^2s_3)^{-1/10}(\log q)^{O(1)}.$$ □

5. Concluding Remarks

At the end of this article, we compare the case of van Dam and Shparlinski with our case. See the following list.

# of variables	Classical	Quantum	ratio (C/Q)
2	1	1/3	3
3	3/2	3/5	5/2

The case of two variables is that of van Dam and Shparlinski and the case of three variables is our case. We notice that the ratio decreases. Does the ratio decrease to 1 when the dimension increase? We can apply the method used in Ref. 3 and this paper to the case of any variables. By roughly calculating, the ratio seems to converge 2, when the number of the variables increases (see Ref. 7 for details). It seems to come from the effect of Grover's algorithm.

Acknowledgments

The authors would like to express their gratitude to Professor Mikio Nakahara for his encouragement. The first author is partially supported by Grant-in-Aid for Scientific Research (C) from JSPS (Grant No. 20540033). The second and third authors are supported by the "Open Research Center" Project for Private Universities: matching fund subsidy from MEXT.

Appendix A. Algebra

In mathematics, sets with binary operations are classified into three main groups, *group, ring* and *field.* As an appendix, we state those notion concisely with some examples. Here, a binary operation $\circ$ on a set G is a map from $G \times G$ to G, for instance addition $+ : G \times G \to G$, $(g_1, g_2) \mapsto g_1 + g_2$, multiplication $\cdot : G \times G \to G$, $(g_1, g_2) \mapsto g_1 \cdot g_2$ and so on.

Appendix A.1. *Group*

A group is an elementary and important notion in algebra. The group axiom is as follows:

Definition A.1 (Group). *A group is a set G together with a binary operation $\circ$ on G such that the following three axioms hold:*

(1) $(a \circ b) \circ c = a \circ (b \circ c)$ *(association) holds for any a, b, c in G.*
(2) *There exists an element e which satisfies $a \circ e = e \circ a = a$ for any a in G. Such element is unique and called the identity element.*
(3) *For each a in G, there exists an element b in G such that $a \circ b = b \circ a = e$, where e is the identity element. Such element is called an inverse element of a.*

Remark A.1. A group G is called *abelian* (or *commutative*), if its operation is commutative (i.e. $a \circ b = b \circ a$ holds for any a and b in G).

Example A.1. The set of all natural numbers $\mathbb{N}$ does not form a group with respect to the addition. Because the identity element does not exist. Hence each element does not have an inverse element.

Example A.2. The set of all integers $\mathbb{Z}$ forms a group with respect to the addition, since the identity element is 0 and each element a has an inverse element $-a$. We can easily see that the addition on $\mathbb{Z}$ is associative. However it is not a group with respect to the multiplication. Note that an element 0 is special, since multiplying 0 by any element gives always 0. Hence we can not expect the existence of the unique identity element. Even if we exclude an element 0, then the identity element is 1, each element does not have an inverse element. Therefore $\mathbb{Z}$ is not group with respect to the multiplication.

Example A.3. Let be

$$M_2(\mathbb{C}) := \left\{ A = \begin{pmatrix} a & b \\ c & d \end{pmatrix} \middle| a, b, c, d \in \mathbb{C} \right\},$$

$$GL_2(\mathbb{C}) := \{ A \in M_2(\mathbb{C}) \mid \det A \neq 0 \}.$$

Note that $GL_2(\mathbb{C})$ is a subset of $M_2(\mathbb{C})$. We see that $GL_2(\mathbb{C})$ forms a group with respect to the matrix multiplication, although $M_2(\mathbb{C})$ does not form a group. The matrix multiplication is associative and there exists the unit matrix which plays a role as the identity element. However each element A in $M_2(\mathbb{C})$ does not have an inverse element, that is an inverse matrix A^{-1}, in general.

Appendix A.2. *Residue class of N*

Let be the residue class of N

$$\mathbb{Z}/N\mathbb{Z} := \{0, 1, \ldots, N-1\} \tag{A.1}$$

for any positive integer N. We define the addition on $\mathbb{Z}/N\mathbb{Z}$ by the following way:

$$a + b \mod N \qquad ({}^{\forall} a, b \in \mathbb{Z}/N\mathbb{Z}).$$

Then $\mathbb{Z}/N\mathbb{Z}$ forms a group with respect to the above addition (The inverse element is 0 and each element a has an inverse element $N-a$). The following table expresses the calculated result of addition for the case of $N = 6$:

Table A1. Addition on $\mathbb{Z}/6\mathbb{Z}$

+	0	1	2	3	4	5
0	0	1	2	3	4	5
1	1	2	3	4	5	0
2	2	3	4	5	0	1
3	3	4	5	0	1	2
4	4	5	0	1	2	3
5	5	0	1	2	3	4

Let us consider the multiplication on $\mathbb{Z}/N\mathbb{Z}$. Similarly, we define the multiplication on $\mathbb{Z}/N\mathbb{Z}$ by

$$a \cdot b \mod N \qquad ({}^{\forall}a, b \in \mathbb{Z}/N\mathbb{Z}).$$

As in the case of $\mathbb{Z}$, 0 must be excluded so that $\mathbb{Z}/N\mathbb{Z}$ forms a group with respect to the above multiplication. However, it is not sufficient to form a group with respect to the multiplication, since there exist elements which induce excluded element 0 by multiplying certain elements. To confirm this phenomenon, we observe the following table which shows the multiplication on $\mathbb{Z}/6\mathbb{Z}$:

Table A2. Multiplication on $(\mathbb{Z}/6\mathbb{Z}) \setminus \{0\}$

$\cdot$	1	2	3	4	5
1	1	2	3	4	5
2	2	4	0	2	4
3	3	0	3	0	3
4	4	2	0	4	2
5	5	4	3	2	1

As can be expected from the above table, elements which are not co-prime to N have the possibility to induce the excluded element 0. In fact, the set

$$(\mathbb{Z}/N\mathbb{Z})^{\times} := \{a \in \mathbb{Z}/N\mathbb{Z} \mid \gcd(a, N) = 1\}$$

forms a group with respect to the multiplication and is called a *modulo multiplication group*. Note that $(\mathbb{Z}/p\mathbb{Z})^{\times} = (\mathbb{Z}/p\mathbb{Z}) \setminus \{0\}$ for any prime number p. It is know that the number of elements of a modulo multiplication group $(\mathbb{Z}/N\mathbb{Z})^{\times}$ is given by the Euler quotient function $\phi(N)$ defined by

$$\phi(N) := N \prod_{\substack{p|N, \\ p:\text{prime}}} \left(1 - \frac{1}{p}\right).$$

We can easily see that $\phi(p) = p - 1$ for any prime number p, since the divisor of p is only itself, and it coincides with the above note.

Appendix A.3. *Ring*

As mentioned above, the set of all integers $\mathbb{Z}$ is closed with respect to the addition and multiplication and forms a group with respect to the addition. Therefore $\mathbb{Z}$ has a richer algebraic structure than the group structure. We call such algebraic structure *ring*. The definition of rings is as follows:

Definition A.2 (Ring). *A ring is a set R together with two binary operations, an addition $(+)$ and a multiplication $(\cdot)$, such that:*

(1) *R is an abelian group with respect to the addition;*
(2) *the multiplication is associative $x \cdot (y \cdot z) = (x \cdot y) \cdot z$;*
(3) *the multiplication is distributive: $(x+y)\cdot z = x\cdot z + y\cdot z$ and $z\cdot(x+y) = z \cdot x + z \cdot y$ for all $x, y, z \in R$.*

Remark A.2. A ring is called *commutative*, if its multiplication is commutative.

Roughly speaking, a ring is an additive group. However it does not require forming a group with respect to the multiplication. We show some examples below.

Example A.4. $\mathbb{Z}/N\mathbb{Z}$ is a ring for any positive integer N. Because $\mathbb{Z}/N\mathbb{Z}$ is an abelian additive group and we can easily confirm the multiplication on $\mathbb{Z}/N\mathbb{Z}$ is associative and distributive.

Example A.5. The set of all integers $\mathbb{Z}$ is a ring. $\mathbb{Q}, \mathbb{R}, \mathbb{C}$ are also rings.

Appendix A.4. *Field*

The aim of this section is to understand an explicit structure of finite fields. For that purpose, we have prepared some notion in the previous sections. We first state the field axiom below.

Definition A.3 (Field). *A field is a commutative ring F such that $F\backslash\{0\}$ is a group under the multiplication, where 0 is the identity element for the addition. In particular, when the number of elements in F is finite, F is called a finite field.*

Example A.6. For any prime number p, $\mathbb{Z}/p\mathbb{Z}$ is a finite field. We often denote $\mathbb{F}_p$ instead of $\mathbb{Z}/p\mathbb{Z}$.

Example A.7. For any composite number N, $\mathbb{Z}/N\mathbb{Z}$ is not a finite field, since $(\mathbb{Z}/N\mathbb{Z})\setminus\{0\}$ does not form a group with respect to the multiplication.

Example A.8. The set of all rational numbers $\mathbb{Q}$ is a field. Let us consider the set

$$\mathbb{Q}(\sqrt{d}) := \{m + n\sqrt{d} \mid m, n \in \mathbb{Q}\},$$

where d is a square-free non-zero integer. We can easily confirm that the above set forms a field. We call such set a *quadratic field.* Note that a quadratic field $\mathbb{Q}(\sqrt{d})$ contains $\mathbb{Q}$ and has a structure as a 2-dimensional vector space over $\mathbb{Q}$.

As in the case of the above example, when a field K contains a field F as a subset, F is called a *subfield* of K. Conversely, K is called an *extension field* of F. An extension field K of F can be identified as a vector space over F. The dimension of K as the vector space over F is called the *degree* of K and denoted by $[K : F]$.

Hereafter we discuss finite fields. It is known that when K is a finite field, there exists a prime number p such that $\mathbb{F}_p$ is contained in K. Hence K is an extension field of $\mathbb{F}_p$. If we put n the degree of K over $\mathbb{F}_p$, then there exists a basis $\alpha_1, \ldots, \alpha_n$ in K and any element of K can be expressed as

$$x_1\alpha_1 + \cdots + x_n\alpha_n \qquad (x_1, \ldots, x_n \in \mathbb{F}_p),$$

since K has a structure as an n-dimensional vector space. Therefore we see that K consists of p^n elements. Further $K^\times (= K \setminus \{0\})$ is an abelian group with respect to the multiplication with order $q-1$. Namely, $x^{q-1} = 1$ holds for any element x in $K^\times$. In fact, we can prove that

$$K^\times = \{a \in \overline{\mathbb{F}}_p \mid a^{q-1} = 1\},$$

where $\overline{\mathbb{F}}_p$ is the algebraic closure of $\mathbb{F}_p$, that is $\overline{\mathbb{F}}_p$ is the set of all the elements that are solutions of $f(x) = 0$ for any $f(x) \in \mathbb{F}_p[x]$. Namely $K^\times$ consists of all $(q-1)$-st root of unity. On the other hand, we can prove that there exists a finite field which consists of p^n elements for any prime number p and any positive integer n. We denote a finite field with p^n elements $\mathbb{F}_{p^n}$.

We deal with a method of representing elements of $\mathbb{F}_q$ $(q = p^n)$ below. There are some methods to represent elements of $\mathbb{F}_q$. Here we introduce a method which employs field extensions.

A polynomial $f \in \mathbb{F}_p[x]$ is said to"irreducible" in $\mathbb{F}_p[x]$, if f has positive degree and $f = bc$ with $b, c \in \mathbb{F}_p[x]$ implies that either b or c is a constant polynomial. We prepare an irreducible polynomial f in $\mathbb{F}_p[x]$ of degree n. The polynomial f has a root α in $\overline{\mathbb{F}}_p$ whereby we consider an extension field $\mathbb{F}_p(\alpha)$ which is a minimal field containing $\mathbb{F}_p$ and α. Then $\mathbb{F}_p(\alpha)$ is exactly $\mathbb{F}_q$ $(q = p^n)$ and $\{1, \alpha, \ldots, \alpha^{n-1}\}$ is a basis of $\mathbb{F}_q$ over $\mathbb{F}_p$. Therefore every element of $\mathbb{F}_q$ can be uniquely expressed as a polynomial in α over $\mathbb{F}_p$ of degree less that n.

Example A.9. Let be $p = 3$. To consider an extension field of $\mathbb{F}_3$ of degree 2, we prepare a polynomial $f(x) = x^2 + 1$ which is irreducible in $\mathbb{F}_3[x]$ of degree 2. If we denote α one of the root of $f(x)$, every element of $\mathbb{F}_{3^2}$ can be expressed as $x_0 + \alpha x_1$ $(x_0, x_1 \in \mathbb{F}_3)$. Hence we have

$$\mathbb{F}_{3^2} = \{0, 1, 2, \alpha, 1 + \alpha, 2 + \alpha, 2\alpha, 1 + 2\alpha, 2 + 2\alpha\},$$

where α is an element in $\mathbb{F}_{3^2}$ satisfying $\alpha^2 = -1$.

See Ref. 5 for further information of finite fields.

References

1. M. Boyer, G. Brassard, P. Høyer and A. Tapp, *Tight bounds on quantum searching*, Fortschritte der Physik, **46** (1998), 493–505.
2. R. Crandall and C. Pomerance, *Prime Numbers: A computational perspective*, Springer-Verlag, Berlin, 2005.
3. W. van Dam and E. Shparlinski, *Classical and Quantum Algorithms for Exponential Congruences*, Proceedings of the Third Workshop on Theory of Quantum Computation, Communication and Cryptography (TQC 2008), Lecture Notes in Computer Science, Vol. 5106, 1–10 (2008)
4. L. Grover, *A fast quantum-mechanical algorithm for database search*, Proceedings of the 28th Annual ACM Symposium on Theory of Computing (STOC '96), 1996, pp. 212–219.
5. R. Lidl and H. Niederreiter, *Finite Fields*, Encyclopedia of Mathematics and Its Applications, Vol. 20, Cambridge Univ. Press, Cambridge, 1997.
6. P. Shor, *Polynomial-time algorithms for prime factorization and discrete logarithms on a quantum computer*, SIAM Journal on Computing **26** (1997), 1484–1509.
7. Y. Sasaki, *On zeros of exponential polynomials and quantum algorithms*, Quantum Inf. Process. **9**, 419–427, 2010.

PART B

Workshop on Matrix Product State Formulation and Density Matrix Renormalization Group Simulations (MPS&DMRG)

APPLICATION OF DENSITY MATRIX RENORMALIZATION GROUP METHOD TO PHOTOINDUCED PHENOMENA IN STRONGLY CORRELATED ELECTRON SYSTEMS

HIROAKI MATSUEDA

Sendai National College of Technology,
Sendai, Miyagi 989-3128, Japan
E-mail: matsueda@sendai-nct.ac.jp

Density matrix renormalization group method (DMRG) is a powerful tool to examine excited and time-dependent states of one-dimensional (1D) strongly correlated systems as well as their ground-state properties. In this review, physical motivation and various numerical techniques for the examination are presented. Then, we summarize our DMRG analysis for photoinduced phenomena in 1D Mott insulators and other transition metal oxides. Close connection between DMRG and key aspects of quantum information such as entanglement is also addressed.

Keywords: Density matrix renormalization group (DMRG); dynamical DMRG; time-dependent DMRG; entanglement entropy; matrix product state (MPS); strongly correlated electron system; spin-charge separation; Hubbard model; Hubbard-Holstein model; double-exchange model; nonlinear optics; photoinduced phase transition (PIPT); pump-probe spectroscopy.

1. Introduction

1.1. *Motivation from condensed matter physics side*

Density matrix renormalization group (DMRG) was introduced by Steven White in 1992 in order to precisely calculate ground-state and low-energy properties of one-dimensional (1D) gapped quantum spin systems.[1] Up to now, DMRG has been applied to various systems and to calculation of various quantities.[2] For instance, it is possible to examine dynamical spectra,[3–5] finite-temperature properties,[6,7] and time-dependent functions.[8–10] DMRG is also extended to classical systems and study of underlying mathematical structure of DMRG,[11,12] bosonic systems with many local degrees of freedom,[13,14] and alternative formulation with symmetry.[15–17]

In viewpoints of application of DMRG to strongly correlated electron systems, calculations of dynamical quantities are necessary. This is because spectroscopic measurements play crucial roles on experimental examination of the systems. Thus, detailed comparison between DMRG data and experiments offers important insights for their electronic properties.[18] A background of this powerfulness is related to development of highly brilliant synchrotron radiation and neutron sources. In addition, laser techniques such as laser photoemission and pump-probe spectroscopy are respectively important for ultrahigh-resolution and time-resolved spectroscopy. For instance, recent active research areas are (1) charge excitation of cuprates measured by resonant inelastic X-ray scattering (RIXS), (2) study of magnetic excitation of high-T_c cuprates by neutron scattering, (3) direct observation of superconducting gaps of f-electrons by laser photoemission, (4) bulk-sensitive photoemission by soft X-ray, and (5) real-time measurements of dynamical coupling and separation between correlated electrons and phonons by pump-probe spectroscopy.

These probes can detect both of low-energy coherent states near the Fermi level and high-energy states beyond an optical gap. The former mainly determines transport properties, whereas the latter local properties. In strongly correlated electron systems, both energy scales mutually interact with each other, and thus the dynamical measurements are essential for deep understanding of the mutual interaction. It has been well-known from the initial stage of study of high-T_c cuprates that high-energy weight of optical absorption spectrum transfers to mid-infrared region for a small amount of hole doping, and further doping induces Drude-like metallic spectral weight.[19] This means global change in band structure due to correlation effects, and in that case a rigid-band picture is not appropriate.

High reliability of recent experiments is originated from detailed research for excitation-energy dependence on spectral shape, atomic selectivity of probe light, and penetration depth. In addition, the reliability is also confirmed by clear observation of spinon and holon excitations in a 1D cuprate that are characterized by edge singularities of angle-resolved photoemission spectra (ARPES).[20] The singularities were not observed definitely 15 years ago.[20]

On the other side, there is a long-standing history for theoretical study of Mott transition.[21] Perhaps, readers know Hubbard's regendary papers, coherent potential approximation, various equation-of-motion techniques for a Green's function, and Gutzwiller approximation. One of recent successful methods is dynamical mean-field theory (DMFT) (or its cluster version) that has been introduced even in standard textbooks.[22] However, their scope is quite limited, and conclusions from them are sometimes incompatible to recent high-resolution experimental data. One of disadvantages in these theoretical treatments is coare-graining of energy scales, and in this case the theory can not describe low-energy magnetic properties. DMFT can take account of local electronic correlation accurately, but loses momentum dependence and spatial dimensionality. Materials of our recent interest have complex crystal structures and electronic orbitals such as 1D chain, layered structure, spinel and pyrochlore. There, dimensionality and geometrical structure of electronic orbitals play dominant roles on their electronic properties.

Because of these reasons, the above-mentioned theoretical approaches and numerical ones for finite-size clusters should be complementary to each other. Two major numerical approaches are exact diagonalization by Lanczos algorithm and quantum Monte Carlo simulation based on Suzuki-Trotter breakup.[23] Taking finite-size clusters would be appropriate as a starting point, since mother compounds of transition metal oxides are insulators. Broadly speaking, they are strong ionic crystals, and a unit cluster composed of a transition metal and oxygens determines basic electronic properties. However, it is fundamentally impossible to access an energy scale lower than an energy gap due to the finiteness of the system. Thus, the systematic size scaling is quite important. DMRG usually treats clusters roughly ten-times larger than those in exact diagonalization without loss of accuracy, and then it is sometimes able to catch low-energy collective behaviors of correlated electrons in the materials.

1.2. *Quantum information perspective*

Matrix product state (MPS) formulation of the original DMRG is important in order to mathematically understand reasons for success of DMRG.[12] On the other side, quantum information theory is based on quantum circuits composed of various quantum devises, and the devises and the circuits are local tensors and their networks. It has been recently recognized that these two are essentially the same, and the concept of 'quantum entanglement' in the information theory flowed into DMRG. The introduction of the entanglement greatly improves DMRG performance for time-dependent calculation, periodic boundary cases, and higher-dimensional models.[8–10,24,25] The crucial new idea for the improvement is to construct tensor-network wave functions to keep basic entanglement structure hidden in the corresponding quantum systems. Now, there are many references that address the roles of entanglement and tensor network on numerical precision of DMRG.[26–30]

2. Concept of DMRG: A Modern Approach

2.1. *Superblock and singular value decomposition*

The most educative articles of the methodology of DMRG would be the White's original papers.[1] In order to aviod redundancy of explanation of details, I present a modern view of DMRG based on entanglement entropy. I hope that this introduction connects smoothly to other research areas that are associated with quantum entanglement.

A basic purpose of DMRG is to precisely calculate the ground-state properties of finite-length 1D quantum systems such as spin models and Hubbard models. One of characteristic features of DMRG is to devide the total system into 'left' and 'right' subsystems/blocks. One of the subsystems is simply called 'system', whereas the other is called 'environment'. The total system is called 'superblock', since the system block and the environment block are combined with each other. On the other hand, quantum coherence over the total system plays an important role on physical properties of quantum systems. Thus, we may be curious about validity of deviding the total system into two subsystems. The key quantity is 'quantum entanglement' between system and environment. Before going to some details of DMRG, let us first study entanglement properties hidden in jointing blocks.

Let us start with the wave function of the superblock, ψ_{ij}. The bases of the system and the environment are respectively represented by $|i\rangle$ and $|j\rangle$. Then, the state of the superblock is given by

$$|\psi\rangle = \sum_{ij} \psi_{ij} |i\rangle |j\rangle \,. \tag{1}$$

Here, we decompose ψ_{ij} into 'singular values'. Then, physical meanings of quantum entanglement emerge quite naturally. Let ψ_{ij} be a matrix. By using unitary matrices, U and V, and non-negative singular values, λ_α, ψ_{ij} can be decomposed into the following form (so called 'singular value decomposition', SVD)

$$\psi_{ij} = \sum_\alpha U_{i\alpha} \lambda_\alpha V_{j\alpha}. \tag{2}$$

These matrices, U and V, do not uniquely determined, but it is mathematically proven that λ_α is uniquely determined. In order to know the physical meaning of λ_α, we substitute Eq. (2) into Eq. (1):

$$|\psi\rangle = \sum_\alpha \lambda_\alpha (\sum_i U_{i\alpha} |i\rangle)(\sum_j V_{j\alpha} |j\rangle) = \sum_\alpha \lambda_\alpha |\alpha\rangle |\alpha'\rangle . \tag{3}$$

Since U, V are unitary matrices, the following two vectors

$$|\alpha\rangle = \sum_i U_{i\alpha} |i\rangle , \tag{4}$$

$$|\alpha'\rangle = \sum_j V_{j\alpha} |j\rangle , \tag{5}$$

are the orthgonal sets of the system and the environment, respectively. Equation (3) is called 'Schmidt decomposition form'. This equation can not be represented by a direct product between system and environment states. Thus, this situation means that the system and the environment are quantum-mechanically 'entangled'. Then, the amount of the entanglement is characterized by λ_α.

The square of the singular value $(\lambda_\alpha)^2$ is the probability of finding the states, $|\alpha\rangle$ and $|\alpha'\rangle$, and the probability is the eigenvalue of the system's density matrix, ρ_A, which is the density submatrix after convention of environmental degrees of freedom

$$\rho_A = tr_B |\psi\rangle \langle\psi| = \sum_\alpha (\lambda_\alpha)^2 |\alpha\rangle \langle\alpha| , \tag{6}$$

where the substript of trace, B, means convention of environmental degrees of freedom. Hereafter, the system and the environment are represented by A and B, respectively. When we exchange the system and the environment, we can also make the density matrix for the enviroment

$$\rho_B = tr_A |\psi\rangle \langle\psi| = \sum_\alpha (\lambda_\alpha)^2 |\alpha'\rangle \langle\alpha'| . \tag{7}$$

Clearly, $(\lambda_\alpha)^2$ is also the eigenvalue of ρ_B.

2.2. *Entanglement entropy*

A physical quantity representing an amount of entanglement is 'entanglement entropy'. The entropy S_A for an observer in A is defined by

$$S_A = -tr_A(\rho_A \log \rho_A) = -\sum_\alpha (\lambda_\alpha)^2 \log(\lambda_\alpha)^2. \tag{8}$$

Thus, S_A is the information entropy for the probability $(\lambda_\alpha)^2$ to find the states, $|\alpha\rangle$ and $|\alpha'\rangle$, in the superblock. Since we have already known that

$$S_A = S_B, \tag{9}$$

the entanglement entropy is an amount of mutual information.

By using the entanglement entropy, we know how to approximately treat Eq. (3). It would be natural to treat $|\psi\rangle$ approximately by taking m-largest values of $(\lambda_\alpha)^2$. The parameter m is called 'truncation number' of the states. In order to make our discussion simple, here we assume that the quantum entanglement between the system and the environment is strongest. This assumption corresponds to a case of the worst numerical cost. In this case,

$$(\lambda_\alpha)^2 = \frac{1}{m}, \tag{10}$$

for all α, and then we obtain

$$S_A = \log m. \tag{11}$$

Thus, the truncation number in the worst case is given by

$$m = e^{S_A}, \tag{12}$$

and the truncation number increases exponentially with the entropy. Generally, the entanglement entropy is an increasing function of the linear system size L of A. However, quantum 1D cases are exceptional. In gapped cases, the entropy obeys well-known 'area-law scaling',

$$S_A \propto \left(\frac{L}{a}\right)^{d-1}, \tag{13}$$

with d and a denoting spatial dimension and lattice constant, respectively, and the entropy saturates for L comparable to the correlation length ξ.[31,32] Thus, in a case of $d = 1$, the entropy is kept constant against the increase in the system size. According to the conformal field theory, the entanglement entropy increases only logarithmically even near critical points

$$S_A = \frac{c}{6}\mathcal{A}\log\left(\frac{\xi}{a}\right), \tag{14}$$

where c is the central charge and $\mathcal{A}$ is the number of the boundary points.[33–35] At criticality, this relation goes to be $S_A = (c/3)\log(L/a)$. Therefore, we obtain

$$m < e^{S_A} = \left(\frac{\xi}{a}\right)^{c\mathcal{A}/6}. \tag{15}$$

This means that m is at most an algebraic function of L, not exponential. We also know that the boundary condition greatly affects DMRG performance, since open and periodic boundary conditions take $\mathcal{A} = 1$ and $\mathcal{A} = 2$, respectively. Our interest is, for example, $S = 1/2$ Heisenberg model, and the central charge of the model is $c = 1$. Thus, even for this critical model, the power of this algebraic function, 1/6, is small in the open boundary condition. The above entropy analysis answers the question why DMRG has been succeeded in 1D quantum systems under the open boundary condition.

2.3. *Sweeping and matrix product state*

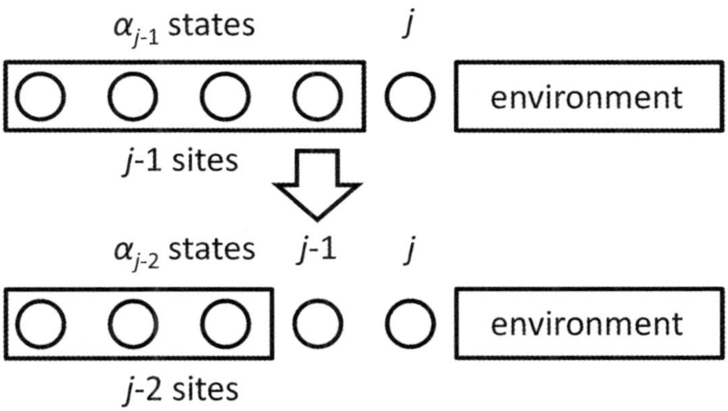

Fig. 1. Orthogonal transformation of the system block.

Let us construct the wave function of 1D quantum systems by conbining SVD with the 'sweeping' procedure. As shown in Fig. 1, we consider the superblock that is devided into two at the right hand side of the j-th site. In order to make clear the position of this site in Eq. (3), we put the subscript j into the index α

$$|\psi\rangle = \sum_{\alpha_j} \lambda_{\alpha_j} |\alpha_j\rangle \, |\, \alpha'_j \,\rangle \,. \tag{16}$$

Here, the system is composed of a block with $j-1$ sites and the j-th site. The states of the block and the site are respectively represented by $|\alpha_{j-1}\rangle$ and $|s_j\rangle$. Then, we can represent $|\alpha_j\rangle$ as

$$|\alpha_j\rangle = \sum_{\alpha_{j-1}, s_j} A^{s_j}_{\alpha_{j-1}\alpha_j} |\alpha_{j-1}\rangle \, |s_j\rangle \,, \tag{17}$$

by using the orthogonal transformation $A^{s_j}_{\alpha_{j-1}\alpha_j}$.[12] Equation (17) corresponds to a procedure to transform $|\alpha_{j-1}\rangle \, |s_j\rangle$ into $|\alpha_j\rangle$, and thus it may be better that the transformation is expressed as $A_{(\alpha_{j-1}s_j)\alpha_j}$. However, we intentionally change the order of the labels, α_{j-1}, α_j, and s_j, so that the labels α_{j-1} and α_j seem to be indices of a matrix. The reason for this special order can be seen by the following iterative transformation (namely 'sweeping'):

$$|\alpha_j\rangle = \sum_{\alpha_{j-1},s_j} A^{s_j}_{\alpha_{j-1}\alpha_j} |s_j\rangle \sum_{\alpha_{j-2},s_{j-1}} A^{s_{j-1}}_{\alpha_{j-2}\alpha_{j-1}} |\alpha_{j-2}\rangle |s_{j-1}\rangle \tag{18}$$

$$= \sum_{s_{j-1},s_j} \sum_{\alpha_{j-2}} \left(A^{s_{j-1}} A^{s_j}\right)_{\alpha_{j-2}\alpha_j} |\alpha_{j-2}\rangle |s_{j-1}s_j\rangle \tag{19}$$

$$= \sum_{s_{j-2},s_{j-1},s_j} \sum_{\alpha_{j-3}} \left(A^{s_{j-2}} A^{s_{j-1}} A^{s_j}\right)_{\alpha_{j-3}\alpha_j} |\alpha_{j-3}\rangle |s_{j-2}s_{j-1}s_j\rangle . \tag{20}$$

Here we use abbreviation $|s_{j-1}\rangle |s_j\rangle = |s_{j-1}s_j\rangle$. Thus, the coefficient is written in the matrix product. Continuing the sweeping procedure from the j-th site to the left edge, we obtain

$$|\alpha_j\rangle = \sum_{s_1,s_2,\ldots,s_j} \sum_{\alpha_0} \left(A^{s_1} A^{s_2} \cdots A^{s_j}\right)_{\alpha_0\alpha_j} |\alpha_0\rangle |s_1 s_2 \cdots s_j\rangle , \tag{21}$$

where α_0 is an auxiliary field. Hereafter, we change the notation (The column vector $|\alpha_0\rangle$ is transposed)

$$\sum_{\alpha_0} \left(A^{s_1} A^{s_2} \cdots A^{s_j}\right)_{\alpha_0\alpha_j} |\alpha_0\rangle \rightarrow \left(\langle\alpha_0| A^{s_1} A^{s_2} \cdots A^{s_j}\right)_{\alpha_j} . \tag{22}$$

The same procedure is done for environmental degrees of freedom. We introduce the orthogonal transformation $B^{s_{j+1}}_{\alpha_j\alpha_{j+1}}$ by defining

$$|\alpha'_j\rangle = \sum_{\alpha_{j+1},s_{j+1}} B^{s_{j+1}}_{\alpha_j\alpha_{j+1}} |\, \alpha'_{j+1} \,\rangle |s_{j+1}\rangle , \tag{23}$$

and continue the sweeping upto the right edge

$$|\alpha'_j\rangle = \sum_{s_{j+1},s_{j+2},\ldots,s_L} \left(B^{s_{j+1}} B^{s_{j+2}} \cdots B^{s_L} |\alpha_n\rangle\right)_{\alpha_j} |s_{j+1}s_{j+2}\cdots s_L\rangle . \tag{24}$$

Substituting Eqs. (21) and (24) into Eq. (16), we finally obtain the following wave function

$$|\psi\rangle = \sum_{s_1,s_2,\ldots,s_L} \psi_{s_1 s_2 \cdots s_L} |s_1 s_2 \cdots s_L\rangle , \tag{25}$$

$$\psi_{s_1 s_2 \cdots s_L} = \langle\alpha_0| A^{s_1} A^{s_2} \cdots A^{s_j} \lambda_j B^{s_{j+1}} B^{s_{j+2}} \cdots B^{s_L} |\alpha_L\rangle . \tag{26}$$

Here λ_j is a diagonal matrix, and the diagonal elements are λ_{α_j}. Therefore, the quantum 1D wave function can be represented by the matrix product, and this is called 'Matrix Product State (MPS)'. An important thing is that this wave function can be factorized by only using local matrices, even though we start with pointing out the importance of the quantum entanglement over the total system.

Up to now, we have considered the decomposition of the total system at site j, but the position of the decomposition can be freely selected. The singular value appears at the position, and thus using an appropriate matrix Γ^{s_j}, we can express the coefficient as[9]

$$\psi_{s_1 s_2 \cdots s_L} = \langle \alpha_0 | \Gamma^{s_1} \lambda_1 \Gamma^{s_2} \lambda_2 \cdots \Gamma^{s_{L-1}} \lambda_{L-1} \Gamma^{s_L} | \alpha_L \rangle . \tag{27}$$

Furthermore, it is also possible to sweep from left to right edges by using one transfmation equation, and we obtain the following expression

$$\psi_{s_1 s_2 \cdots s_L} = \langle \alpha_0 | A^{s_1} A^{s_2} \cdots A^{s_L} | \alpha_L \rangle . \tag{28}$$

In mathematical formulation of MPS, Eq. (28) is the most common form.

Here, we briefly touch on the boundary condition. In a periodic boundary condition, $\alpha_0 = \alpha_L$ and then

$$\psi_{s_1 s_2 \cdots s_L} = tr(A^{s_1} A^{s_2} \cdots A^{s_L}), \tag{29}$$

where the translational invariance due to the periodic boundary condition is represented by the rotational invariance of the trace. In an open boundary condition,

$$\psi_{s_1 s_2 \cdots s_L} = tr(A^{s_1} A^{s_2} \cdots A^{s_L} Q), \tag{30}$$

and Q is the boundary matrix defined by

$$Q = |\alpha_L\rangle \langle \alpha_0| . \tag{31}$$

Equation (30) indicates that appropriate boundary deformation may reduce finite-size effects. Actually, it has been known that the boundary matrix plays a crucial role on periodicity of ordered states.[36]

The transformation by the matrix $A^s_{\alpha\beta}$ should conserve an orthonormal basis set, $\langle \alpha | \alpha' \rangle = \delta_{\alpha\alpha'}$, and then the following relation is hold

$$\delta_{\alpha\alpha'} = \sum_{\beta,\beta',s,s'} \left(A^{s'}_{\alpha'\beta'} \right)^* A^s_{\alpha\beta} \langle s' | s \rangle \langle \beta' | \beta \rangle = \sum_s \left(A^s \left(A^s \right)^\dagger \right)_{\alpha\alpha'} . \tag{32}$$

2.4. *Finite-system sweep of DMRG*

An introduction of general MPS formulation has been finished in the previous sections. The remaining problem is to determine a set of the orthogonal matrices, $A^{s_j}_{\alpha_j \alpha_{j+1}}$. DMRG is a method for the determination. In particular, the sweeping discussed previously is well matched onto the finite-system sweep of DMRG.[1] In the following, we explain outline of this algorithm, and make clear the relation among the DMRG sweep algorithm, SVD in Eq. (2), and MPS in Eq. (17).

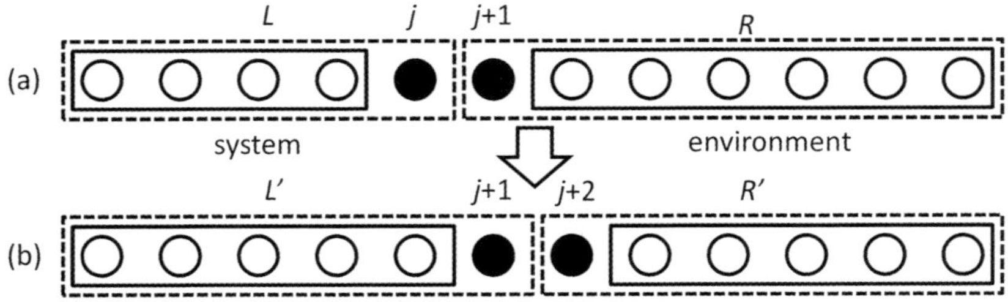

Fig. 2. Sweep algorithm of DMRG.

We start with the following state (see Fig. 2(a))

$$|\psi\rangle = \sum_{l,s_j,s_{j+1},r} \psi_{ls_js_{j+1}r} |l\rangle |s_j\rangle |s_{j+1}\rangle |r\rangle , \tag{33}$$

where two sites at j and $j+1$ and two blocks, L and R, are combined into the total system. The two local sites with dimension s are exactly treated, while the blocks with many sites are treated approximately. The blocks, L and R, are represented by the states $|l\rangle$ and $|r\rangle$, respectively. The block L and the site j are in the system block, and the block R and the site $j+1$ are in the enviromental block. First, we diagonalize the superblock Hamiltonian, and obtain the ground-state wave function, $\psi_{ls_js_{j+1}r}$. However, the calculation is impossible for large L values, and thus we assume an initial condition where two blocks L and R are approximately represented by m states. We trace over enviromental degrees freedom to make the density matrix

$$\rho_{l,s_j;l',s'_j} = \sum_{s_{j+1},r} \psi_{ls_js_{j+1}r} \psi^*_{l's'_js_{j+1}r}. \tag{34}$$

The density matrix has sm dimension, but after diagonalization of ρ we ratain only m states with large eigenvalues. Then by using these m states, the new block L' in Fig. 2(b) is represented. When these procedures are matched onto Eq. (17), we know that the matrix diagonalizing ρ is corresponding to $A^{s_j}_{\alpha_j\alpha_{j+1}}$. The auxilliary field α is the state with large probability of realization, and the eigenstate of the density submatrix with large eigenvalue. This formulation is also consistent with Eq. (2). Next, the same procedures are operated to the new superblock $(L', j+1, j+2, R')$. We continue to sweep from left to right, then reverse the sweeping direction at the edges, until the ground-state energy is converged.

3. Extention of DMRG for Excited States

3.1. *Targetting excited states into density matrix*

The basic DMRG algorithm restricts the exponentially increasing Hilbert space to the relevant DMRG bases using the ground-state density matrix. However, the DMRG basis is optimized only for the ground state. The state defining the density matrix is called 'target state'. To make informations about excited states reflect on the density matrix, we need to target appropriate excited states into the density matrix. The new density matrix for the system block is given by

$$\rho_{ii'} = \sum_{\alpha} p_{\alpha} \sum_{j} \psi^{*}_{\alpha,ij} \psi_{\alpha,i'j}, \tag{35}$$

where

$$\sum_{\alpha} p_{\alpha} = 1, \tag{36}$$

with α denoting each target state.[3,4] This is the density matrix for a mixed ensemble. Let us remember the thermal density matrix

$$\rho = \sum_{\alpha} p_{\alpha} \left|\alpha\right\rangle \left\langle\alpha\right|, \tag{37}$$

where we do not trace out the enviromental degrees of freedom. Here, p_{α} is the Boltzmann factor defined by

$$p_{\alpha} = \frac{e^{-\beta E_{\alpha}}}{\sum_{\alpha} e^{-\beta E_{\alpha}}} \geq 0, \tag{38}$$

with $H\left|\alpha\right\rangle = E_{\alpha}\left|\alpha\right\rangle$. In DMRG, p_{α} is regulated by hand, since the target states are the states that we need to know. The factor, p_{α}, gives us a criterion of precision of the corresponding target state, ψ_{α}. As we will discussed in detail, we calculate the one-particle spectral function

$$A(k,\omega) = -\frac{1}{\pi}\mathrm{Im}\left\langle 0\right| c^{\dagger}_{k,\sigma} \frac{1}{E_0 - \omega - H + i\gamma} c_{k,\sigma}\left|0\right\rangle, \tag{39}$$

by the following targets (with normalization factors)

$$\{\left|\alpha\right\rangle\} = \left\{\left|0\right\rangle, c_{k,\sigma}\left|0\right\rangle, (E_0 - \omega - H + i\gamma)^{-1} c_{k,\sigma}\left|0\right\rangle\right\}. \tag{40}$$

When we represent normalized target states as matrices ψ_α (row and column represent system and environmental degrees of freedom), the density matrix for the mixed ensemble is given by

$$\rho_A = \sum_\alpha p_\alpha \psi_\alpha \psi_\alpha^*, \tag{41}$$

where the environmental degrees of freedom are trace out by the matrix multiplication. Furthermore, we have

$$tr_A\ \rho_A = \sum_\alpha p_\alpha tr_A\ (\psi_\alpha \psi_\alpha^*) = \sum_\alpha p_\alpha \sum_{i,j} (\psi_\alpha)_{i,j} (\psi_\alpha^*)_{ij} = \sum_\alpha p_\alpha = 1. \tag{42}$$

After defining the density matrix for the mixed ensemble, the numerical procedures are the same as those of the ground-state DMRG. We diagonalize Eq. (41) by the unitary matrix U, and remain the eigenstates with the large eigenvalues. Note that it might be problematic that the unitary matrix, U, does not diagonalize each matrix, $\psi_\alpha \psi_\alpha^*$, with different α simultaneously.

The mixed-ensemble form should be distinguished from the following phase coherent state (pure state)

$$\psi = \sum_\alpha \sqrt{p_\alpha} e^{i\theta_\alpha} \psi_\alpha. \tag{43}$$

This is because the corresponding density matrix contains quantum interference between different states

$$\tilde{\rho} = \psi\psi^* = \rho + \sum_{\alpha \neq \beta} \sqrt{p_\alpha p_\beta} e^{i(\theta_\alpha - \theta_\beta)} \psi_\alpha \psi_\beta^*. \tag{44}$$

It is in general difficult to prove whether the second term vanishes due to the sign change of the phase factor, $e^{i(\theta_\alpha - \theta_\beta)}$.

There are various versions of calculating spectra. The most sophisticated numerical algorithm is called dynamical DMRG (DDMRG) that is based on a variational method.[5] In the following, we examine various numerical techniques based on the correction vector DMRG that is one of pre-DDMRG techniques, since this is physically clear.[4]

Before closing this subsection, we consider validity of dynamical calculation by DMRG in a viewpoint of the entanglement-entropy scaling. Targetting appropriate states into the density matrix is certainly the main idea for succesive optimization of the DMRG basis states. At the same time, the entanglement entropy and its scaling relation are key ingredients for the MPS formalism of DMRG. Usually, the entropy is defined by using the ground-state wave function, and in general the entropy scaling seems to be valid only for the pure state. However, the ground state is factorized by a simple product of spin and charge sectors, and a one-photon photoexcited state is composed of only charge degree of freedom. Thus, the basic properties of the ground state still remain in the photoexcited state. This is also the reason for the success of the dynamical calculation of DMRG.

3.2. *Optical spectra*

3.2.1. *Linear optical absorption and optically forbidden spectrum*

Let us first derive the target states of the current-current correlation function

$$\chi_j(\omega) = \langle 0| \, j \frac{1}{\omega + E_0 - H + i\gamma} j^\dagger \, |0\rangle \,, \tag{45}$$

where $|0\rangle$ is the ground state with energy E_0 and $\gamma \to 0+$. Here, the current operator j is defined by

$$j = -ita \sum_{i,\sigma} (c^\dagger_{i,\sigma} c_{i+1,\sigma} - \text{H.c.}), \tag{46}$$

where $c^\dagger_{i,\sigma}$ and $c_{i,\sigma}$ are creation and annihilation operators for an electron at site i with spin σ, t is electron hopping matrix element, and a is a lattice constant. We use the natural unit. Hereafter, we introduce the notation

$$\chi_j(\omega) = \chi'_j(\omega) + i\chi''_j(\omega), \tag{47}$$

where $\chi'_j(\omega)$ and $\chi''_j(\omega)$ are real and imaginary parts of $\chi_j(\omega)$, respectively. The current-current correlation function is a fundamental physical quantity to examine optical properties of the system. A regular real part of the complex optical conductivity is given by

$$\sigma'_{reg}(\omega) = -\frac{1}{L\omega} \chi''_j(\omega). \tag{48}$$

For insulating ground states, this is directly the optical conductivity. In cases of metallic ground states, there exists an irregular part called Drude part in addition to $\sigma'_{reg}(\omega)$

$$\sigma'_{sing}(\omega) = -\frac{\pi}{L}(\langle \tau \rangle + \chi'_j(\omega))\delta(\omega) = 2\pi D \delta(\omega), \tag{49}$$

where D is the Drude weight characterizing metallic properties of this system. Here τ is defined by

$$\tau = -ta^2 \sum_{i,\sigma} (c^\dagger_{i,\sigma} c_{i+1,\sigma} + \text{H.c.}). \tag{50}$$

This is called stress energy tensor.[37] We introduce the following spectrum in addition to $\chi_j(\omega)$

$$\chi_\tau(\omega) = \langle 0| \tau \frac{1}{\omega + E_0 - H + i\gamma} \tau^\dagger |0\rangle . \quad (51)$$

Since j and τ have odd and even parities under the space inversion, respectively, $\chi_j(\omega)$ detects the odd-parity states while $\chi_\tau(\omega)$ does the even-parity states. The latter detects optically-forbidden states.

The optical spectrum $\chi_j(\omega)$ is composed of the following states

$$|\psi_1\rangle = |0\rangle , \quad (52)$$

$$|\psi_2\rangle = j^\dagger |0\rangle , \quad (53)$$

$$|\Psi_\gamma(\omega)\rangle = (\omega + E_0 - H + i\gamma)^{-1} j^\dagger |0\rangle . \quad (54)$$

Thus, we need to target them into the density matrix. Our strategy is quite simple. We just target the states we need in order to optimize them by the DMRG procedures. The first state, $|\psi_1\rangle$, is the ground state of the superblock made by the Lanczos method. The next state, $j^\dagger |0\rangle$, can be easily prepared by operating j to the ground state directly, since j does not change any quantum numbers. Here, j is represented by the bases that also represents the ground state. In order to generate j, what we do is only to change the sign of the hopping term in the Hubbard Hamiltonian. The last state, $|\Psi_\gamma(\omega)\rangle$, is called 'correction vector'. The correction vector picks up a particular photoexcited state, when we fix the value of photon energy ω

$$\lim_{\gamma \to 0+} \left(-\frac{1}{\pi} \right) \mathrm{Im} |\Psi_\gamma(\omega)\rangle = \sum_n |n\rangle \langle n| j^\dagger |0\rangle \, \delta(\omega + E_0 - E_n). \quad (55)$$

In order to introduce the last state, we must generate the ground state in advance. For this reason, we target the above three states, although we only need the last state for examination of the photoexcited state. Finally, we must solve the following linear equation

$$(\omega + E_0 - H + i\gamma) |\Psi_\gamma(\omega)\rangle = j^\dagger |0\rangle . \quad (56)$$

There are many solvers for this equation, but they would take much time.

3.2.2. *Kramers-Kronig relation*

It should be noted that we need to treat both of the real and the imaginary parts of $|\Psi_\gamma(\omega)\rangle$, even though we only need the imaginary part for the examination of optical physics. This is due to the Kramers-Krönig relation

$$\chi''(\omega) = -\frac{1}{\pi} P \int_{-\infty}^{\infty} d\omega' \frac{\chi'(\omega')}{\omega' - \omega} \tag{57}$$

$$\chi'(\omega) = \frac{1}{\pi} P \int_{-\infty}^{\infty} d\omega' \frac{\chi''(\omega')}{\omega' - \omega}, \tag{58}$$

with P denoting the principle part. According to the linear response theory, they are causal relations, and thus essentially important. If we target only the imaginary part of the correction vector, the numerical precision becomes considerably worse particularly at high energy region. In order to aviod to treat complex valuables, we target the following four states, $|\psi_1\rangle$, $|\psi_2\rangle$, $|\psi_3\rangle = \mathrm{Re}\,|\Psi_\gamma(\omega)\rangle$, and $|\psi_4\rangle = \mathrm{Im}\,|\Psi_\gamma(\omega)\rangle$. It is noted that the numerical precision of the ground state is quite important, since the other states are basically made by operating appropriate operators to the ground state. Usually we take $p_1 = 0.5$. The spectral intensity for a particular ω value is given by an inner product between ψ_2 and ψ_4, $\langle\psi_2|\,\psi_4\rangle$. In actual numerical simulation, we stop sweeping after the spectral weight converges.

3.2.3. *Two correction vectors*

It is clear that the basis for the photoexcited states is optimized for a particular ω value. Thus, we need many-time calculation by changing ω to make full spectral shape. The energy resolution of the present calculation is roughly γ, and thus we must repeat calculation at reast $(\omega_2 - \omega_1)/\gamma$ times for energy range (ω_1, ω_2). In order to reduce this intricate situation, we also target $|\Psi_\gamma(\omega + 2\gamma)\rangle$ as well as the correction vector $|\Psi_\gamma(\omega)\rangle$. Then, the spectrum for $\omega < \epsilon < \omega + 2\gamma$ is calculated after one DMRG run. In order to understand the validity of this method, we take an inner product between $|\Psi_\gamma(\omega)\rangle$ and $|\Psi_\gamma(\omega + 2\gamma)\rangle$, leading to

$$\langle\Psi_\gamma(\omega)\,|\Psi_\gamma(\omega+2\gamma)\rangle = \sum_n \frac{\left|\langle n|\, j^\dagger\,|0\rangle\right|^2}{(\omega+\gamma+E_0-E_n)^2 + 2\gamma(\omega+E_0-E_n)i}, \tag{59}$$

for small γ values. This means that the inner product is relatively large for $E_n - E_0 \sim \omega + \gamma$. Thus, the information for $|\epsilon\rangle$ around $\epsilon \sim \omega + \gamma$ can be well interpolated by the two targets. For sufficiently large m, it is possible to increase the number of the target states $|\Psi_\gamma(\omega + 4\gamma)\rangle$, ... and scan wide energy range in one DMRG run. However, it is not so easy to keep numerical accuracy. It is technically useful to note that the numerical convergence is obtained when $I(\epsilon \sim \omega)$ for the target range $\omega - 2\gamma \leq \epsilon \leq \omega$ is smoothly connected with $I(\epsilon \sim \omega)$ for the target range $\omega \leq \epsilon \leq \omega + 2\gamma$.

3.3. *Angle resolved photoemission spectroscopy*

3.3.1. *DMRG calculation of ARPES spectra*

Angle resolved photoemission spectroscopy (ARPES) can directly observe one-particle spectral function $A(k,\omega)$. Experimental techniques of ARPES have been continuously developed after the discovery of high-T_c cuprates. It is thus indispensable to develop a method for calculating $A(k,\omega)$ by DMRG in order to extract important information from such high resolution data. The spectral function is the imaginary part of the Green's function $G_\sigma(k,\omega)$

$$A(k,\omega) = -\frac{1}{\pi}\mathrm{Im}G_\sigma(k,\omega), \tag{60}$$

and thus we would like to know how to calculate the Green's function. The Green's function at absolute zero temperature is defined by

$$G_\sigma(k,\omega) = \langle 0| c^\dagger_{k,\sigma} \frac{1}{E_0 - \omega - H + i\gamma} c_{k,\sigma} |0\rangle . \tag{61}$$

Thus, the target states (we need normalization) are given by

$$|\psi_1\rangle = |0\rangle , \tag{62}$$

$$|\psi_2\rangle = c_{k,\sigma} |0\rangle , \tag{63}$$

$$|\Psi_\gamma(\omega)\rangle = (E_0 - \omega - H + i\gamma)^{-1} c_{k,\sigma} |0\rangle . \tag{64}$$

In this case, we need to generate the state $c_{k,\sigma}|0\rangle$. However, the quantum number of this state is different from that of the ground state, since the electron operator changes momentum and spin by k and σ, respectively. Usually, the ground state is spanned by the restricted Hilbert space with electron number N. Here, we enlarge the Hilbert space so that the electron number is N or $N-1$. This is reasonable extention, since the block states L and R in Fig. 2(a) have particle-number fluctuation.

3.3.2. *Pseudo-momenta*

The numerical precision of DMRG is better for the open boundary condition in comparison with the periodic boundary condition. As we have already discussed, the reason for the better performance of the open boundary condition comes from a fact that the entanglement entropy for the open boundary condition is two-times larger than that for the periodic boundary condition. Rough estimation shows that

$$m_{periodic} \sim (m_{open})^2 . \tag{65}$$

Thus, we frequently use the open boundary condition. In this case, the standard momenta for the periodic boundary condition are not good quantum numbers. Then, we regard our system as an electron system inside of a potential well, and introduce a quantum number called pseudo-momentum. When the electrons feel infinitely large potential outside of the sites from 1 to L (we call the two boundaries as site 0 and site $L+1$), then the pseudo-mementun is given by simple qunatum mechanics as

$$k = \frac{n\pi}{L+1}, \tag{66}$$

for $n = 1, 2, ..., L$. Then, we have the following discrete Fourier expansion

$$c_{k,\sigma} = \sqrt{\frac{2}{L+1}} \sum_{l} \sin(kl) c_{l,\sigma}. \tag{67}$$

The ramaining process is to detemine $c_{i,\sigma}$ at each site. How to take Fermion signs is a somehow delicate problem. The detail has been discussed in the reference.[7]

4. Time-dependent DMRG

4.1. *Direct approach similar to dynamical DMRG*

Time-dependent calculation can be done by directly extending the multiple targetting that is the basic concept of dynamical DMRG. Let us examine a time-dependent quantity, $\langle \tau | O | \tau \rangle$, where $|\tau\rangle$ is the formal solution of the time-dependent Schrödinger equation, $i(\partial/\partial\tau)|\tau\rangle = H(\tau)|\tau\rangle$. Here, the Hamiltonian contains time-dependent vector potential of the pump light field. The solution is given by

$$|\tau\rangle = T \exp\left\{-i \int_0^\tau H(\tau')d\tau'\right\} |0\rangle \tag{68}$$

with T being the time ordering operator. Then, the target states are $|0\rangle$ and $|\tau\rangle$. In contrast to targetting photoexcited states, the time-dependent state, $|\tau\rangle$, is far from the eigenstate of the Hamiltonian. Thus, we need relatively large truncation number. The outline of the numerical technique is almost the same as the dynamical calculation for spectra, but we sometimes change targets, depending on complexity of models treated.

When we consider one-phonon pumping with energy ω_0, the photogenerated state may be expressed by a simpler form given by

$$|\psi(\tau)\rangle \propto e^{-iH\tau} \mathrm{Im} \left(\frac{1}{\omega_0 + E_0 - H + i\gamma_0} \right) j^\dagger |0\rangle, \tag{69}$$

where $1/\gamma_0$ corresponds to duration time of the pump light. In this case, the target states are $|0\rangle$, $j^\dagger |0\rangle$, $(\omega_0 + E_0 - H + i\gamma_0)^{-1} j^\dagger |0\rangle$, and $|\psi(\tau)\rangle$, for a fixed τ.

4.2. *MPS and time evolution*

An advantage of MPS formalism is that the time-dependent calculation is efficient. A point for the efficiency is to proceed a time step by local update of the MPS-type wave function continuously. On the other hand, the time evolution operator, e^{-iHt}, for the total Hamiltonian, H, is not described by a product of local matrices. Thus, we would like to consider the following steps.

Let us start with the Hamiltonian with the nearest-neighbor interaction

$$H = \sum_{j=1}^{L-1} H_j, \tag{70}$$

where the local Hamiltonian, H_j, is defined on the bond between the sites j and $j+1$. H_j also contains local terms at site j. We devide H into sum of even and odd bonds

$$H = H_o + H_e = \sum_{j:odd} H_j + \sum_{j:even} H_j. \tag{71}$$

The two operators, H_o and H_e, do not commute, but the operators inside of H_o (or H_e) commute. Actually we have the following relation

$$e^{-iH_o\delta t} = \prod_{j:odd} e^{-iH_j\delta t}. \tag{72}$$

We apply the Suzuki-Trotter breakup to the time evolution operator

$$e^{-i(H_o+H_e)t} = \lim_{n\to\infty} \left(e^{-iH_o\delta t} e^{-iH_e\delta t}\right)^n, \tag{73}$$

where $\delta t = t/n$.[23] The time evolution has been done by operating first $e^{-iH_o\delta t}$, then $e^{-iH_e\delta t}$, again $e^{-iH_o\delta t}$, until all of the terms are operated to the state before time evolution.

The above concept can be also incorporated into the original DMRG sweep algorithm mentioned previously. This is called 'adaptive' time-dependent DMRG method.[10] In the j-th step of DMRG, it is possible to exactly apply a local operator O including H_j to the state $|\psi\rangle$

$$(O\psi)_{ls_js_{j+1}r} = \sum_{s'_js'_{j+1}} O_{s_js_{j+1};s'_js'_{j+1}} \psi_{ls'_js'_{j+1}r}. \tag{74}$$

In order to combine this operation with sweeping, the time evolution operator for the total system is decomposed into the following form

$$e^{-iH\delta t} = e^{-iH_1\delta t/2}e^{-iH_2\delta t/2}\cdots e^{-iH_L\delta t}\cdots e^{-iH_2\delta t/2}e^{-iH_1\delta t/2}. \quad (75)$$

The main idea is to apply $e^{-iH_1\tau/2}$ at the first step of DMRG, then $e^{-iH_2\tau/2}$ at the next step, forming the usual left-to-right sweep, then reverse at the right edge in order to make the right-to-left sweep.

4.3. *Wave function transformation*

In order to do all of the local operations, we need so-called 'wave function transformation'.[38] Now we are in the step $j+1$. In the previous step j, we have already obtained the state, $|\tau, j\rangle$, defined by

$$|\tau, j\rangle = \prod_{i=1}^{j} e^{-iH_i\tau/2}\,|\psi\rangle\,. \quad (76)$$

From the step j to the step $j+1$, we operate $e^{-iH_{j+1}\tau/2}$ to the previously obtained state, $|\tau, j\rangle$, but because of the sweeping the basis states have changed. Thus, the state must be rearranged by the basis in the step $j+1$. Now we write the basis-transformation operator as U_j. Then, what we actually do at each step is the following calculation

$$|\tau, j+1\rangle = e^{-iH_{j+1}\tau/2}U_j\,|\tau, j\rangle\,. \quad (77)$$

(In that sense, Eq. (76) is not correct.) By continuing this process, we apply all of the local time evolution oparators to the corresponding state. When we have obtained the wave function at a time τ, $|\tau\rangle = e^{-iH\tau}\,|\psi\rangle$, we next go to the calculation of $|2\tau\rangle = e^{-iH\tau}\,|\tau\rangle$, $|3\tau\rangle = e^{-iH\tau}\,|2\tau\rangle$, $\cdots$.

In realistic simulations, numerical precision of the wave function transformation is crucial in comparison with the error from the Suzuki-Trotter breakup. There are various trial methods to recover numerical accuracy.[39–41]

5. Optical Physics of 1D Mott Insulators

5.1. *Introduction*

There are solid theoretical approaches to attack the ground-state and low-energy properties of 1D quantum systems such as Bethe ansatz, conformal field theory, and bosonization. However, when we go to optically-excited states, reliable theoretical techniques are quite limited. There, numerical approaches are indispensable. Recently, various fascinating optical phenomena have been found in 1D Mott insulators and other transition metal oxides. Dynamical and time-dependent DMRG are applied to analysis of such phenomena.

Among these phenomena, gigantic third-order nonlinear optics and ultrafast phase control by intense laser light are interesting topics in strongly correlated 1D electron systems. Here, the target materials are 1D cuprates, $SrCuO_2$, Sr_2CuO_3, and Ca_2CuO_3, halogen-bridged Ni halides, $[Ni(chxn)_2Br]Br_2$, $[Ni(chxn)_2Cl]Cl_2$, and $[Ni(chxn)_2Cl](NO_3)_2$, and TCNQ salts, K-TCNQ, Rb-TCNQ, and ET-F_2TCNQ. As for the TCNQ salts, maretial dependence is reflected on the strength of the spin-Peierls distortion. Development of nonlinear optics is also essential for next-generation opto-electronics. The third-order nonlinear susceptibilities in the typical 1D Mott insulators are thousand-times larger than those of band insulators or Peierls insulators.[42,43] Ultrafast transient optics is an approach exploring new functionalities of materials and observing properties hidden in equilibrium conditions. The phase control by laser light is called 'photoinduced phase transition' (PIPT).[44] As for the PIPT phenomena of the 1D Mott insulators, the ultrafast Mott transition after irradiation and its subsequent recovery of the order of picoseconds have been observed.[45,46]

Even apart from these 1D systems, transient optics and ultrafast electronic-state control are current hot topics in a wide area of strongly correlated electron systems and others. Let us remember phase diagrams of high-T_c cuprates, manganites, and organic compounds where various electronic phases are competing with each other. The origin of the various phases comes from sattle balance among charge, spin, orbital, and lattice degrees of freedom. The phases near the boundaries are not stable. There, magnetic, transport, and optical properties are entangled. Thus, only tiny amount of external perturbations can dramatically change the electronic properties. This is the basic strategy of PIPT. It is obvious that photoexcitation into insulators generates metallic carriers irrespective of the strength of electron correlation. Thus, it is important to make clear the difference between doped semiconductor and PIPT in strongly correlated systems. Furthermore, ultrafast manipulation of femtosecond pulse laser is characteristic of PIPT, and would provide a striking difference between PIPT and the other phase transitions. When we observe the system in time domain, various excitations can be separately treated because of their different energy scales. This is crucial for understanding detailed mechanism of dynamical change in electronic structure.

5.2. *Fundamental concepts in photoexcited 1D Mott insulators*

5.2.1. *Photo-carrier doping into 1D systems*

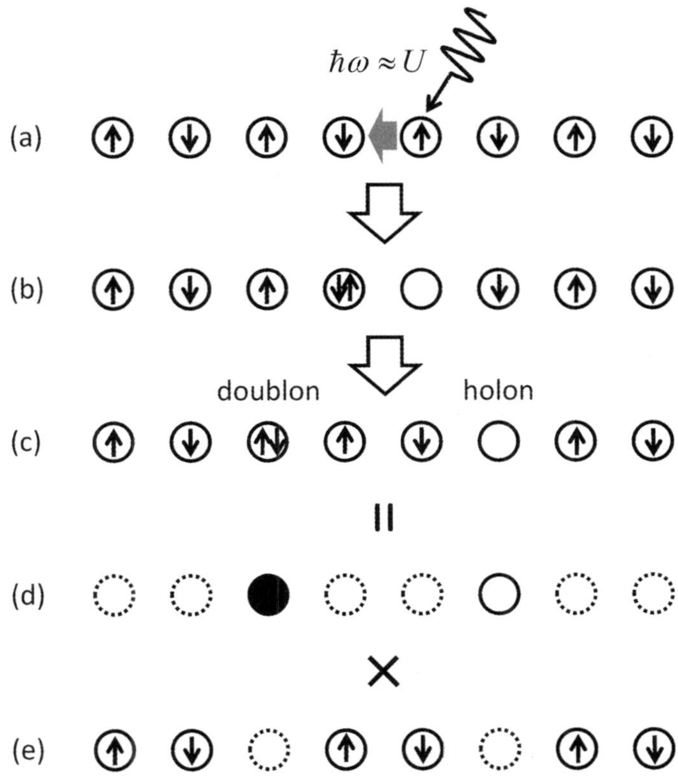

Fig. 3. Schematice viewgraph of wave-function factorization in the photoexcited 1D Mott insulator. Two charge modes, called 'holon' and 'doublon', do not disturb background spin configuration. The spin configuration is still antiferromagnetic even after the holon and the doublon move. Thus, the spin and the charge collective excitations are decoupled with each other.

As already mentioned, there exists a group of transition metal oxides and organic materials that are ideal 1D Mott insulators. In the Mott insulator, each site is occupied by one electron, but because of large on-site Coulomb repulsion, U, the double occupancy at each site is prohibited (Fig. 3(a)). This is the origin of the insulating ground state. The electron spins align antiferromagnetically. Experimentally, it is usually impossible to do chemical doping into these 1D Mott insulators in contrast to high-T_c cuprates in 2D where the carrier doping induces d-wave superconductivity. The absence of methods for carrier doping is disadvantageous, since the carrier doping into several types of insulators (band insulator, Mott insulator, topological insulator, ...) is one of long-standing topics in condensed matter physics. Photodoping is a promising way to overcome this difficulty.

5.2.2. *Physics of spin-charge separation, holon-doublon picture*

It is well known that the wave function of the antiferromagnetic Mott insulator is 'factorized' into a direct product of spin and charge parts.[47] This phenomenon is called 'spin-charge separation'. As shown in Fig. 4, in the

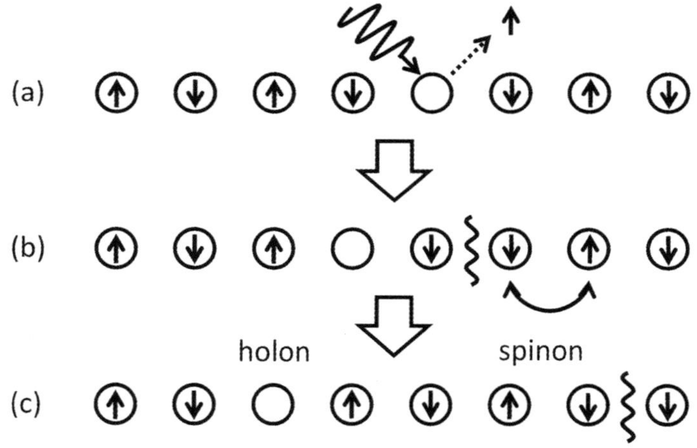

Fig. 4. Schematice viewgraph of single-particle excitation and generation of spinon and holon excitations.

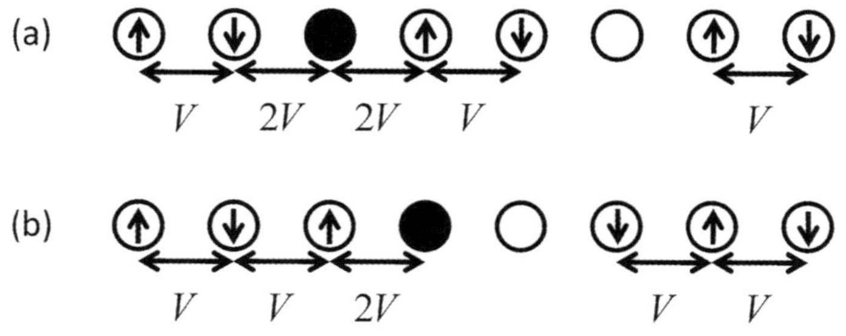

Fig. 5. Energy gain by V due to the formation of exciton.

single-particle excitation process, spin defect, called 'spinon', and photogenerated empty site, called 'holon', are separately moving. In the photoexcited states shown in Fig. 3, only the charge degree of freedom survives. Then, the system after one-photon excitation is well described by an effective two particle model, called the 'holon-doublon model', which is composed of a 'holon' (solid open circle in Fig. 3(d)) and a 'doublon' representing one doubly occupied site (filled black circle in Fig. 3(d)). These carriers attract with each other by the nearest-neighbor Coulomb interaction V to form an excitonic bound state (see Fig. 5). The bound state is generated for $V = 2t$ in the large-U limit. These carriers can not exchange each other in the large-U limit. This special condition offers essential difference between photocarriers in band and 1D Mott insulators.

5.2.3. *Coupling between photocarriers and phonons*

In the ground state of the Mott insulator, the on-site Coulomb repulsion makes charge distribution uniform, and then electron-phonon (EP) interactions are irrelevant. However, the holon and the doublon induce charge fluctuation. Thus, the EP interaction is expected to play an important role in various physical quantities. In praticular, to take account of phonon dynamics quantum mechanically may be neccesary for examination of relaxation dynamics. In addition, we are interested in whether many theoretical predictions without phonon degrees of freedom are robust against the EP coupling. For instance, a problem is to examine stability of the spin-charge separation in the presence of the EP coupling in real materials. It is also important to understand the effect of the EP interaction on holon and doublon excitations and to examine U-dependence on the phonon effect.

5.3. *Relevant models for optically-excited 1D Mott insulators*

5.3.1. *Hubbard models*

The single-band Hubbard model is the most fundamental model of cuprates. The t-J model is also frequently used, since the Hilbert-space dimension is much smaller than that of the Hubbard model. However, the t-J model can not describe the photoexcited states beyond the Mott gap. The Hubbard Hamiltonian is defined by

$$H = -t \sum_{i,\sigma} (c^{\dagger}_{i,\sigma} c_{i+1,\sigma} + \text{H.c.}) - U \sum_{i} n_{i,\uparrow} n_{i,\downarrow}, \tag{78}$$

where $c_{i,\sigma}^\dagger$ is the creation operator for an electron with spin σ at site i, $n_i = n_{i,\uparrow} + n_{i,\downarrow}$, t is the hopping integral along the chain axis, and $U \gg t$ is the on-site Coulomb interaction. When we further consider the effect of the nearest-neighbor Coulomb interaction V on the electronic states, we use the following Hamitonian

$$H = -t \sum_{i,\sigma} (c_{i,\sigma}^\dagger c_{i+1,\sigma} + \text{H.c.}) + U \sum_i n_{i,\uparrow} n_{i,\downarrow} + V \sum_i n_i n_{i+1}, \quad (79)$$

and this is called extended Hubbard model. Sometimes we redefine the V-term as

$$\sum_i n_i n_{i+1} \to \sum_i (n_i - 1)(n_{i+1} - 1), \quad (80)$$

since each site is occupied by one electron. This notation does not induce a bad boundary effect which is localization of a doublon at the edge. When we use the standard notation $V \sum_i n_i n_{i+1}$, we need the boundary potential, $\tilde{H} = H + V(n_1 + n_L)$. Recently, it has been examined to systematically deform spatial curvature and interaction strength near the boundary.[48] One of techniques, called 'hyperbolic deformation', controls energy scales of edge excitations, and in that sense is resemble to the role of the boundary potential on the doublon excitation.

5.3.2. *Holstein and Peierls couplings*

There are mainly two types of EP couplings. One is described by the Hubbard-Holstein model

$$\begin{aligned} H = &-t \sum_{i,\sigma} (c_{i,\sigma}^\dagger c_{i+1,\sigma} + \text{H.c.}) \\ &+U \sum_i n_{i,\uparrow} n_{i,\downarrow} + V \sum_i (n_i - 1)(n_{i+1} - 1) \\ &+\omega_0 \sum_i b_i^+ b_i - g \sum_i (b_i^+ + b_i)(n_i - 1), \end{aligned} \quad (81)$$

where b_i^+ (b_i) is the creation (annihilation) operator for an Einstein phonon at site i, ω_0 is the phonon frequency, and g is the EP coupling constant.

The other is the Hubbard-Peierls coupling given by

$$H = -t\sum_{i,\sigma}\left\{\left(b^{\dagger}_{i+1/2} + b_{i+1/2}\right) c^{\dagger}_{i,\sigma}c_{i+1,\sigma} + \text{H.c.}\right\}$$
$$+U\sum_{i} n_{i,\uparrow}n_{i,\downarrow} + V\sum_{i}(n_i - 1)(n_{i+1} - 1)$$
$$+\omega_0\sum_{i} b^{+}_{i+1/2}b_{i+1/2}, \qquad (82)$$

where the phonon operator, $b_{i+1/2}$, is defined on the link between sites i and $i+1$.

They have different roles on the electronic properties of transition metal oxides and organic compounds. The Peierls type interaction is important when alternating lattice deformation occurs. However, in cases of 1D cuprates and ET-F_2TCNQ, the deformation does not play a role on optically-excited states. The Peierls type changes the value of exchange coupling $J = 4t^2/U$ through the electron hopping, but in the 1D cases the spin and the charge excitations are decoupled. In the following, we focus on the Holstein-type coupling.

5.3.3. *Pseudo-phonon method*

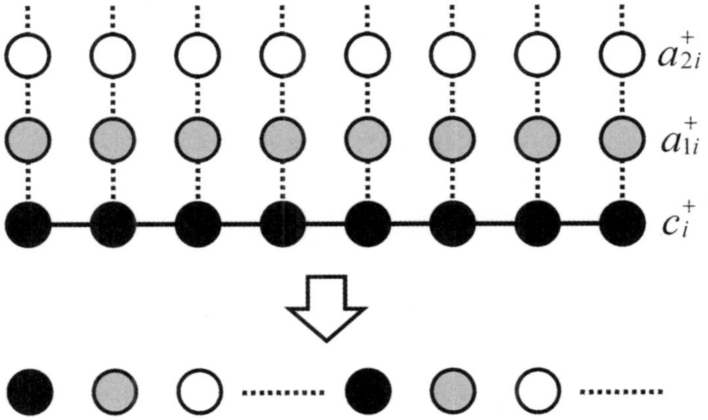

Fig. 6. Pseudo-site method for the Hubbard-Holstein model.

In DMRG calculation, phononic degrees of freedom at each site are truncated up to $M = 2^N$, and then exactly transformed into a set of N hard-core bosons. The bosonic and electronic degrees of freedom are treated as different 'sites'. This method is called 'pseudo-site' method.[13] For instance, the restricted boson operator with 4 states (0 phonon, 1 phonon, 2 phonons, and 3 phonons) is transformed into

$$b^\dagger = a_1^\dagger + \sqrt{2}a_2^\dagger a_1 + (\sqrt{3}-1)a_1^\dagger a_2^\dagger a_2, \tag{83}$$

with use of two hard-core boson operators a_1 and a_2 (see Fig. 6) satisfying the anti-commutation relation $\{a_i, a_j^\dagger\} = \delta_{ij}$. Then, step-by-step renormalization of each boson is carried out, although the interactions are long-ranged. This transformation enables us to calculate the spectra for large values of g.

5.3.4. *Holon-doublon model*

In order to study the optical spectra in the large-U limit, we can introduce a holon-doublon model due to the spin-charge separation. This model describes the charge sector of the photoexcited states. Dropping the spin degrees of freedom greatly reduces the Hilbert-space dimension. The effective Hamiltonian of the extended Hubbard-Holstein model is defined by

$$\begin{aligned} H = &-t\sum_i (h_i^\dagger h_{i+1} + d_i^\dagger d_{i+1} + \text{H.c.}) + \frac{U}{2}\sum_i h_i^\dagger h_i \\ &+\frac{U}{2}\sum_i d_i^\dagger d_i - V\sum_i (h_{i+1}^\dagger h_{i+1} + h_{i-1}^\dagger h_{i-1}) d_i^\dagger d_i \\ &+\omega_0 \sum_i b_i^\dagger b_i + g\sum_i (b_i^\dagger + b_i)(h_i^\dagger h_i - d_i^\dagger d_i), \end{aligned} \tag{84}$$

where $h_i^\dagger$ (h_i) and $d_i^\dagger$ (d_i) are creation (annihilation) operators of holon and doublon at site i, respectively. The holon and doublon obey the local constraint

$$h_i^\dagger h_i + d_i^\dagger d_i < 1. \tag{85}$$

The current-current correlation function can be described by

$$\chi_j(\omega) = -\frac{2}{\pi L}\text{Im}\left(1\left|\frac{1}{\omega + E_0 - H + i\gamma}\right|1\right), \tag{86}$$

where $|1)$ is a state with one holon-doublon pair. This state is created by applying the current operator

$$j = -it\sum_i (d_i^\dagger h_{i+1}^\dagger - h_i^\dagger d_{i+1}^\dagger) \tag{87}$$

to the classical Neel-ordered state. Here, we introduce the prefactor 2 in order to take account of the effect of quantum spin fluctuation on the spectrum.

5.3.5. *Introduction of vector potential of pump laser light*

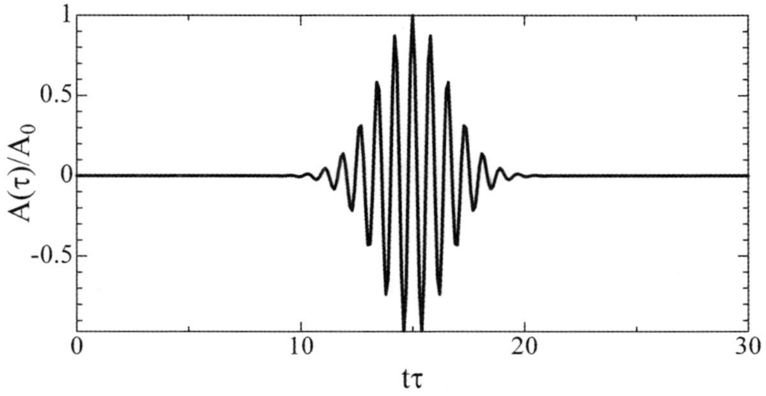

Fig. 7. Vector potential of pump laser. $t\tau_d = 5$, $t\tau_0 = 15$, and $\omega_{\text{pump}} = U - V = 8t$.

When we examine PIPT, classical vector potential of pump light is introduced into the hopping term of the Hamiltonian as a Peierls phase

$$-t\sum_{i,\sigma}(c_{i,\sigma}^{\dagger}c_{i+1,\sigma} + \text{H.c.}) \to -t\sum_{i,\sigma}(e^{iA(\tau)}c_{i,\sigma}^{\dagger}c_{i+1,\sigma} + \text{H.c.}). \tag{88}$$

The time-dependent vector potential, $A(\tau)$, is defined by

$$A(\tau) = A_0 e^{-(\tau-\tau_0)^2/\tau_d}\cos(\omega_{\text{pump}}(\tau - \tau_0)), \tag{89}$$

where τ_d is the duration time and ω_{pump} is the frequency of pump laser. Figure 7 is an example of the pulse shape.

5.4. *Photoexcited and single-particle excited states of 1D Mott insulators*

First, we brielfy summarize DMRG results for optically-excited states before going to study of ultrafast relaxation dynamics.

5.4.1. *Degenerate photoexcited states with even and odd parity*

The third-order nonlinear susceptibility is proportional to transition dipole moments between photoexcited states. The third-order nonlinear susceptibilities are anomalously large in the 1D Mott insulators. The anomalous enhancement of the nonlinear susceptibilities originates in the large transition dipole moment between an optically-allowed state with odd parity and an optically-forbidden (two-photon allowed) state with even parity due to the degeneracy of these states.[49,50] The degeneracy comes from prohibition of exchange of a holon and a doublon.

In order to analyze the degeneracy systematically, we employ DMRG to the calculation of one-photon absorption spectrum $\chi_j(\omega)$ and optically-forbidden spectrum $\chi_\tau(\omega)$ in the extended Hubbard model.[51] The DMRG calculation for $U \geq 10t$ has shown that the even-parity and the odd-parity photoexcited states are almost degenerate for $V < 2t$. For finite-U cases, the holon and the doublon can exchange their positions, and in this case the complete dengeneracy is formally broken. However, in realistic U-cases ($U = 8t \sim 10t$), nearly degenerate feature still remains. On the other hand, for $V \geq 2t$ where the holon and the doublon form an exciton, the degeneracy is lift. In the 1D cuprates and the halogen-bridged Ni compounds, the value of V is expected to be at most $V \sim 2t$, and thus they are good candidates for the degenerate photoexcited states.

The reason why the exciton is not degenerate is interpreted as follows. For $V > 2t$, a probability of the exchange is large, since these particles approach each other to form the bound states. Therefore, the lift of the degeneracy becomes remarkable for $V > 2t$.

5.4.2. *Effect of on-site and nearest-neighbor Coulomb interactions on Holstein-type electron-phonon coupling*

Regarding the interplay of the electron-electron interactions and the Holstein-type EP interaction, a particular interest is whether the phonon effect is enhanced or reduced by increasing the values of on-site and nearest-neighbor Coulomb repulsions. We have examined the one-phonon absorption spectrum in the extended Hubbard-Holstein model and its effective holon-doublon model by DMRG.[52–54] First, we have found that the holon-doublon exciton is strongly broadened by phonons. Thus, even if we experimentally observe broad absorption spectrum without an exciton-like peak, the degeneracy important for the large nonlinear susceptibility might be broken. The next finding is that the increase in U enhances the effect of the EP coupling on the spectrum. In contrast to the ground state, the EP coupling is quite important for detailed examination of the photoexcited Mott insulators.

5.4.3. *Stability of spin-charge separation in the presence of Holstein-type electron-phonon coupling*

We have examined the single-particle excitation spectrum in the 1D Hubbard-Holstein model at half-filling.[55,56] The DMRG calculation reveals that both of the spinon and holon singularities are smeared out. Particularly, in strong EP coupling regime, the phonon-charge (Holstein) coupling strongly affects the spinon as well as the holon branch. Nevertheless, the spin-charge separation seems to be robust against the EP coupling.

Because of the spin-charge separation, the effective model of ARPES spectrum in the Hubbard model without phonon degrees of freedom is basically given by the superposition formula

$$A_{\text{eff}}(k,\omega) = \sum_{q=-\pi/2}^{\pi/2} A_h\left(k-q, \omega+2t+\varepsilon_s\left(q+\frac{\pi}{2}\right)\right), \tag{90}$$

where A_h represents the holon spectrum

$$A_h(p,\varepsilon) = \delta(\varepsilon - 2t\cos p), \tag{91}$$

and ε_s represents the spinon dispersion

$$\varepsilon_s(q) = -(\pi J/2)|\sin(q)|, \tag{92}$$

When we take account of non-local phase-string effects on the spectrum, more precise singular structure at the holon branchs can be reproduced.[57] In the presence of the EP coupling, each of the holon branches couples with phonons independently. Namely, $A_h(p,\varepsilon)$ is given by the spectra for a spinless carrier dressed with phonons. Actually, this model well reproduces the DMRG results, and thus the spin-charge separation is robust.

5.5. *Ultrafast relaxation dynamics*

5.5.1. *Motivations and purposes*

The photocarriers of the 1D Mott insulators are the holon and the doublon. They recombine by emitting energy to other elementary excitations. The relaxation time is very fast. However, these charge carriers do not couple strongly with any quantum degrees of freedom. Since our system is in 1D, the concept of the spin-charge separation is quite robust, as we have often pointed out. Furthermore, in systematic study of a TCNQ family, the relaxation time for a material with smaller EP coupling is much faster that that with larger EP coupling. Thus, our problem is to clarify a pass way of energy dissipation due to the recombination.

5.5.2. *Time-dependent quantities*

Photodoping rate due to pumping, δ, which is controlled by the amplitude of laser pulse A_0 in (89), is defined by

$$\delta = \lim_{\tau \to \infty} \frac{E(\tau) - E_0}{L\omega_{\text{pump}}}, \tag{93}$$

where $E(\tau) = \langle \tau | H | \tau \rangle$. We also calculate the time evolution of doublon number, $N_{\text{d}}(\tau)$, phonon number, $N_{\text{ph}}(\tau)$, and local spin correlation, $C_{\text{s}}(\tau)$. They are defined by

$$N_{\text{d}}(\tau) = \sum_i \langle \tau | n_{i,\uparrow} n_{i,\downarrow} | \tau \rangle, \tag{94}$$

$$N_{\text{ph}}(\tau) = \sum_i \langle \tau | b_i^\dagger b_i | \tau \rangle, \tag{95}$$

and

$$C_{\text{s}}(\tau) = \sum_i \langle \tau | \vec{S}_i \cdot \vec{S}_{i+1} | \tau \rangle. \tag{96}$$

5.5.3. *Numerical results and perspectives*

We examine relaxation dynamics of the photocarriers doped into the extended Hubbard-Holstein model. Figure 8 shows photodoping rate δ as a function of pulse intensity A_0 in the extended Hubbard model ($g = 0$). Hereafter, the common parameters are taken to be $L = 12$, $U = 10t$, $V = 2t$, and $\omega_0 = 0.1t$. The pump energy is set to be the exitation energy of the holon-doublon exciton appearing in the linear optical absorption spectrum. It is noted that the rate of the recombination of the holon and the doublon is large for $V \geq 2t$. In order to clearly look at the recombination dynamics, we take $V = 2t$. The doping rate δ oscillates due to the finite-size effect, but globaly we see that the rate tends to increase with A_0. For very small A_0 region, the data may be fit to the condition, $\delta \propto A_0^2$. It is noted that the doping rate for a fixed A_0 value increases with introducing the EP coupling g. Because of the energy gain of the EP coupling term in the presence of the photocarriers, the system absorbs much energy and generates more carriers in comparison with the pure Hubbard model.

Figure 8 shows relaxation dynamics in the extended Hubbard-Holstein model after the pump pulse has been turned off at $t\tau \sim 10$. Here, we compare two cases with and without the EP coupling. We take a small

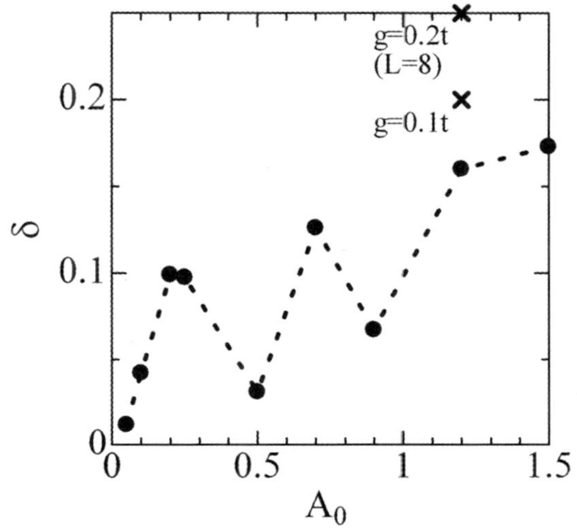

Fig. 8. Photodoping rate δ as a function of pulse intensity A_0 in the extended Hubbard model. Due to the finite-size effect, the doping rate oscillates, but tends to increase as A_0.

value of the EP coupling, $g = 0.1t$, for which the absorption spectrum is almost the same as that without g. Just after photoirradiation, the number of the phonon bases per site required for accurate numerical simulation is around 12, although in this case we take 6 phonons per site. In the absence of the EP coupling, the spin degrees of freedom absorb the photon energy, but the amount of the energy is too small. This tendency is consistent with the exact-diagonalization study.[58] This is because the spin and charge sectors are almost decoupled, though in this case high-energy states are mixed by the strong photoexcitation. Thus, the spin sector does not work as an energy reservior. On the other hand, in the presence of the EP coupling, we clearly see that the phonon emission dominates the initial stage of the relaxation against the spin degrees of freedom. Actually, the increase in the energy of the spin sector seen in the case of the Hubbard model is suppressed, when we introduce only a small amount of the EP coupling. In Fig. 9(c), the rapid increase in $N_{ph}(\tau)$ is determined by the inverse of the EP coupling. We conclude that the roles of the EP coupling on the ground and photoinduced states are quite different. In the photoinduced states, the EP coupling is quite important even if the value is small.

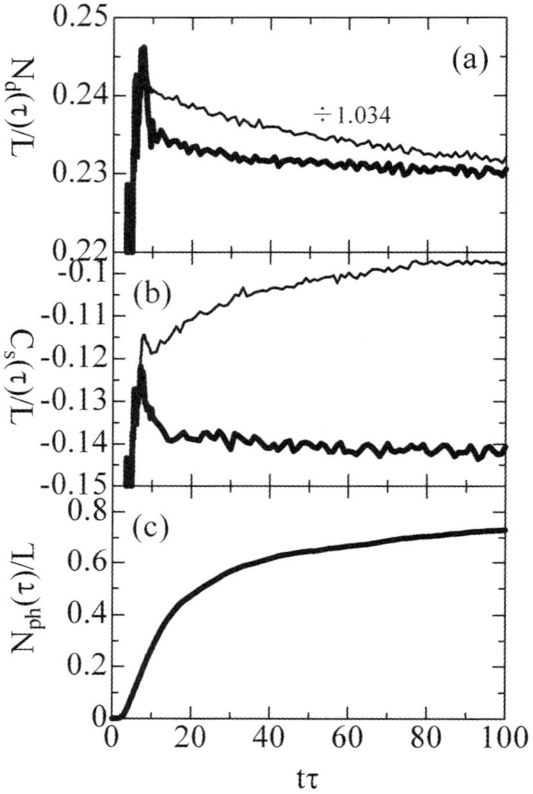

Fig. 9. Relaxation dynamics of photocarriers, localized spins, and phonons after the pump pulse has been turned off. Bold and solid lines represents results for the Hubbard-Holstein ($g = 0.1t$) and Hubbard models, respectively.

6. What is Effective Spatial Dimension of Fermion-Boson Coupled System?

In a viewpoint of the area-law scaling of the entanglement entropy and its logarithmic violation, quantum systems with bosonic degrees of freedom should be taken carefully. Accroding to the pseudo-site method for the Hubbard-Holstein model, the phonon operator is expanded by a set of the hard-core boson operators. If we regard the set as effective sites, this expansion enlarges an effective dimension of our model space (again see Fig. 4). Here, the infinitely large phononic degrees of freedom are decomposed into more compact local states in the enlarged space. Each phononic chain seems to be independent, but the electron hopping indirectly connects neighboring chains with each other. Thus, we expect that the area law is satisfied in this 2D space. In this case, MPS inherent in DMRG may be insufficient, and we need TPS formalism. In the present stage, it is hard to treat large phonon systems quite accurately with only using standard DMRG. The reason for the hardness would come from inconsistency between DMRG and TPS. It is a challenging task to confirm the entanglement-entropy scaling in the Hubbard-Holstein model.

7. Photoinduced States of Multi-Orbital Systems

Photoinduced ultrafast dynamics is also extensively examined in perovskite manganites and cobaltites. In the following, we summarize results of our DMRG examinations.[59–62]

7.1. *Manganites*

The manganites $R_{1-x}A_xMnO_2$ (R: a rare-earth ion, A: an alkaline-earth ion) have various electronic phases due to the presence of spin, charge, and orbital degrees of freedom, and thus provide good playgrounds for elucidating effects of these multiple degrees of freedom on PIPT. In particular, we focus on transition from charge-ordered insulating to ferromagnetic metallic phases. In contrast to the 1D cuprates, spin textures of localized t_{2g} electrons play important roles on conductivity of photocarriers doped into e_g orbitals, because of double-exchange coupling between the photocarriers and the localized spins.

Some important properties of PIPT on the manganites can be studied by the extended double-exchange model in a 1D chain. Thus, DMRG is applicable to the system. The Hamiltonian is given by

$$H = -t \sum_{\langle i,j\rangle,\sigma} c^\dagger_{i,\sigma} c_{j,\sigma} + U \sum_i n_{i,\uparrow} n_{i,\downarrow} + V \sum_{\langle i,j\rangle} n_i n_j \\ -J_H \sum_{i,s,s'} \vec{S}_i \cdot c^\dagger_{i,s} (\vec{\sigma})_{ss'} c_{i,s'} + J_s \sum_{\langle i,j\rangle} \vec{S}_i \cdot \vec{S}_j, \tag{97}$$

where $\vec{S}$ is the t_{2g} spin. We assume that $S = 1/2$ and the mobile carriers are in a single band for simplicity, although in realistic cases $S = 3/2$ and the e_g orbitals have two different degrees of freedom. Now, the single band is quarter-filled, and because of large V, the system is in a charge-ordered state. In 1D interacting models, the spin-charge separation is quite robust. Thus, the photocarriers were basically decoupled to spin degree of freedom in the Mott insulators. However, the double-exchange interaction is in some sense exceptional. The spin and the charge degrees of freedom on the single carrier band may be decoupled, but the energy-gain due to the double-exchange coupling occurs locally, depending on the local spin texture. Local electron hopping is necessarily affected by the texture. This situation is rather different from the separation of collective spin and charge excitations. The transient optical absorption spectrum is defined by

$$X_j(\omega,\tau) = -\frac{1}{\pi}\mathrm{Im}\,\langle\psi(\tau)|\, j\frac{1}{\omega + E_1 - H + i\gamma} j^\dagger\,|\psi(\tau)\rangle\,, \qquad (98)$$

where the state $|\psi(\tau)\rangle$ has been defined in Eq. (69), $E_1 = \langle\psi(\tau)|\,H\,|\psi(\tau)\rangle$, and $\tau > \gamma_0^{-1} + \gamma^{-1}$.

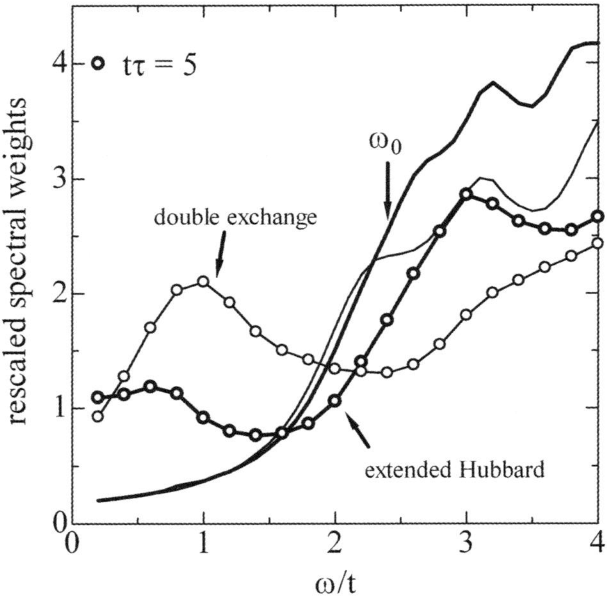

Fig. 10. One-photon absorption spectra (solid lines) and transient spectra with $t\tau = 5$ (solid lines with marks). Fine (bold) solid lines are for the double-exchange (extended Hubbard) model. Pump energy is denoted by $\omega_0 = 2.4t$. Model parameters are taken to be $L = 13$, $N = 7$, $U = 10t$, $V = 4t$, $J_H = 2t$, $J_s = 0.2t$ for the double-exchange model, while $L = 33$, $N = 17$, $U = 10t$, and $V = 6t$ for the Hubbard model in order to set almost the same value of the optical gap. Note that the absolute spectral weights for the double-exchange model are normalized for the comparison.

We have found that high-energy spectral weight of the optical conductivity before pumping transfers to the inside of the charge gap in the transient optical absorption spectrum.[59–61] This means that the system becomes metallic just after photoirradiation. Although the in-gap spectral weight is located at finite energy range, we have numerically confirmed that the peak position approaches to zero as L increases. In addition, the low-energy spectral weight evolves as time increases. This spectral-weight transfer is accompanied with remarkable suppression of antiferromagnetic correlations of the localized spins. If we drop the localized spin degrees of freedom (extended Hubbard model), the evolution of the low-energy metallic weight is suppressed as shown in Fig. 10. Thus, we see that it is crucial for understanding the metallic properties after photoexcitation to look at what kinds of quantum degrees of freedom are coupled to the photocarriers. Furthermore, we see that the largest total-S_z space is most stable in the photoexcited state. This result also supports highly conductive nature of the photoexcited state.

Detailed calculations have demonstrated that photocarrier motion govens a short-time scale where the spin and the charge degrees of freedom are combined strongly, while in a long-time scale controlled by spin relaxation, the charge sector does not follow the remarkable change in the spin sector.[60,61] This is in sharp contrast to the double-exchange physics in equillibrium states.

7.2. *Cobaltites*

In the cobaltites $R_{1-x}A_xCoO_3$, photo-control of spin state transition is one of topics. A single photon doped to the low-spin band insulator induces a bound state composed of a high-spin site and a photogenerated hole.[62] This bound state produces a remarkable peak structure in the pump-probe spectrum. This result explains recent experimental data for $RBaCo_2O_{6-\delta}$ (R=Sm, Gd, and Tb).

8. Summary

In this article, we have first presented recent viewpoints of DMRG based on quantum entanglement, and then various numerical methods for dynamical and time-dependent quantities are reviewed. After that, optical physics of 1D Mott insulators are examined, and PIPT of multi-orbital systems are brielfy summarized. DMRG is powerful for the dynamical study of strongly correlated 1D electron systems. Unfortunately, in the present stage, we are not using full performance of tenser network formalism in vaious parts of PIPT studies. This research direction is quite necesary for theoretical development of PIPT physics.

Acknowledgement

I would like to thank many colleagues for discussions and hospitality of Yukawa Institute for Theoretical Physics. This work was supported by the Next Generation Supercomputing Project of Nanoscience Program and Grant-in-Aid for Scientific Research from MEXT (21740268). A part of numerical calculations was performed in the supercomputing facilities in ISSP, University of Tokyo, YITP, Kyoto University, and IMR, Tohoku University.

References

1. S. R. White, Phys. Rev. Lett. **69**, 2863 (1992); Phys. Rev. **B48**, 10345 (1993).
2. I. Peschel, X. Wang, M. Kaulke, and K. Hallberg, eds., "Density Matrix Renormalization: A New Numerical Method in Physics", Springer (1999).
3. K. Hallberg, Phys. Rev. **B52**, R9827 (1995).
4. T. D. Kühner and S. R. White, Phys. Rev. **B60**, 335 (1999).
5. E. Jeckelmann, Phys. Rev. **B66**, 045114 (2002).
6. X. Wang and T. Xiang, Phys. Rev. **B56**, 5061 (1997).
7. N. Shibata, J. Phys. A: Math. Gen. **36**, R381 (2003); N. Shibata and H. Tsunetsugu, J. Phys. Soc. Jpn. **68**, 744, 3138 (1999).
8. M. A. Cazalilla and J. B. Marston, Phys. Rev. Lett. **88**, 256403 (2002); **91**, 049702 (2003); H. G. Luo, T. Xiang, and H. Q. Wang, Phys. Rev. Lett. **91**, 049701 (2003).
9. G. Vidal, Phys. Rev. Lett. **91**, 147902 (2003); **93**, 040502 (2004).
10. S. R. White and A. E. Feiguin, Phys. Rev. Lett. **93**, 076401 (2004).
11. T. Nishino and K. Okunishi, J. Phys. Soc. Jpn., **64**, 4084 (1995); **65**, 891 (1996); **66**, 3040 (1997); **67**, 3066 (1998).
12. S. Östlund and S. Rommer, Phys. Rev. Lett. **75**, 3537 (1995); S. Rommer and S. Östlund, Phys. Rev. **B55**, 2164 (1997).
13. E. Jeckelmann and S. R. White, Phys. Rev. **B57**, 6376 (1998).
14. C. Zhang, E. Jeckelmann, and S. R. White, Phys. Rev. Lett. **80**, 2661 (1998); Phys. Rev. **B60**, 14092 (1999).
15. S. Ramasesha, S. K. Pati, H. R. Krishnamurthy, Z. Shuai, and J. L. Bredas, Phys. Rev. **B54**, 7598 (1996).
16. G. Sierra and T. Nishino, Nucl. Phys. **B456**, 505 (1997).
17. I. P. McCulloch and M. Gulacsi, Europhys. Lett. **57**, 852 (2002).
18. S. Maekawa, T. Tohyama, S. E. Barnes, S. Ishihara, W. Koshibae, and G. Khaliullin, "Physics of Transition Metal Oxides", Springer (2004).
19. S. Uchida, T. Ido, H. Takagi, T. Arima, Y. Tokura, and S. Tajima, Phys. Rev. **B43**, 7942 (1991).

20. C. Kim, A. Y. Matsuura, Z. X. Shen, N. Motoyama, H. Eisaki, S. Uchida, T. Tohyama, and S. Maekawa, Phys. Rev. Lett. **77**, 4054 (1996); C. Kim, Z. X. Shen, N. Motoyama, H. Eisaki, S. Uchida, T. Tohyama, and S. Maekawa, Phys. Rev. **B56**, 15589 (1997); B. J. Kim, H. Koh, E. Rotenberg, S.-J. Oh, H. Eisaki, N. Motoyama, S. Uchida, T. Tohyama, S. Maekawa, Z.-X. Shen, and C. Kim, Nat. Phys. **2**, 397 (2006).
21. M. Imada, A. Fujimori, and Y. Tokura, Rev. Mod. Phys. **70**, 1039 (1998).
22. M. J. Rozenberg, X. Y. Zhang, and G. Kotliar, Phys. Rev. Lett. **69**, 1236 (1992); A. Georges and G. Kotliar, Phys. Rev. **B45**, 6479 (1992).
23. M. Suzuki, Prog. Theor. Phys. **56**, 1454 (1976).
24. F. Verstraete and J. I. Cirac, arXiv:cond-mat/0407066.
25. F. Verstraete, D. Porras, and J. I. Cirac, Phys. Rev. Lett. **93**, 227205 (2004).
26. A. J. Daley, C. Kollath, U. Schollwöck, and G. Vidal, J. Stat. Mech., 04005 (2004).
27. U. Schollwöck, Rev. Mod. Phys. **77**, 259 (2005); U. Schollwöck and S. R. White, arXiv:cond-mat/0606018.
28. E. Jeckelmann, Prog. Theor. Phys. Suppl. **176**, 143 (2008).
29. F. Verstraete, J. I. Cirac, and V. Murg, Adv. Phys. **57**, 143 (2008); arXiv:0907.2796.
30. G. Sierra and M. A. Martin-Delgado, arXiv:cond-mat/9811170.
31. L. Bombelli, R. K. Koul, J. Lee, and R. D. Sorkin, Phys. Rev. D **34**, 373 (1986).
32. M. Srednicki, Phys. Rev. Lett. **71**, 666 (1993).
33. C. Holzhey, F. Larsen, F. Wilczek, Nucl. Phys. B **424**, 443 (1994).
34. P. Calabrese and J. Cardy, J. Stat. Mech. **0406**, P002 (2004) [note added: arXiv:hep-th/0405152].
35. P. Calabrese and J. Cardy, J. Phys. A **42**, 504005 (2009).
36. H. Ueda, I. Maruyama, and K. Okunishi, J. Phys. Soc. Jpn. **80**, 023001 (2011).
37. B. S. Shastry and B. I. Shraiman, Phys. Rev. Lett. **65**, 1068 (1990).
38. S. R. White, Phys. Rev. Lett. **77**, 3633 (1996).
39. K. Ueda, C. Jin, N. Shibata, Y. Hieida, and T. Nishino, arXiv:cond-mat/0612480.
40. A. E. Feiguin and S. R. White, Phys. Rev. **B72**, 020404(R) (2005).
41. A. E. Feiguin and S. R. White, Phys. Rev. **B72**, 220401(R) (2005).
42. H. Kishida, H. Matsuzaki, H. Okamoto, T. Tanabe, M. Yamashita, Y. Taguchi and Y. Tokura, Nature **405**, 929 (2000).
43. H. Kishida, M. Ono, K. Miura, H. Okamoto, M. Izumi, T. Manako, M. Kawasaki, Y. Taguchi, Y. Tokura, T. Tohyama, K. Tsutsui, and S. Maekawa, Phys. Rev. Lett. **87**,177401 (2001).
44. M. Kuwata-Gonokami and S. Koshihara, eds., "SPECIAL TOPICS Photo-Induced Phase Transitions and their Dynamics", J. Phys. Soc. Jpn. **75**, 011001-011008 (2006).
45. S. Iwai, M. Ono, A. Maeda, H. Matsuzaki, H. Kishida, H. Okamoto, and Y. Tokura, Phys. Rev. Lett. **91**, 057401 (2003).

46. H. Okamoto, H. Matsuzaki, T. Wakabayashi, Y. Takahashi, and T. Hasagawa, Phys. Rev. Lett. **98**, 037401 (2007); H. Uemura et al., J. Phys. Soc. Jpn. **77**, 113714 (2008).
47. M. Ogata and H. Shiba, Phys. Rev. **B41**, 2326 (1990).
48. H. Ueda and T. Nishino, J. Phys. Soc. Jpn. **78**, 014001 (2009).
49. T. Ogasawara, M. Ashida, N. Motoyama, H. Eisaki, S. Uchida, Y. Tokura, H. Ghosh, A. Shukla, S. Mazumdar and M. Kuwata-Gonokami, Phys. Rev. Lett. **85**, 2204 (2000).
50. Y. Mizuno, K. Tsutsui, T. Tohyama and S. Maekawa, Phys. Rev. **B62**, R4769 (2000).
51. H. Matsueda, T. Tohyama, and S. Maekawa, Phys. Rev. **B70**, 033102 (2004).
52. H. Matsueda, A. Ando, T. Tohyama, and S. Maekawa, Phys. Rev. B **77**, 193112 (2008).
53. H. Matsueda, A. Ando, T. Tohyama, and S. Maekawa, J. Phys. Chem. Solids **69**, 3070 (2008).
54. T. Tohyama and H. Matsueda, Prog. Theor. Phys. Suppl. **176**, 165 (2008).
55. H. Matsueda, T. Tohyama, and S. Maekawa, Phys. Rev. B **74**, 241103(R) (2006).
56. H. Matsueda, T. Tohyama, and S. Maekawa, J. Mag. Mag. Mater. **310**, 975 (2007).
57. H. Suzuura, H. Yasuhara, A. Furusaki, N. Nagaosa, Y. Tokura: Phys. Rev. Lett. **76**, 2579 (1996).
58. A. Takahashi, H. Gomi, and M. Aihara, Phys. Rev. Lett. **89**, 206402 (2002); Phys. Rev. **B69**, 075116 (2004).
59. H. Matsueda and S. Ishihara, J. Phys. Soc. Jpn. **76**, 083703 (2007).
60. Y. Kanamori, H. Matsueda, S. Ishihara, Phys. Rev. Lett. **103**, 267401 (2009).
61. Y. Kanamori, H. Matsueda, S. Ishihara, Phys. Rev. B **82**, 115101 (2010).
62. Y. Kanamori, H. Matsueda, S. Ishihara, Phys. Rev. Lett. **107**, 167403 (2011).

DENSITY-MATRIX RENORMALIZATION GROUP METHOD FOR TOMONAGA-LUTTINGER LIQUID

TOSHIYA HIKIHARA

Department of Physics, Hokkaido University,
Sapporo 060-0810, Japan
E-mail: hikihara@phys.sci.hokudai.ac.jp

The application of the density-matrix renormalization group (DMRG) method to the one-dimensional quantum critical state, the Tomonaga-Luttinger liquid (TLL), is discussed. I review a combined scheme of the DMRG calculation and the bosonization technique, which has proven to be a powerful tool in studying the TLL. The scheme provides a full, *quantitative* description of TLL and enables us to make accurate predictions on various observables in thermodynamic limit. I then present examples of studies in which the combined scheme has been used successfully in reproducing and explaining experimental results.

Keywords: Density-matrix renormalization group method; Tomonaga-Luttinger liquid; one-dimensional quantum system.

1. Introduction

Since the development by White,[1,2] the density-matrix renormalization group (DMRG) method has been applied to many studies in condensed-matter physics and has proven its efficiency.[3] Remarkable advantages of the DMRG method are that it can achieve extremely accurate (almost exact) data for large systems (typically with a few thousand sites) and can be applicable to, in principle, any 1d models including frustrated systems for which the quantum Monte Carlo method suffers from the minus-sign problem. Another feature of the DMRG method is that it is extremely powerful in studying one-dimensional (1d) systems while the calculation becomes much less accurate for higher dimensions. From these characteristics, the DMRG method is now widely accepted as the leading numerical tool to study 1d quantum many-body problem. The reason why the DMRG has a forte for 1d systems is understood from the fact that DMRG is a variational method using the matrix-product state as a variational function.[4–6]

In this paper, we discuss the application of the DMRG method to a typical universality class in the 1d quantum systems, i.e., Tomonaga-Luttinger liquid (TLL). The TLL is characterized by gapless excitations of collective modes and an algebraic decay of correlation functions which exhibit a quasi long-range order (LRO) at zero temperature. The TLL is commonly seen in several 1d systems and plays an essential role in the physics of 1d systems as a counterpart of the Fermi liquid in three dimensions.

It is well known that the low-energy properties of TLL are well described by the bosonization approach.[7] In the bosonization, the TLL is described as a system of free bosons of the gapless collective excitations. The approach enables us to obtain analytic forms of correlation functions in the thermodynamic limit and to predict parameter- and temperature-dependences of various observables at low temperatures. Nevertheless, there are also limitations in the bosonization method; Similar to most of analytic methods, the bosonization relies on some approximations. As a result, although the bosonization approach can describe TLL successfully, it can not tell, in principle, whether or not the TLL is indeed realized in the microscopic model considered. Therefore, one must check the emergence of the TLL in the model by other approaches such as the Bethe ansatz and numerical methods. Furthermore, the bosonization results include nonuniversal parameters which are not determined within the approach. The latter issue prevents us from deriving quantitative results from the bosonization and would be problematic when one wants to compare the theory with experiments.

On the numerical side, the DMRG method has been used in many studies of TLLs and been successful in obtaining accurate data of various observables in the large 1d systems. However, DMRG analysis also has difficulties; Since the correlation length is divergent in TLLs, the finite-size effect can not be negligible even in the large systems which can be treated by DMRG. Furthermore, as the DMRG calculation is usually performed under the open boundary condition, one must manage the open boundary effects, which are also amplified by the divergent correlation length. For extracting accurate and reliable results from the DMRG data, it is necessary to take account of these finite-size and boundary effects.

In order to overcome the difficulties and construct a strong approach to analyze TLLs, my collaborator and I have proposed a combined scheme of the bosonization and DMRG methods.[8–10] In the scheme, we determine the nonuniversal parameters in the bosonization theory using the DMRG data. The resultant theory enables us to aquire even quantitatively accurate results for various observables of the microscopic 1d models in the thermodynamic limit. The method has been applied to several 1d models in the

TLL state successfully[11–13] and good agreements with experimental results have been achieved.[14–16]

The paper is organized as follows. In Sec. 2, we review the combined scheme we have proposed. We consider the spin-1/2 XXZ chain under magnetic field as an example of TLL and explain how the scheme is applied to the problem. In Sec. 3, we discuss the applications of our scheme to analyses of various (quasi) 1d compounds modeled by the spin-1/2 XXZ chain, two-leg ladder, and frustrated zigzag ladder systems under magnetic field. We will see that the method works well to obtain quantitative results for the systems and make comparisons and predictions to experiments. Section 4 is devoted to summary and concluding remarks.

2. Combined Scheme of Bosonization and DMRG Methods

In this section, I explain our scheme of the combination of analytic bosonization approach and the numerical DMRG calculation to study 1d quantum systems of TLL. The scheme provides full, not only qualitative but quantitative, description of low-energy properties of the systems in the thermodynamic limit. In the following, we focus on the 1d spin-1/2 XXZ model under magnetic field as a typical example of TLL. The Hamiltonian is given by

$$\mathcal{H} = J \sum_{l} \left(S_l^x S_{l+1}^x + S_l^y S_{l+1}^y + \Delta S_l^z S_{l+1}^z \right) - h \sum_{l} S_l^z, \tag{1}$$

where $\mathbf{S}_l$ is the spin-1/2 operator at lth site, $J > 0$ is the exchange constant, $\Delta > -1$ is the anisotropy parameter, and h is the external magnetic field. We assume that the magnetic field h is in the critical regime $h_{c1} \leq h \leq h_{c2}$, where h_{c1} is the lower critical field ($h_{c1} = 0$ for $-1 < \Delta \leq 1$ and $h_{c1} > 0$ for $\Delta > 1$) and h_{c2} is the saturation field. Applications of the scheme to other 1d systems are discussed in Sec. 3.

In Sec. 2.1, I review the bosonization approach to the model (1) and derive analytic forms of spin correlation functions in the finite system with open boundaries. The results are used in Sec. 2.2 to fit the DMRG data and then to determine the nonuniversal parameters in the bosonization results to complete the quantitative bosonization theory.

2.1. *Bosonization approach*

It has been established[7] that the low-energy states of the XXZ chain (1) are described by the bosonization approach. The low-energy excitations of the model are free massless bosons, whose dynamics are governed by the Gaussian model,

$$\widetilde{\mathcal{H}}_0 = \frac{v}{2}\int dx \left[K\left(\frac{d\theta}{dx}\right)^2 + \frac{1}{K}\left(\frac{d\phi}{dx}\right)^2 \right], \tag{2}$$

where v is the spin-wave velocity and $\phi(x)$ and $\theta(x)$ are the boson fields which obeys the commutation relation $[\phi(x), \partial_y\theta(y)] = i\delta(x-y)$. Here, I have taken the lattice constant $\alpha = 1$ and identified the site index l with the continuous coordinate x. K is called the TLL parameter, which is the most important parameter to characterize the TLL but can not be determined within the bosonization approach. I note that for the XXZ chain (1), K can be obtained exactly from the Bethe ansatz,[17–19] while for the other models, in general, K is a parameter to be determined by our scheme. (See Sec. 3.2 and 3.3.) The spin operators $\mathbf{S}_l$ are expressed in terms of the boson fields as

$$S_l^z = M + \frac{1}{\sqrt{\pi}}\frac{d\phi(x)}{dx} - (-1)^l a \sin[2\pi M l + \sqrt{4\pi}\phi(x)] + \cdots, \tag{3}$$

$$S_l^+ = (-1)^l b e^{i\sqrt{\pi}\theta(x)} + b' e^{i\sqrt{\pi}\theta(x)} \sin[2\pi M l + \sqrt{4\pi}\phi(x)] + \cdots, \tag{4}$$

where M is the magnetization per site. The coefficients a, b, and b' are the nonuniversal parameter to be determined as well as K.

Equations (2), (3), and (4) represent the TLL theory for the XXZ chain (1); In principle, one can derive all the low-energy properties of the system from the relations. For example, the asymptotic behaviors of the ground-state equal-time spin correlation functions in the thermodynamic limit are obtained as

$$\langle S_0^z S_r^z \rangle = M^2 - \frac{\eta}{4\pi^2 r^2} + A_1^z \frac{(-1)^r \cos(2\pi M r)}{|r|^\eta} + \cdots, \tag{5}$$

$$\langle S_0^x S_r^x \rangle = A_0^x \frac{(-1)^r}{|r|^{1/\eta}} - A_1^x \frac{\cos(2\pi M r)}{|r|^{\eta+1/\eta}} + \cdots, \tag{6}$$

where $A_1^z = a^2/2$, $A_0^x = b^2/2$, $A_1^x = b'^2/4$ (with appropriate short-distance regularization), and the decay exponent η is related to the TLL parameter K by $\eta = 2K$. Therefore, once the TLL parameter K and the coefficients a, b, b', ..., are determined, we can evaluate quantitatively any observables in the model (1). Our strategy is to determine those parameters by fitting the DMRG data of correlation functions to the bosonization forms.

Before proceeding to the fitting procedure, we carry out a pretreatment for the bosonization forms, which is essential for obtaining accurate estimates of the parameters. As mentioned in Sec. 1, the DMRG method works most efficiently when applied to a finite system with open boundaries.[20,21] Under the open boundary condition, the finite-size and boundary effects is prominent and we observe large deviation in the spin correlation functions $\langle S_l^z S_{l'}^z \rangle$ and $\langle S_l^x S_{l'}^x \rangle$ from the asymptotic behaviors in Eqs. (5) and (6). Therefore, it is necessary to prepare the analytic formulas of the correlators in the finite open chains for accomplishing the accurate fitting. I also note that the open boundary effect can be even useful to aquire the information of low-energy properties of TLL; For example, in the periodic systems, the local spin polarization is constant, $\langle S_l^z \rangle = M$, and not informative. On the other hand, in the chains with open boundaries, $\langle S_l^z \rangle$ around the boundaries exhibits the Friedel oscillation, from which we can extract various results such as the values of K and a. [See Eqs. (7) and (13) and Sec. 2.2 below.]

Let us consider the finite open chain consisting of L spins, $\mathbf{S}_l$ ($l = 1, 2, ..., L$). The presence of the open boundaries can be represented by assigning vanishing spins at the ghost sites $l = 0$ and $L+1$, i.e., $\mathbf{S}_0 = \mathbf{S}_{L+1} = 0$. In the bosonization approach, the condition is reproduced by imposing the Dirichlet boundary condition on the boson field $\phi(x)$ as $\phi(0) = \phi(L+1) = 0$.[22] Under the condition, one can calculate the ground-state expectation values of the local spin polarization and the spin correlation functions in the finite open chain. (See Ref. 10 for the details of the calculation.) The resultant forms are

$$\langle S_l^z \rangle = z(l;q) \equiv \frac{q}{2\pi} - a\frac{(-1)^l \sin(ql)}{f_{\eta/2}(2l)}, \tag{7}$$

$$\begin{aligned} \langle S_l^z S_{l'}^z \rangle &= Z(l,l';q) \\ &\equiv \left(\frac{q}{2\pi}\right)^2 - \frac{\eta}{4\pi^2}\left[\frac{1}{f_2(l-l')} + \frac{1}{f_2(l+l')}\right] - \frac{qa}{2\pi}\left[\frac{(-1)^l \sin(ql)}{f_{\eta/2}(2l)} + \frac{(-1)^{l'} \sin(ql')}{f_{\eta/2}(2l')}\right] \\ &\quad + \frac{(-1)^{l-l'} a^2}{2f_{\eta/2}(2l)f_{\eta/2}(2l')}\left\{\cos[q(l-l')]\frac{f_\eta(l+l')}{f_\eta(l-l')} - \cos[q(l+l')]\frac{f_\eta(l-l')}{f_\eta(l+l')}\right\} \\ &\quad - \frac{\eta a}{2\pi}\left\{\frac{(-1)^l \cos(ql)}{f_{\eta/2}(2l)}[g(l+l')+g(l-l')] + \frac{(-1)^{l'} \cos(ql')}{f_{\eta/2}(2l')}[g(l+l')-g(l-l')]\right\}, \end{aligned} \tag{8}$$

$$\begin{aligned} \langle S_l^x S_{l'}^x \rangle &= X(l,l';q) \\ &\equiv \frac{f_{1/2\eta}(2l)f_{1/2\eta}(2l')}{f_{1/\eta}(l-l')f_{1/\eta}(l+l')} \\ &\quad \times \left\{\frac{(-1)^{l-l'} b^2}{2} + \frac{\mathrm{sgn}(l-l')bb'}{2}\left[\frac{(-1)^{l'} \cos(ql)}{f_{\eta/2}(2l)} - \frac{(-1)^l \cos(ql')}{f_{\eta/2}(2l')}\right]\right. \\ &\quad \left. - \frac{b'^2}{4f_{\eta/2}(2l)f_{\eta/2}(2l')}\left[\cos[q(l+l')]\frac{f_\eta(l-l')}{f_\eta(l+l')} + \cos[q(l-l')]\frac{f_\eta(l+l')}{f_\eta(l-l')}\right]\right\}, \end{aligned} \tag{9}$$

where

$$q = \frac{2\pi LM}{L+1}, \tag{10}$$

and

$$f_\alpha(x) = \left[\frac{2(L+1)}{\pi}\sin\left(\frac{\pi|x|}{2(L+1)}\right)\right]^\alpha, \tag{11}$$

$$g(x) = \frac{\pi}{2(L+1)}\cot\left[\frac{\pi x}{2(L+1)}\right]. \tag{12}$$

I note that in the thermodynamic limit $L \to \infty$ with $|l-L/2|, |l'-L/2| \ll L$, Eqs. (8) and (9) reduce to the asymptotic forms Eqs. (5) and (6), respectively. On the other hand, in the limit $L \to \infty$ with $l \ll L$, Eq. (7) reduces to

$$\langle S_l^z \rangle = M - \frac{(-1)^l a}{(2l)^{\eta/2}}\sin(2\pi M l). \tag{13}$$

The presence of open boundaries results in the Friedel oscillation in $\langle S_l^z \rangle$, which oscillates with the wavenumber $Q = 2\pi M$ and decays algebraically with the exponent $\eta/2 = K$.

2.2. *DMRG results*

In this subsection, I show the results of the fitting of DMRG data to the bosonization forms in the finite open chain. Using the DMRG method, we have calculated the local magnetization $\langle S_l^z \rangle$ and the spin correlations $\langle S_l^z S_{l'}^z \rangle$ and $\langle S_l^x S_{l'}^x \rangle$ in the lowest-energy state in the subspace characterized by the magnetization per site $M = (1/L)\sum_l S_l^z$. From the fitting, we can achieve accurate estimates of the nonuniversal parameters in the bosonization approach such as the coefficients a, b, and b', which enable us to complete the quantitative TLL theory for the model (1).

Figure 1 shows the DMRG data of the local magnetization $\langle S_l^z \rangle$, longitudinal spin correlation $\langle S_l^z S_{l'}^z \rangle$, and transverse spin correlation $\langle S_l^x S_{l'}^x \rangle$ and their fits to Eqs. (7), (8), and (9), respectively. We see that the DMRG data and the fits are in excellent agreement with each other; The deviations of the correlators from the asymptotic behaviors, such as the Friedel oscillation in $\langle S_l^z \rangle$ and the deviations around the open boundaries in $\langle S_l^z S_{l'}^z \rangle$ and $\langle S_l^x S_{l'}^x \rangle$, are reproduced well by the bosonization forms for the finite open chain. I note that there are only one (two) free parameters a (b, and b') in the fitting of $\langle S_l^z \rangle$ and $\langle S_l^z S_{l'}^z \rangle$ ($\langle S_l^x S_{l'}^x \rangle$). The excellent success of the fitting procedure demonstrates that the low-energy physics of the model (1) is indeed described by the TLL theory defined by Eqs. (2), (3), and (4).

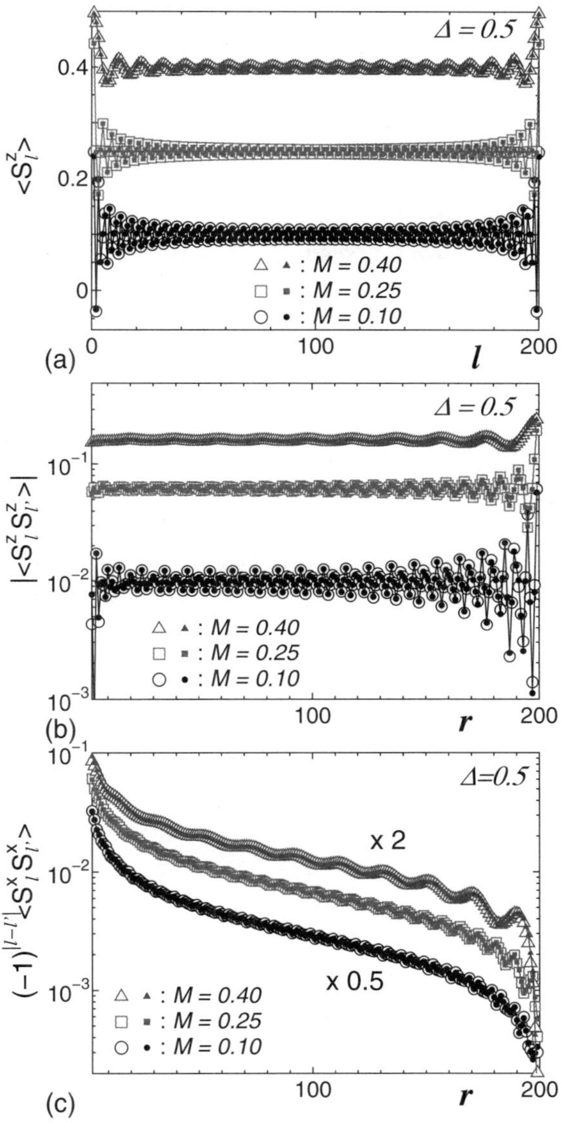

Fig. 1. Correlation functions in the spin-1/2 XXZ chain (1) with $\Delta = 0.5$ and $L = 200$; (a) local spin polarization $\langle S_l^z \rangle$, (b) longitudinal spin correlation function $\langle S_l^z S_{l'}^z \rangle$, and (c) transverse spin correlation function $\langle S_l^x S_{l'}^x \rangle$. The open symbols represent the DMRG data and the solid symbols are fits to Eqs. (7), (8), and (9). In (b) and (c), the data for $l = L/2 - [r/2]$ and $l' = L/2 + [(r+1)/2]$ are shown as a function of $r = |l - l'|$. For the clarity, the data for $M = 0.10$ and $M = 0.40$ in (c) are multiplied by a factor of 0.5 and 2, respectively.

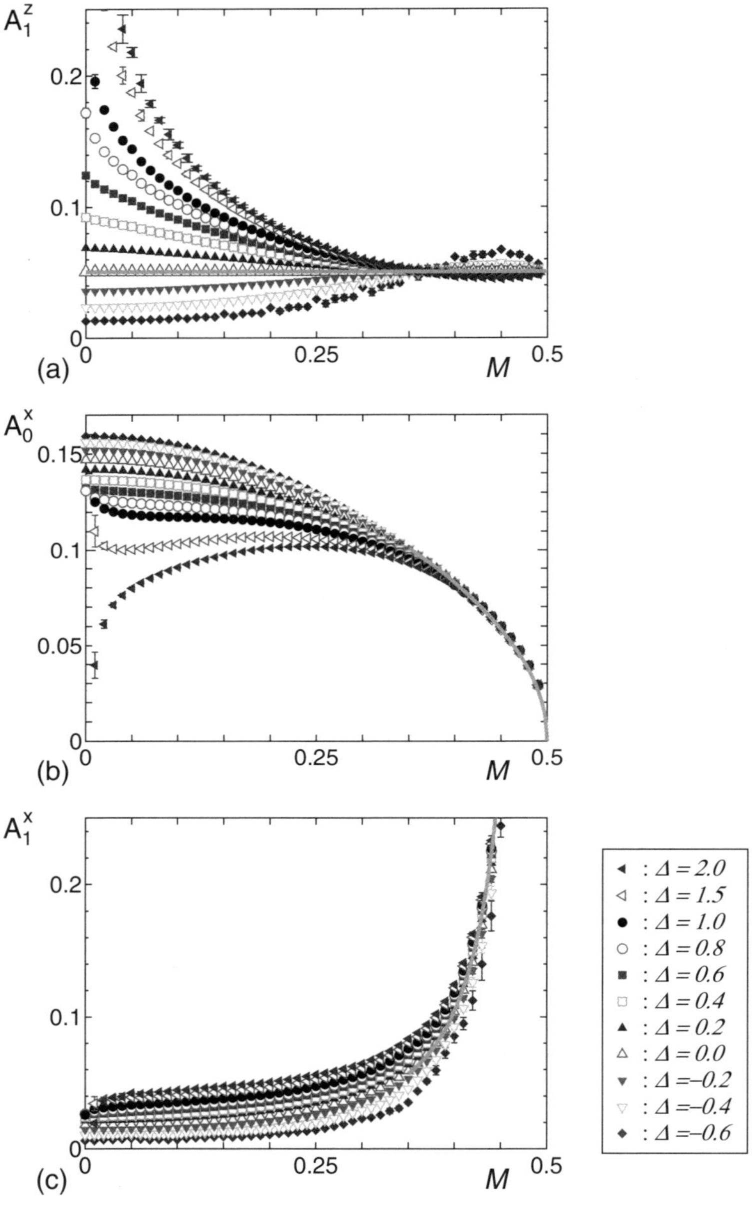

Fig. 2. Correlation amplitudes as functions of M for several typical values of Δ; (a) A_1^z, (b) A_0^x, and (c) A_1^x. The solid line in (a) represents the exact result $A_1^z = 1/(2\pi^2)$ for $\Delta = 0$ while the solid curves at large M in (b) and (c) represent the asymptotic behaviors at $M \to 1/2$ [Eq. (14)].

In Fig. 2, I present the estimates of the correlation amplitudes $A_1^z(= a^2/2)$, $A_0^x(= b^2/2)$, and $A_1^x(= b'^2/4)$ as functions of Δ and M. For the amplitudes, there are exact results available for some limiting cases;

(i) At $M = 0$, the expressions of the correlation amplitudes A_1^z, A_0^x, and A_1^x are obtained analytically for $-1 < \Delta < 1$.[23–25] (See Eqs. (32)-(34) in Ref. 10 for the expression.)

(ii) At the saturation limit $M \to 1/2$, A_0^x and A_1^x should obey the universal behaviors irrespective of Δ as[26–28]

$$A_0^x = \frac{\rho}{2\sqrt{\pi}}\left(\frac{1}{2} - M\right)^{1/2}, \quad A_1^x = \frac{\rho}{16\pi^{5/2}}\left(\frac{1}{2} - M\right)^{-3/2}, \tag{14}$$

with $\rho = 0.92418\ldots$. Furthermore, A_1^z must converge to a universal value $A_1^z \to 1/(2\pi^2)$ at $M \to 1/2$ since $\langle S_l^z S_{l'}^z \rangle \to 1/4$ and $\eta \to 2$ at the limit [see Eq. (5)].

(iii) For $\Delta = 0$, the model is a free fermion system and we can calculate exactly $\langle S_l^z S_{l'}^z \rangle$, from which it follows that $A_1^z = 1/(2\pi^2)$ for arbitrary M.

The estimates of A_1^z, A_0^x, and A_1^x obtained from the fitting procedure above agree with these analytic results, verifying the accuracy of the estimates. We will see in Sec. 3.1 that the TLL theory with these estimates of the coefficients a, b, and b' is applied successfully to quantitative analyses of various observables, which can be compared to experimental results on real compounds.

3. Applications

In the preceding section, we have seen that combining the bosonization approach and DMRG calculation yields the complete TLL theory, which is able to derive *quantitatively* low-energy properties of the microscopic model in the thermodynamic limit. In this section, I review the applications of the scheme to other 1d models as well as real quasi-1d compounds.[11,14–16,29]

3.1. *XXZ chain under magnetic field*

We have shown in Sec. 2 that the TLL parameter K and the coefficients a, b, and b' in the XXZ chain (1) are obtained accurately as a function of Δ and M. The results have been used to evaluate various observables and succeeded in explaining experimental results on real compounds.

First, let us discuss the application of the scheme to Cu benzoate. The material is a quasi-1d spin-1/2 antiferromagnet but known to exhibit an unexpected excitation gap under uniform magnetic field. This field-induced gap is explained by the weak staggered magnetic field induced by the uniform field via a staggered g-tensor and a Dzyaloshinskii-Moriya interaction.[30] The model to describe the material under field is thus found to be

$$\mathcal{H}_{\mathrm{CuB}} = J \sum_l \mathbf{S}_l \cdot \mathbf{S}_{l+1} - h \sum_l S_l^z + h_s \sum_l (-1)^l S_l^x, \tag{15}$$

where $h_s = \gamma h$ with $\gamma \sim 0.06$ for Cu benzoate.

In Ref. 14, Essler *et al.* used the combined scheme of the TLL theory and the DMRG results to explain the neutron scattering experiments of Cu benzoate. They mapped the model (15) to the Sine-Gordon model by the bosonization and derived analytic forms of energy gaps of soliton and breather excitations and of the longitudinal/transverse dynamical structure factors in the model. Then, substituting the exact values of parameters such as K and spin velocity v as well as the numerical values of the coefficients a and b into the analytic forms, they evaluated the ratio of the spectral weights of the first breather and the soliton modes. The result was compared to the neutron scattering experiment[31] and found to be in a good agreement with the experimentally observed value. The agreement demonstrated that the physics of the field-induced gap system of Cu benzoate was described successfully within the low-energy effective theory of the Sine-Gordon model.

The combined scheme of quantitative bosonization has also been used to study a quasi-1d compound $BaCo_2V_2O_8$, which is believed to be modeled by the weakly-coupled spin-1/2 XXZ chains (1) with Ising-type anisotropy, $\Delta = 2.17$. For the compound, it has been observed experimentally that at intermediate magnetic field the compound undergoes a transition at $T = T_N$ between high-temperature paramagnetic phase and low-temperature incommensurate magnetic ordered phase.[32] In Ref. 15, Okunishi and Suzuki applied the quantitative TLL theory of the XXZ spin chain discussed above to the analysis of the experimental result. They described each XXZ chains using the TLL theory and incorporated the effect of weak interchain couplings J' within the mean-field approximation to derive the analytic form of the transition temperature T_N. Then, using the exact values of K and v as well as the correlation amplitudes of A_1^z and A_0^x, they evaluated T_N as a function of only J' and the field h. The resultant T_N with optimized $J' \sim 0.0014J$ succeeded in reproducing the experimentally observed field dependence of T_N with remarkable accuracy (see Fig. 3 in Ref. 15).

These two successes in the comparison of the TLL theory with the experiments elucidate the ability of the combined scheme in quantitative analyses.

3.2. *Two-leg ladder under magnetic field*

Let us turn to the application of the combined scheme to a system other than the XXZ chain (1). Here, we consider the spin-1/2 two-leg ladder under magnetic field,

$$\mathcal{H}_{2\mathrm{leg}} = J \sum_{n=1,2} \sum_{l} \mathbf{S}_{n,l} \cdot \mathbf{S}_{n,l+1} + J_{\perp} \sum_{l} \mathbf{S}_{1,l} \cdot \mathbf{S}_{2,l} - h \sum_{l} S_l^z, \tag{16}$$

where $\mathbf{S}_{n,l}$ is the spin-1/2 operator at the lth site of nth chain ($n = 1, 2$) and we consider the case of antiferromagnetic leg and rung couplings, $J, J_{\perp} > 0$.

With varying the field h, the system exhibits three different phases; the singlet phase with a spin gap for $h < h_{c1}$, a gapless phase with partial magnetization $0 < M < 1/2$ for $h_{c1} < h < h_{c2}$, and the fully polarized phase for $h > h_{c2}$. It has been shown[9,33–35] that in the gapless phase the low-energy physics of the model is described by the TLL theory with the Gaussian model (2). In this phase, the operators exhibiting the quasi LROs are the longitudinal component of the total rung spins, $S^z_{+,l} = S^z_{1,l} + S^z_{2,l}$, and the transverse component of the staggered rung spins, $S^x_{-,l} = S^x_{1,l} - S^x_{2,l}$. The asymptotic forms of the correlation functions are found to be

$$\langle S^z_{+,0} S^z_{+,r} \rangle = M_{\mathrm{lad}}^2 - \frac{\eta}{4\pi^2 r^2} + A_1^z \frac{\cos(2\pi M_{\mathrm{lad}} r)}{|r|^{\eta}} + \cdots, \tag{17}$$

$$\langle S^x_{-,0} S^x_{-,r} \rangle = 2A_0^x \frac{(-1)^r}{|r|^{1/\eta}} - 2A_1^x \frac{(-1)^r \cos(2\pi M_{\mathrm{lad}} r)}{|r|^{\eta+1/\eta}} + \cdots, \tag{18}$$

where $M_{\mathrm{lad}} = (1/L_{\mathrm{rung}}) \sum_{l=1}^{L_{\mathrm{rung}}} (S^z_{1,l} + S^z_{2,l})$ is the magnetization per rung and L_{rung} the number of rungs. The TLL parameter K and the amplitudes A_1^z, A_0^x, A_1^x, ..., are functions of $J_{\perp}/J$ and h (or M). Once these parameters are determined, we can evaluate various observables in the two-leg ladder model in the critical regime by the bosonization theory. The correlation functions of the transverse total rung spin $S^x_{+,l} = S^x_{1,l} + S^x_{2,l}$ and the longitudinal staggered rung spin $S^z_{-,l} = S^z_{1,l} - S^z_{2,l}$ decay exponentially.

The estimates of K and the correlation amplitudes are obtained in a similar manner to the case of the XXZ chain discussed in Sec. 2. In the finite open ladder, the analytic forms of the local magnetization and the correlation functions are obtained as

$$\langle S^z_{+,l}\rangle = \frac{1}{2} + z(l; q_{\rm lad}), \tag{19}$$

$$\langle S^z_{+,l} S^z_{+,l'}\rangle = \frac{1}{4} + \frac{1}{2}\left[z(l; q_{\rm lad}) + z(l'; q_{\rm lad})\right] + Z(l, l'; q_{\rm lad}), \tag{20}$$

$$\langle S^x_{-,l} S^x_{-,l'}\rangle = 2X(l, l'; q_{\rm lad}), \tag{21}$$

with

$$q_{\rm lad} = \frac{2\pi L}{L+1}\left(M_{\rm lad} - \frac{1}{2}\right). \tag{22}$$

By fitting the DMRG data of the correlators to these forms, we can determine K and the correlation amplitudes accurately.[9] Figure 3 shows the

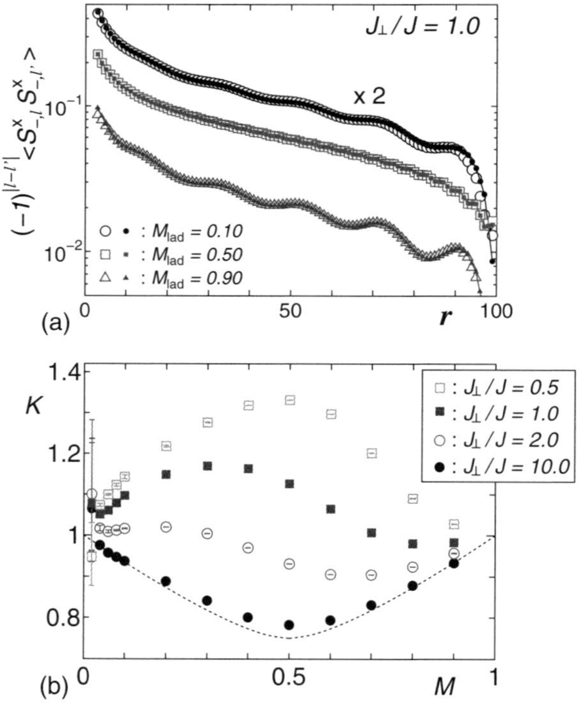

Fig. 3. (a) The transverse staggered spin correlation $\langle S^x_{-,l} S^x_{-,l'}\rangle$ in the two-leg ladder model (16) with $L_{\rm rung} = 100$ rungs and $J_\perp/J = 1.0$. The open symbols represent the DMRG data and the solid symbols are fits to Eqs. (21). The data for $l = L_{\rm rung}/2 - [r/2]$ and $l' = L_{\rm rung}/2 + [(r+1)/2]$ are shown as a function of $r = |l - l'|$. For the clarity, the data for $M = 0.10$ are multiplied by a factor of 2. (b) The TLL parameter K estimated from the fitting of $\langle S^x_{-,l} S^x_{-,l'}\rangle$ for $J_\perp/J = 10.0, 2.0, 1.0, 0.5$. The dotted curve represents the exact result at the strong rung-coupling limit $J_\perp/J \to \infty$.[35–37]

results of the fitting of $\langle S^x_{-,l} S^x_{-,l'} \rangle$. The excellent agreement between the DMRG data and the fits suggests that our scheme works well also for the two-leg ladder model. The resultant estimates of K are also shown in Fig. 3. It is found that K changes non-monotonically with M between the universal value $K = 1$ at the limits $M \to 0$ and $M \to 1/2$, where the system is equivalent to the dilute limit of hard-core bosons. The data for $J_\perp/J = 10.0$ almost coincide with the exact result obtained at the strong rung-coupling limit $J_\perp/J \to \infty$,[35–37] suggesting the correctness of the estimates.

The TLL theory with the estimates of K and correlation amplitudes has been used to analyze experimental results of the two-leg ladder compound $CuBr_4(C_5H_{12}N)_2$. In Ref. 16, using the values of K and A_0^x obtained from the fitting and the spin velocity v from the susceptibility, Klanjšek *et al.* calculated the field dependence of the nuclear spin-lattice relaxation rate $1/T_1$ in the nuclear magnetic resonance (NMR) measurement. Furthermore, including the effect of weak interladder coupling within the mean-field approximation, they also evaluated the Néel temperature to the three-dimensional (3d) ordering and the amplitude of the magnetic moment in the 3d ordered phase. The results was then compared to the experimental data and proved to be successful in reproducing the data accurately. (See Fig. 3 of Ref. 16 for the results.) The remarkable agreement between the theory and the experiments demonstrates that the combined scheme is also efficient in studying the TLL in the two-leg ladder system.

3.3. *Zigzag ladder under magnetic field*

The combined scheme has also been applied to the spin-1/2 chain with competing nearest- and next-nearest-neighbor exchange couplings under magnetic field,

$$\mathcal{H}_{J_1-J_2} = J_1 \sum_l \mathbf{S}_l \cdot \mathbf{S}_{l+1} + J_2 \sum_l \mathbf{S}_l \cdot \mathbf{S}_{l+2} - h \sum_l S_l^z. \tag{23}$$

The model is called the $J_1 - J_2$ chain or zigzag ladder and includes frustration when J_2 is antiferromagnetic. It has been shown that the model exhibits several exotic phases, including critical and spin-gapped phases, in the ground-state phase diagram in the J_1/J_2 versus h plane.[38]

The scheme was used to reveal the nature of the critical phases appearing in the magnetic phase diagram and successful in identifying the low-energy effective theory for the phases.[11,12] Furthermore, the field and exchange-coupling dependences of the TLL parameter K, which directly relates to the decay exponent of the dominant (and sub-dominant) correlation functions, were determined by the analysis. The results were used to discuss the field-induced phase transitions in the multiferroic spin-1/2 chain cuprate $LiCuVO_4$.[39–41] We can also make predictions on the magnetic-field dependence of the NMR relaxation rate $1/T_1$ in the compounds, such as $LiCu_2O_2$, $LiCuVO_4$, $PbCuSO_4(OH)_2$, and $Rb_2Cu_2Mo_3O_{12}$, which will be reported elsewhere.[29]

4. Summary

In this paper, I have reviewed the combined scheme of the DMRG method and bosonization approach for studying the TLL. By fitting the DMRG data of correlation functions to the analytic forms obtained from the bosonization for the finite system with open boundaries, we determine nonuniversal parameters in the bosonization description of the system. We can thereby construct a full TLL theory which describes the low-energy state of the 1d model considered accurately. The theory enables us to make quantitative predictions on various observables in the system. The comparison of the results of the TLL theory to experiments has been performed successfully for several real compounds, such as a field-induced gap system Cu benzoate, the spin-1/2 XXZ chain compound $BaCo_2V_2O_8$, the two-leg ladder compound $CuBr_4(C_5H_{12}N)_2$, and the J_1-J_2 chain cuprate $LiCuVO_4$.

While we have discussed the spin-1/2 chain and ladder models under magnetic field as examples, the combined scheme is applicable to any other critical models which is described as a TLL. Once the relation between the spin or fermion/boson operators in the original microscopic model and the boson-field operators in the effective Hamiltonian of the Gaussian model is established, we can derive the analytic forms of correlation functions and carry out the fitting analysis of the numerical data. The application of the scheme to 1d interacting fermion models such as the 1d Hubbard model would be promising.

Acknowledgments

It is my pleasure to acknowledge the collaboration with A. Furusaki, on which all of the present work was based. I would also like to thank F. H. L. Essler, H. Kawamura, S. Lukyanov, T. Momoi, K. Okunishi, and M. Sato for their collaboration and discussions. This work was supported in part by Grants-in-Aid for Scientific Research from the Ministry of Education, Culture, Sports, Science and Technology (MEXT) of Japan (Grants No. 21740277 and No. 22014016).

References

1. S. R. White, *Phys. Rev. Lett.* **69**, 2863 (1992).
2. S. R. White, *Phys. Rev. B* **48**, 10345 (1993).
3. U. Schollwöck, *Rev. Mod. Phys.* **77**, 259 (2005).
4. Stellan Östlund and Stefan Rommer, *Phys. Rev. Lett.* **75**, 3537 (1995).
5. Stefan Rommer and Stellan Östlund, *Phys. Rev. B* **55**, 2164 (1997).
6. T. Nishino, T. Hikihara, K. Okunishi, and Y. Hieida, *Int. J. Mod. Phys. B* **13**, 1 (1999).
7. T. Giamarchi, *Quantum Physics in One Dimension* (Clarendon Press, Oxford, 2004).
8. T. Hikihara and A. Furusaki, *Phys. Rev. B* **58**, R583 (1998).
9. T. Hikihara and A. Furusaki, *Phys. Rev. B* **63**, 134438 (2001).
10. T. Hikihara and A. Furusaki, *Phys. Rev. B* **69**, 064427 (2004).
11. T. Hikihara, L. Kecke, T. Momoi, and A. Furusaki, *Phys. Rev. B* **78**, 144404 (2008).
12. T. Hikihara, T. Momoi, A. Furusaki, and H. Kawamura, *Phys. Rev. B* **81**, 224433 (2010).
13. S. Lukyanov, A. Furusaki, and T. Hikihara, unpublished.
14. F. H. L. Essler, A. Furusaki, and T. Hikihara, *Phys. Rev. B* **68**, 064410 (2003).
15. K. Okunishi and T. Suzuki, *Phys. Rev. B* **76**, 224411 (2007).
16. M. Klanjšek, H. Mayaffre, C. Berthier, M. Horvatić, B. Chiari, O. Piovesana, P. Bouillot, C. Kollath, E. Orignac, R. Citro, and T. Giamarchi, *Phys. Rev. Lett.* **101**, 137207 (2008).
17. N. M. Bogoliubov, A. G. Izergin, and V. E. Korepin, *Nucl. Phys. B* **275**, 687 (1986).
18. S. Qin, M. Fabrizio, L. Yu, M. Oshikawa, and I. Affleck, *Phys. Rev. B* **56**, 9766 (1997).
19. D. C. Cabra, A. Honecker, and P. Pujol, Phys. Rev. B **58**, 6241 (1998).
20. The fact that the DMRG works most efficiently under the open boundary condition can be understood from the system-size and boundary-condition dependence of the entanglement entropy; See, for example, Ref. 21.
21. P. Calabrese and J. Cardy, *J. Stat. Mech.: Theory Exp.* (**2004**) P06002.
22. To be precise, there is no strict reason to set the ghost sites at $x = 0$ and $L+1$; we can put them at $x = \delta$ and $L+1-\delta$ and take δ as a free parameter to be optimized in the following fitting procedure. In this paper, I set $\delta = 0$ for simplicity. For the treatment of $\delta \neq 0$, see Ref. 12.
23. S. Lukyanov and A. Zamolodchikov, *Nucl. Phys. B* **493**, 571 (1997).
24. S. Lukyanov, *Phys. Rev. B* **59**, 11163 (1999).
25. S. Lukyanov and V. Terras, *Nucl. Phys. B* **654**, 323 (2003).
26. H. G. Vaidya and C. A. Tracy, Phys. Rev. Lett. **42**, 3 (1979).
27. H. G. Vaidya and C. A. Tracy, *J. Math. Phys.* **20**, 2291 (1979).
28. D. B. Creamer, H. B. Thacker, and D. Wilkinson, *Phys. Rev. D* **23**, 3081 (1981).
29. M. Sato, T. Hikihara, and T. Momoi, unpublished.

30. For the studies of the field-induced gap in Cu benzoate, see M. Oshikawa and I. Affleck, *Phys. Rev. Lett.* **78**, 1984 (1997), I. Affleck and M. Oshikawa, *Phys. Rev. B* **60**, 1038 (1999), and references therein.
31. D. C. Dender, P. R. Hammar, D. H. Reich, C. Broholm and G. Aeppli, *Phys. Rev. Lett.* **79**, 1750 (1997).
32. S. Kimura, T. Takeuchi, K. Okunishi, M. Hagiwara, Z. He, K Kindo, T. Taniyama, and M. Itoh, *Phys. Rev. Lett.* **100**, 057202 (2008).
33. R. Chitra and T. Giamarchi, *Phys. Rev. B* **55**, 5816 (1997).
34. T. Giamarchi and A. M. Tsvelik, *Phys. Rev. B* **59**, 11398 (1999).
35. A. Furusaki and S. C. Zhang, *Phys. Rev. B* **60**, 1175 (1999).
36. F. Mila, Eur. Phys. J. B **6**, 201 (1998).
37. K. Totsuka, Phys. Rev. B **57**, 3454 (1998).
38. For the phase diagram, see Refs. 11 and 12 and references therein.
39. M. G. Banks, F. Heidrich-Meisner, A. Honecker, H. Rakoto, J.-M. Broto, and R. K. Kremer, *J. Phys.: Condens. Matter* **19**, 145227 (2007).
40. N. Büttgen, H.-A. Krug von Nidda, L. E. Svistov, L. A. Prozorova, A. Prokofiev, and W. Aßmus, *Phys. Rev. B* **76**, 014440 (2007).
41. Y. Naito, K. Sato, Y. Yasui, Y. Kobayashi, Y. Kobayashi, and M. Sato, *J. Phys. Soc. Jpn.* **76**, 023708 (2007).

SUPERSYMMETRIC VALENCE-BOND SOLID MODELS — HIDDEN ORDER AND DYNAMICS

KEISUKE TOTSUKA

Yukawa Institute for Theoretical Physics,
Kyoto University, Kitashirakawa, Kyoto 606-8502, Japan

KAZUKI HASEBE

Kagawa National College of Technology, Mitoyo, Kagawa 769-1192, Japan

Generalized valence bond solid states with super Lie group symmetries are discussed in the context of hole-pair doped valence bond solid states. Depending on the hole-doping parameter, they exhibit both insulating and superconducting properties in the charge sector. Though supersymmetric valence bond solid states are disordered spin liquids, there still exists a generalized hidden order detected by non-local string order parameter. The supersymmetric extension of the matrix product state formalism is also developed for the purpose of calculating such physical quantities as spin- and superconducting correlations. The spectra of gapped excitations are investigated too.

1. Introduction

It has been a long-sought desire to write down the explicit wavefunction of non-trivial many-body states. Obviously, two of the well-known examples of such wavefunctions are the BCS wavefunction[1] of superconductivity and the Laughlin wavefunction of fractional quantum Hall effect.[2] In quantum magnetism, the valence bond solid (VBS) state introduced by Affleck, Kennedy, Lieb and Tasaki[3,4] plays a similar role as the paradigmatic state for topologically-ordered (gapped) spin liquid in one dimension. Remarkably, it has been pointed out[5] some time ago that there is an intimate connection between the VBS states and the Laughlin wavefunction (on a sphere). Moreover, recent development in quantum information has revealed intriguing new applications of the VBS states (e.g. quantum computation[6] and the matrix-product representation[7–9] which approximates weakly-entangled gapped states[10]).

Guided by the analogy to the supersymmetric extension of the quantum Hall effect, Arovas, Hasebe, Qi and Zhang have recently proposed[11] an extension of the VBS state, which contains doped hole-pair degrees of freedom, with exact supersymmetry: *supersymmetric valence bond solid (SVBS) states.* Interestingly, it has been shown[11] that the VBS state resembles the other non-trivial many-body wavefunction –the BCS wavefunction– when doped with hole pairs. Later, with the use of the supersymmetric matrix-product formalism, physical properties (hidden order, low-lying excitation, etc.) of the SVBS states have been extensively explored by Hasebe and Totsuka.[12] For arbitrary amount of hole doping, the SVBS states are invariant under supersymmetric transformations. The SVBS states mathematically realize the Anderson's scenario of high-T_c superconductivity[13] and Zhang's idea of symmetry unification of superconductivity and antiferromagnetism.[14] The SVBS states generally interpolate the VBS states and the resonating valence bond solid states as the two extreme limits of the hole doping (zero and infinity), and exhibit superconducting property for arbitrary finite amount of hole doping.

In the first part of the paper, we will review some basic properties of the original VBS state and its supersymmetric extension. In the second part, we will report our results on the hidden order and the dynamical properties of the SVBS states. We will also develop a supersymmetric generalization of the MPS (SMPS).[9]

2. Valence Bond Solid Models

We begin with briefly recapitulating the ordinary VBS states. The VBS models have exactly solvable ground states that, in 1D, realize the celebrated Haldane's scenario on gapped antiferromagnetic spin liquids. In terms of the Schwinger boson operators

$$|\uparrow\rangle = a^{\dagger}|0\rangle, \quad |\downarrow\rangle = b^{\dagger}|0\rangle, \tag{1}$$

representing the spin degrees of freedom, the ground states of the VBS models generically read:

$$|\mathrm{VBS}\rangle = \prod_{<ij>} (a_i^{\dagger}b_j^{\dagger} - b_i^{\dagger}a_j^{\dagger})^M |0\rangle, \tag{2}$$

where $<ij>$ represents pairs of nearest neighbor sites and M is an integer characterizing the VBS states. Since the state $|\mathrm{VBS}\rangle$ is made up of products of spin-singlet bond $a_i^\dagger b_j^\dagger - b_i^\dagger a_j^\dagger$, it describes an entangled antiferromagnetic quantum state, which is invariant under the $SU(2)$ transformation generated by $S_a = \frac{1}{2}(a^\dagger\ b^\dagger)\sigma_a(a\ b)^t$. The spin magnitude S at each site is given in terms of the lattice coordination number (the number of nearest neighbor sites) z and the number M of valence bonds connecting adjacent sites as:

$$S = \frac{1}{2} zM \,. \tag{3}$$

For instance, for 1D chain $z = 2$, for 2D square lattice $z = 4$, and $z = 3$ for 2D hexagonal lattice. Since $|\mathrm{VBS}\rangle$ accommodates spin S at each site, the total spin J of any neighboring spin pair can take $J = S \otimes S = 0, 1, 2, \cdots, 2S$. In the VBS case, however, the highest value of J is smaller than $2S$ and the spin-pair, when decomposed into irreducible representations, does not contain the following J-values

$$J_{\max} < J \leq 2S \ \ (J_{\max} = 2S - M) \tag{4}$$

since there are already M singlet valence bonds on the pair. From this, one readily sees that the state $|\mathrm{VBS}\rangle$ is annihilated by the spin-J projection operator P_J

$$\begin{aligned} P_J(i,j) &= \prod_{J' \neq J} \frac{\{\mathbf{S}(i) + \mathbf{S}(j)\}^2 - J'(J'+1)}{J(J+1) - J'(J'+1)} \\ &= \prod_{J' \neq J} \frac{2\mathbf{S}(i)\cdot\mathbf{S}(j) + 2S(S+1) - J'(J'+1)}{J(J+1) - J'(J'+1)} \end{aligned} \tag{5}$$

with $J_{\max} < J \leq 2S$. Therefore, the parent Hamiltonian realizing $|\mathrm{VBS}\rangle$ as its exact zero-energy ground state is written as

$$\mathcal{H} = \sum_{<ij>} \sum_{J > J_{max}} V_J P_J(i,j), \tag{6}$$

where V_J are arbitrary positive constants to guarantee the energy eigenvalues to be positive semi-definite. It is noticed that this construction applies on any lattice and in any dimensions. In 1D ($z = 2$), the parent

Hamiltonians for $M = 1, 2, 3$ respectively read:[9]

$$\mathcal{H}_{S=1} = \sum_i \left\{ \mathbf{S}(i)\cdot\mathbf{S}(i+1) + \frac{1}{3}(\mathbf{S}(i)\cdot\mathbf{S}(i+1))^2 + \frac{2}{3} \right\}, \tag{7a}$$

$$\mathcal{H}_{S=2} = \sum_i \left\{ \mathbf{S}(i)\cdot\mathbf{S}(i+1) + \frac{2}{9}(\mathbf{S}(i)\cdot\mathbf{S}(i+1))^2 + \frac{1}{63}(\mathbf{S}(i)\cdot\mathbf{S}(i+1))^3 + \frac{10}{7} \right\}, \tag{7b}$$

$$\mathcal{H}_{S=3} = \sum_i \left\{ \mathbf{S}(i)\cdot\mathbf{S}(i+1) + \frac{59}{358}(\mathbf{S}(i)\cdot\mathbf{S}(i+1))^2 + \frac{19}{1611}(\mathbf{S}(i)\cdot\mathbf{S}(i+1))^3 + \frac{1}{3222}(\mathbf{S}(i)\cdot\mathbf{S}(i+1))^4 + \frac{396}{179} \right\}, \tag{7c}$$

where we have chosen the coefficients V_J in such a way that $\mathcal{H}_S$ consists only of $(S+1)$-th order polynomials and the coefficient of $\mathbf{S}(i)\mathbf{S}(i+1)$ is unity.

The above 1D VBS state have several important features:

- The spin correlation is short-ranged and exponentially dumped with the correlation length $\xi = 1/\ln\left(\frac{S+2}{S}\right)$. In this sense, the VBS states are disordered spin liquids.
- Gapped triplon (*crackion*) excitations. The exponential decay of spin correlations suggests gapped excitations over the VBS ground states. A model of such excitations, sometimes dubbed *crackions*, may be found by replacing one of the singlet valence bonds to a triplet. Remarkably, the crackions are equivalent to the Feynman states created by (the Fourier-transform of) the spin operators $S^a(k)$.[9,15] It has been shown[9] that the gap of these crackions (or equivalently, the Feynman frequencies of the single-mode approximation) decreases exponentially as spin S is increased.
- Edge states and ground-state degeneracy. The finite-length VBS chain has effective spin-$S/2$ degrees of freedom at each of the two edges, which lead to the $(S+1)\times(S+1)$-fold degenerate ground states.[16] These degenerate ground states converge to a unique ground state in the infinite length limit.[4]
- Hidden (string) order. The exponential decay of spin correlation also implies that the VBS state is a disordered spin liquid with no local order parameter. Still, however, there is a kind of order in the spin-S VBS state as will be explained in section 6. In the $S = 1$

case, it is well-known that this 'hidden' order is measured by the non-local string operator parameter[17]

$$\langle S_z(i)\exp\Big(i\pi\sum_{k=i}^{j-1}S_z(k)\Big)S_z(j)\rangle. \tag{8}$$

3. Supersymmetric Valence Bond Solid Models

Now let us add hole degrees of freedom to the VBS state. An extension of the VBS model including spinless (bosonic) holes has been considered in Ref. 18. A more realistic model with a spin-1/2 hole has been introduced[19] in the context of the midgap state found in nickelates. Another route to adding holes is to extend the bosonic SU(2) symmetry to something else which include fermionic states as well. In the first case, we replace the SU(2) symmetry of the VBS states with UOSp(1|2) supersymmetry to construct the supersymmetric VBS (SVBS) states.

3.1. *UOSp(1|2) supersymmetry*

The supersymmetry is a transformation between bosons and fermions. For UOSp(1|2) algebra, we have five generators, three of which are bosonic S_a ($a=1,2,3$), and the other two are fermionic K_μ ($\mu=1,2$). They satisfy the following (anti)commutation relations

$$[S_a,S_b]=i\epsilon_{abc}S_c,\quad [S_a,K_\mu]=\frac{1}{2}(\sigma_a)_{\nu\mu}K_\nu,\quad \{K_\mu,K_\nu\}=\frac{1}{2}(\epsilon\sigma_a)_{\mu\nu}S_a, \tag{9}$$

where $\epsilon=i\sigma_2$. The UOSp(1|2) Casimir given by

$$\mathcal{S}_A\mathcal{S}_A=S_aS_a+\epsilon_{\mu\nu}K_\mu K_\nu, \tag{10}$$

has the eigenvalue $\widetilde{S}(\widetilde{S}+\frac{1}{2})$ with the superspin index $\widetilde{S}=0,1/2,1,3/2,2,\cdots$. The corresponding irreducible representation is $(4\widetilde{S}+1)$-dimensional. The Clebsch-Gordan decomposition of the product of two identical irreducible representations is given by

$$\widetilde{S}\otimes\widetilde{S}=0\oplus\frac{1}{2}\oplus1\oplus\frac{3}{2}\oplus2\oplus\cdots\oplus2\widetilde{S}\,. \tag{11}$$

Note that the difference of superspins is not 1 but 1/2, unlike the SU(2) case. The three-dimensional fundamental representation of the UOSp(1|2) generators are

$$s_a=\frac{1}{2}\begin{pmatrix}\sigma_a & 0\\ 0 & 0\end{pmatrix},\qquad \kappa_\mu=\frac{1}{2}\begin{pmatrix}\mathbf{0} & \frac{1}{x^2}\tau_\mu\\ -(\epsilon\tau_\mu)^t & 0\end{pmatrix}, \tag{12}$$

where $\tau_1 = (1,0)^t$ and $\tau_2 = (0,1)^t$ (t denotes transposition); x is an arbitrary parameter*. A convenient way of realizing the fully symmetric representation is to adopt the oscillator representation:

$$S_a = \hat{\psi}^\dagger s_a \hat{\psi}, \qquad K_\mu = \hat{\psi}^\dagger \kappa_\mu \hat{\psi}, \tag{13}$$

where $\hat{\psi} = (a, b, x\, f)^t$ is the Schwinger superoperators whose first two components are the standard Schwinger bosons and last one is fermion:

$$[a, a^\dagger] = [b, b^\dagger] = \{f, f^\dagger\} = 1. \tag{14}$$

Then, the UOSp(1|2) Casimir operator (10) is expressed as

$$C = S_A S_A = \sum_{a=1,2,3} S_a S_a + \sum_{\mu=1,2} \epsilon_{\mu\nu} K_\mu K_\nu = \frac{1}{4} N(N+1), \tag{15}$$

where N is the total particle number:

$$N = n_a + n_b + n_f = a^\dagger a + b^\dagger b + f^\dagger f\ , \tag{16}$$

which is related to the superspin as $\widetilde{S} = \frac{1}{2}N$. Since the SU(2) spin magnitude is given by $S = \frac{1}{2}(a^\dagger a + b^\dagger b)$, the UOSp(1|2) multiplet consists of two SU(2) multiplets of $S = \frac{1}{2}N = \widetilde{S}$ (when $n_f = 0$) and $S = \frac{1}{2}(N-1) = \widetilde{S} - 1/2$ ($n_f = 1$),

$$\frac{1}{\sqrt{m_1! m_2!}} a^{\dagger m_1} b^{\dagger m_2} |0\rangle, \qquad \frac{1}{\sqrt{n_1! n_2!}} a^{\dagger n_1} b^{\dagger n_2} f^\dagger |0\rangle, \tag{17}$$

where $m_1 + m_2 = n_1 + n_2 + 1 = N$, and thus the dimension of UOSp(1|2) multiplet is $(N+1) + (N) = 2N + 1 = 4\widetilde{S} + 1$. For $N = 1$, the states spanned by (17) are given by

$$\underbrace{a^\dagger|0\rangle, \quad b^\dagger|0\rangle}_{S=1/2}, \quad \underbrace{f^\dagger|0\rangle}_{S=0}\ . \tag{18}$$

Similarly, for $N = 2$, the $\widetilde{S} = 1$ (five-dimensional) irreducible representation is spanned by (see Fig. 1)

$$\underbrace{\frac{1}{\sqrt{2}} a^{\dagger 2}|0\rangle,\ a^\dagger b^\dagger|0\rangle,\ \frac{1}{\sqrt{2}} b^{\dagger 2}|0\rangle}_{S=1}, \quad \underbrace{a^\dagger f^\dagger|0\rangle,\ b^\dagger f^\dagger|0\rangle}_{S=1/2} \tag{19}$$

consists of two SU(2) representations ($S = 1$ and $S = 1/2$), which are connected with each other by the fermionic generators $K_{1,2}$. In the next

*In the usual matrix realization of the UOSp(1|2) generators, x is taken as 1, such that κ_μ satisfy $\kappa_\mu^\ddagger = \epsilon_{\mu\nu}\kappa_\nu$ under the super-adjoint $\ddagger$: $\begin{pmatrix} B_1 & F_1 \\ F_2 & B_2 \end{pmatrix}^\ddagger = \begin{pmatrix} B_1^\dagger & F_2^\dagger \\ -F_1^\dagger & B_2^\dagger \end{pmatrix}$.

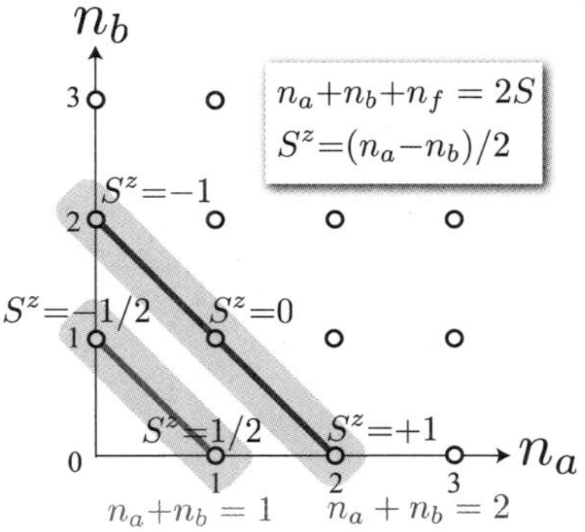

Fig. 1. The numbers (n_a, n_b) of Schwinger bosons and states of the five-dimensional $\widetilde{S}=1$ representation. The fermionic K operators connect $S=1/2$ $(n_a+n_b=1)$ and $S=1$ $(n_a+n_b=2)$.

subsection, we will use the $S=1/2$- and the $S=1$ subspace as the states with and without a hole f.

3.2. *UOSp(1|2) SVBS states*

Having defined the five states which transform under UOSp(1|2), let us construct the supersymmetric version of the VBS state. As the original VBS state is constructed by placing singlet valence bonds between adjacent sites, a natural extension would be to replace the SU(2)-singlet valence bonds with the singlet combination of the Schwinger *super* operators:

$$|\mathrm{SVBS}\rangle = \prod_{<ij>} (a_i^\dagger b_j^\dagger - b_i^\dagger a_j^\dagger - r f_i^\dagger f_j^\dagger)^M |0\rangle, \tag{20}$$

where r should satisfy the condition $r=x^2$ in order for $|\mathrm{SVBS}\rangle$ to be invariant under the UOSp(1|2) transformation (13). One may find that the above expression is antisymmetric under the interchange $i \leftrightarrow j$ (note that f_i and f_j anti-commute). Physically, the parameter r represents the concentration of hole pairs $f_i^\dagger f_j^\dagger$ doped into the standard VBS state†. To see the physical meaning of SVBS states, we consider the spin-hole coherent state representation[20] of the SVBS states. In the spin-hole coherent representation, the Schwinger superoperators $(a^\dagger, b^\dagger, f^\dagger)$ are replaced by the set of numbers (u, v, θ), in which u and v are the usual complex numbers, while

†For early attempts at including the hole degrees of freedom into the VBS state, see, for instance, Refs. 18,19.

θ is Grassmann odd quantity. Then, the 'wave function' of $|\text{SVBS}\rangle$ reads

$$\Psi_{\text{SVBS}}(\{u,v,\theta\};r) = \prod_i (u_i\, v_{i+1} - v_i\, u_{i+1} - r\,\theta_i\theta_{i+1})^M. \tag{21}$$

Carrying out a formal expansion[‡] of Ψ_{SVBS} with respect to the parameter r, we obtain a superposition of many different VBS states doped with hole pairs:

$$\Psi_{\text{SVBS}} = \Phi_{\text{VBS}} - r\sum_i \frac{\theta_i\theta_{i+1}}{u_iv_{i+1}-v_iu_{i+1}}\Phi_{\text{VBS}} + \frac{1}{2}r^2\sum_i\left(\frac{\theta_i\theta_{i+1}}{u_iv_{i+1}-v_iu_{i+1}}\right)^2\Phi_{\text{VBS}}$$
$$+\cdots+(-r)^N\theta_1\theta_2\cdots\theta_N\left(\Phi_{\text{MG}}^{\text{even}} - \Phi_{\text{MG}}^{\text{odd}}\right), \tag{22}$$

where Φ_{VBS} is the coherent representation of the original VBS state obtained by setting $r=0$ in eq.(21)

$$\Phi_{\text{VBS}} = \prod_i (u_iv_{i+1} - v_iu_{i+1})^M, \tag{23}$$

and $\Phi_{\text{MG}}^{\text{even (odd)}}$ is (the coherent representation of) the Majumdar-Ghosh dimer state:[21,22]

$$\Phi_{\text{MG}}^{\text{even(odd)}} = \prod_{i=\text{even(odd)}} (u_iv_{i+1} - v_iu_{i+1}). \tag{24}$$

In the expansion (22), the leading term is the original VBS state. In the second- and the third term, the VBS state added with one- and two hole pairs appear, respectively. Repeating this procedure, we finally obtain the dimer states in which the hole pairs appear on every other bond (see Fig. 2).

It would be interesting to see the analogy between $|\text{SVBS}\rangle$ and the BCS state of superconductivity[§]. The BCS state is given by

$$|\text{BCS}\rangle = \prod_k (1 + g_k c^\dagger_{k\uparrow}c^\dagger_{-k\downarrow})|0\rangle \tag{25}$$

and is expanded as[1]

$$|\text{BCS}\rangle = |0\rangle + \sum_k g_kc^\dagger_{k\uparrow}c^\dagger_{-k\downarrow}|0\rangle + \frac{1}{2}(\sum_k g_kc^\dagger_{k\uparrow}c^\dagger_{-k\downarrow})^2|0\rangle + \cdots + \prod_k g_kc^\dagger_k|0\rangle, \tag{26}$$

---

‡This expansion can also be regarded as the Grassmann coordinate expansion in superfield formalism

§There also exist interesting analogies between the VBS state and the Laughlin wavefunction in quantum Hall effect.[5] The original AKLT VBS state corresponds to the Laughlin-Haldane wavefunction on a two-sphere, and the SVBS states to the supersymmetric Laughlin-Haldane wavefunction on a supersphere.[23]

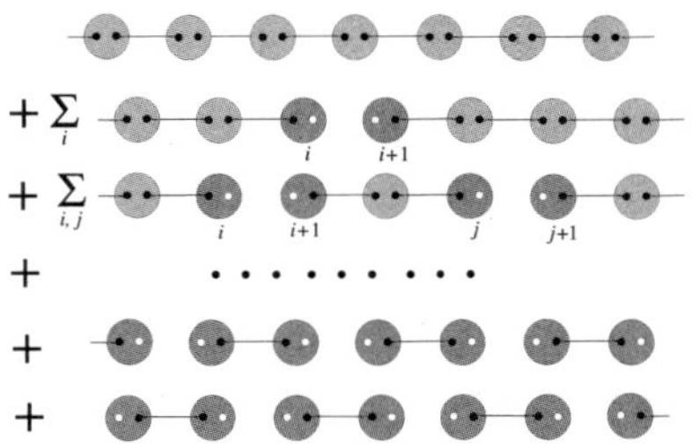

Fig. 2. The expansion of the SVBS states in the number of doped holes. The black circles represent bosons, while white circles holes. The black lines between neighboring sites signify the VB, *i.e.*, spin singlet bond.

where g_k is the coherence factor and $c^\dagger_{k\sigma}$ is the creation operator for electron with momentum k and spin σ. The BCS state is a superposition of the electron pairs, representing uncertainty of the electron number particular to the superconducting state. One may find correspondence: $r \leftrightarrow g_k$ and $\theta_i\theta_{i+1} \leftrightarrow c^\dagger_{k\uparrow}c^\dagger_{-k\downarrow}$. Then, the SVBS chain is expected to exhibit the superconducting property with finite amount of the hole doping. To summarize, SVBS states have the following properties:

- $r = 0$: the original VBS state (insulating)
- $r =$finite : superconducting, hole number fluctuates
- $r = \infty$: the Majumdar-Ghosh dimer state (insulating).

This realizes the property mentioned in Introduction: Hole-doping renders phase transition from antiferromagnetic insulating state to superconducting phase.

3.3. *Supersymmetric parent Hamiltonian*

The parent Hamiltonians for the SVBS states are constructed analogously to those of the original VBS states. First we note from eq.(11) that the total superspin $\widetilde{J}$ of a pair of two superspins $\widetilde{S}$ can take half-integer values between 0 and $2\widetilde{S}$. Again, the existence of M SUSY-singlet valence bonds restricts the maximal value of $\widetilde{J}$ to $\widetilde{J}_{\max} = 2\widetilde{S} - M$ with $\widetilde{S} = \frac{1}{2}zM$.

The strategy used in constructing the parent Hamiltonian of the usual VBS states in section 2 is carried over to the SUSY case; all we have to do is to replace the SU(2) operators in the projection operators (5) with the UOSp(1|2) operators. Specifically, the SUSY projection operator onto the total spin $\widetilde{J}$ is explicitly given by

$$\mathcal{P}_{\widetilde{J}}(i,j) = \prod_{\widetilde{J}'\neq\widetilde{J}} \frac{(\mathcal{S}_A(i)+\mathcal{S}_A(j))^2 - \widetilde{J}'(\widetilde{J}'+\frac{1}{2})}{\widetilde{J}(\widetilde{J}+\frac{1}{2}) - \widetilde{J}'(\widetilde{J}'+\frac{1}{2})}$$
$$= \prod_{\widetilde{J}'\neq\widetilde{J}} \frac{2\mathcal{S}_A(i)\mathcal{S}_A(j) + 2\widetilde{S}(\widetilde{S}+\frac{1}{2}) - \widetilde{J}'(\widetilde{J}'+\frac{1}{2})}{\widetilde{J}(\widetilde{J}+\frac{1}{2}) - \widetilde{J}'(\widetilde{J}'+\frac{1}{2})}, \tag{27}$$

where $\mathcal{S}_A(i)\mathcal{S}_A(j) \equiv S_a(i)S_a(j) + \epsilon_{\mu\nu}K_\mu(i)K_\nu(j)$ and $\widetilde{J}'$ runs over all half-integer values between 0 and $2\widetilde{S}$. By using them, the parent Hamiltonian for SVBS chain is constructed as¶

$$\mathcal{H} = \sum_{<ij>} \sum_{\widetilde{J}>\widetilde{J}_{\max}} V_{\widetilde{J}} \mathcal{P}_{\widetilde{J}}(i,j). \tag{29}$$

For instance, the parent Hamiltonian for the $\widetilde{S}=1$ SVBS chain reads

$$\mathcal{H}_{\widetilde{S}=1} = \sum_i \left\{ V_{3/2}\mathcal{P}_{3/2}(i,i+1) + V_2\mathcal{P}_2(i,i+1) \right\}$$
$$= \sum_i \left\{ \frac{1}{35}(5V_2 + 63V_{3/2})(\mathcal{S}_A(i)\mathcal{S}_A(i+1)) + \frac{2}{45}(9V_2 - 7V_{3/2})(\mathcal{S}_A(i)\mathcal{S}_A(i+1))^2 \right.$$
$$+ \frac{16}{45}(V_2 - 5V_{3/2})(\mathcal{S}_A(i)\mathcal{S}_A(i+1))^3$$
$$\left. + \frac{32}{315}(V_2 - 7V_{3/2})(\mathcal{S}_A(i)\mathcal{S}_A(i+1))^4 + V_{3/2} \right\}. \tag{30}$$

4. Matrix Product Representation

It is well-known[8,9] that the 1D spin-S VBS states are conveniently represented as products of $(S+1)$-dimensional matrices, whose elements are local spin-S states. One of the greatest merits of using the matrix-product (MP) representation is that the concept of edge states and a certain kind of hidden order inherent in the VBS states are naturally incorporated. In this section, we construct the supersymmetric MP representation of the SUSY

¶The Hamiltonian (29) is *not* Hermitian because of the non-Hermitian fermion generators. A natural Hermitian Hamiltonian can be constructed as

$$H_{\widetilde{S}=1} = \sum_i \sum_{\widetilde{J}>\widetilde{J}_{max}} V_{\widetilde{J}} \mathcal{P}^\dagger_{\widetilde{J}}(i,i+1)\mathcal{P}_{\widetilde{J}}(i,i+1). \tag{28}$$

If the projection operators were Hermitian and satisfied $\mathcal{P}^\dagger_J\mathcal{P}_J = \mathcal{P}^2_J = \mathcal{P}_{\widetilde{J}}$, $H_{\widetilde{S}=1}$ would reduce to the original parent Hamiltonian (29). In this sense, $H_{\widetilde{S}=1}$ is a natural Hermitian counterpart of $\mathcal{H}_{\widetilde{S}=1}$.

VBS state $|\text{SVBS}\rangle$. In the rest of this paper, we will restrict ourselves to the 1D case, although the (S)VBS states can be defined on any lattice and in any dimension.

4.1. *Matrix-product formalism*

There are several ways to introduce the matrix-product states (MPS) (see Ref. 10 for a recent review on MPS). In our case, the simplest way is to use the Schwinger-superoperator representation (20).

Before going to the SUSY case, let us illustrate how the MP representation of the usual VBS states is obtained from their Schwinger-boson representation.[9] Defining short-hand notations $\hat{\phi}^t = (a^\dagger, b^\dagger)$ and $\epsilon = i\sigma^2$, the valence bond can be compactly written as $a_i^\dagger b_j^\dagger - b_i^\dagger a_j^\dagger = \hat{\phi}_i^t \epsilon \hat{\phi}_j$ and the VBS state can be written as[9]

$$\begin{aligned}
|\text{VBS}\rangle_{\alpha\beta} &= (\epsilon\phi_1)^\alpha \prod_{i=1}^{L-1} (\phi_i^t \epsilon \phi_{i+1}) \phi_L^\beta |0\rangle \\
&= (\epsilon\phi_1)^\alpha (\phi_1^t \epsilon \phi_2) \cdots (\phi_{i-1}^t \epsilon \phi_i) \cdot (\phi_i^t \epsilon \phi_{i+1}) \cdots (\phi_{L-1}^t \epsilon \phi_L) \phi_L^\beta \\
&= (\epsilon\phi_1)^\alpha \phi_1^t \cdots (\epsilon\phi_i \cdot \phi_i^t) \cdots \cdot (\epsilon \phi_L \phi_L^\beta) \\
&= (\epsilon\phi_1)^\alpha \phi_1^t \left(\prod_{i=2}^{L-1} g_i \right) (\epsilon \phi_L \phi_L^\beta) |0\rangle \\
&= (g_1 g_2 \cdots g_{L-1} g_L)_{\alpha\beta} ,
\end{aligned} \tag{31}$$

where the local matrix state on the site-i is defined as

$$g_i = \epsilon\, \phi_i \cdot \phi_i^t |0\rangle = \begin{pmatrix} a_i^\dagger b_i^\dagger |0\rangle & {b_i^\dagger}^2 |0\rangle \\ -{a_i^\dagger}^2 |0\rangle & -a_i^\dagger b_i^\dagger |0\rangle \end{pmatrix} \tag{32}$$

and the Schwinger bosons on the left (right) edge $\phi_{1(L)}^\alpha$ ($\alpha = 1, 2$) represent the effective $S = 1/2$ degrees of freedom. Thus, the VBS states are represented by a product of 2×2 matrices each matrix elements of which corresponds to one of the four-fold degenerate ground states stemming from the two effective spin-1/2 degrees of freedom on both edges.

A similar argument shows that for the periodic chain the VBS chain is expressed as $((a_{L+1}, b_{L+1}) = (a_1, b_1))$

$$\begin{aligned}
|\text{VBS}\rangle_{\text{PBC}} &= \prod_{j=1}^{L} (a_j^\dagger b_{j+1}^\dagger - b_j^\dagger a_{j+1}^\dagger) \\
&= \operatorname{tr}(g_1 g_2 \cdots g_L) .
\end{aligned} \tag{33}$$

With such matrix-product expressions, calculation of norms and various expectation values reduces to calculating strings of the following 4×4 'transfer matrix':[8–10]

$$G = g_i^\dagger \otimes g_i \ , \ \mathcal{O}_m = g_m^\dagger \otimes (O_m g_m) \ . \tag{34}$$

4.2. *SMPS formalism*

Now let us generalize the above formalism for the bosonic MPS to the supersymmetric MPS (SMPS).[12] The SMPS is formulated, basically by replacing trace with supertrace‖ in MPS:

$$\mathrm{tr} \to \mathrm{str} \ . \tag{36}$$

With $\psi^t = (a^\dagger, b^\dagger, \sqrt{r} f^\dagger)$ and

$$\mathcal{R} = \begin{pmatrix} 0 & 1 & 0 \\ -1 & 0 & 0 \\ 0 & 0 & -1 \end{pmatrix} , \tag{37}$$

the SVB is concisely expressed as

$$a_i^\dagger b_j^\dagger - b_i^\dagger a_j^\dagger - r f_i^\dagger f_j^\dagger = \psi_i^t \mathcal{R} \psi_j , \tag{38}$$

and the SVBS state is

$$\begin{aligned} |\mathrm{SVBS}\rangle_{\alpha\beta} &= (\mathcal{R}\psi_1)^\alpha \prod_{i=1}^{L-1} (\psi_i^t \mathcal{R} \psi_{i+1}) \psi_L^\beta |0\rangle = (\mathcal{R}\psi_1)^\alpha \psi_1^t \cdot (\prod_i g_i) \cdot (\mathcal{R}\psi_L \psi_L^\beta)|0\rangle \\ &= (g_1 g_2 \cdots g_L)_{\alpha\beta} , \end{aligned} \tag{39}$$

where

$$g_i = \mathcal{R}\psi_i \cdot \psi_i^t |0\rangle = \begin{pmatrix} a_i^\dagger b_i^\dagger |0\rangle & {b_i^\dagger}^2 |0\rangle & \sqrt{r} b_i^\dagger f_i^\dagger |0\rangle \\ -{a_i^\dagger}^2 |0\rangle & -a_i^\dagger b_i^\dagger |0\rangle & -\sqrt{r} a_i^\dagger f_i^\dagger |0\rangle \\ -\sqrt{r} a_i^\dagger f_i^\dagger |0\rangle & -\sqrt{r} b_i^\dagger f_i^\dagger |0\rangle & 0 \end{pmatrix} . \tag{40}$$

The edge operators $\mathcal{R}\psi_1 = (b_1^\dagger, -a_1^\dagger, -\sqrt{r} f_1^\dagger)$ and ψ_L appearing respectively on the left- and the right edges represent the three possible edge states

‖The supertrace is defined as

$$\mathrm{str}\left\{ \begin{pmatrix} B & F \\ F' & B' \end{pmatrix} \right\} \equiv \mathrm{tr}(B) - \mathrm{tr}(B') , \tag{35}$$

where B and B' are bosonic block components and F and F' are fermionic block components of supermatrix.

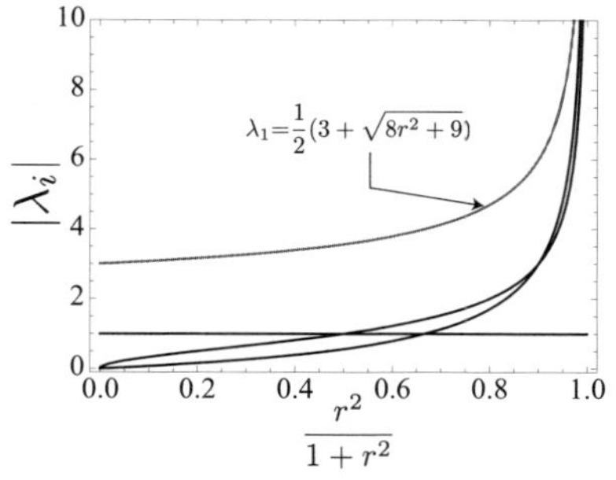

Fig. 3. Plot of the absolute values of the eigenvalues of G.

(spin-up/down and hole). From eqs.(39) and (40), one can readily see the nine-fold degeneracy of the ground state.

For a periodic chain, the fermion sign has to be treated carefully. Using the identity $\psi_L^t \mathcal{R}\psi_1 = \mathrm{str}(\mathcal{R}\psi_1\psi_L^t)$, we can express the SVBS state as

$$\begin{aligned}|\mathrm{SVBS}\rangle_{\mathrm{PBC}} &= \underbrace{\prod_{i=1}^{L-1}(\psi_i^t \mathcal{R}\psi_{i+1})\,\psi_L^\alpha(\mathcal{R}\psi_1)^\alpha}_{\text{Grassmann even}}|0\rangle \\ &= \mathrm{str}\left\{(\mathcal{R}\psi_1)\prod_{i=1}^{L-1}(\psi_i^t\mathcal{R}\psi_{i+1})\psi_L^t\right\} = \mathrm{str}(g_1 g_2 \cdots g_L).\end{aligned} \tag{41}$$

As in the case of the usual (bosonic) VBS states, the inner products and the expectation values of the SVBS states are calculated in terms of the c-numbered transfer matrix

$$G = g_i^\dagger \otimes g_i. \tag{42}$$

as

$$\langle \mathrm{SVBS}|\mathrm{SVBS}\rangle \sim \prod_i G_i = G^L\ . \tag{43}$$

The (moduli of the) eigenvalues of the transfer matrix G are shown in Fig. 3. The calculation of two point functions of SVBS states is performed almost in the same manner as in the original VBS states except that special care about anticommuting fermionic components has to be taken.

5. Magnetic- and Superconducting Properties: "Bosons v.s. Fermions"

The SVBS state constructed in sections 3 and 4 contains both bosonic- and fermionic degrees of freedom; the "bosons" represent spin degrees of

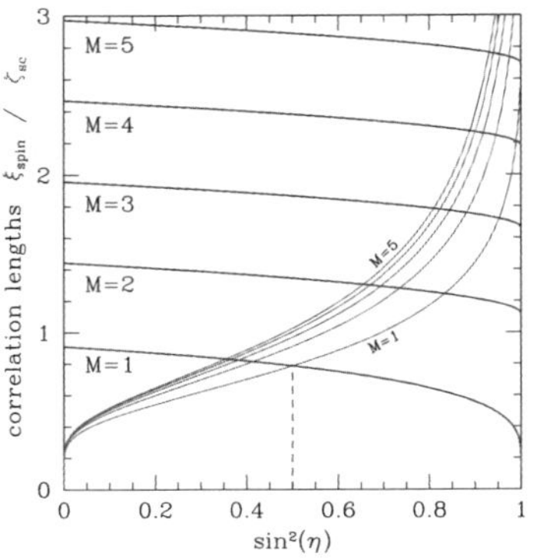

Fig. 4. The spin correlation and the superconducting correlation exactly coincide at the "supersymmetric point" $r = 1$. The spin correlation decreases with increase of hole doping r, since hole doping subtracts the spin degrees of freedom. For $M = 1$ SVBS states, at $r = \infty$, the resulting state is the Majumdar-Ghosh dimer state in which the spin correlation completely vanishes. For $M \geq 2$, spin correlation is finite even at $r = \infty$, since VBS chains still exist at the limit.

freedom and the "fermions" holes. Therefore, intuitively, the spin correlation reflects coherence of bosons, while the hole correlation does that of fermions. In this section, we investigate the hole concentration dependence of the two different properties of bosons and fermions, *i.e.* spin correlation and superconducting correlation.

We begin with the spin correlation function for $\widetilde{S} = 1$ SVBS chain:

$$\langle S_z(i)S_z(i+n)\rangle = \begin{cases} \frac{2\left(r^2+\sqrt{8r^2+9}+3\right)}{\sqrt{8r^2+9}\left(\sqrt{8r^2+9}+3\right)} & \text{for } n=0 \\ \frac{13r^2+24+\left(r^2+8\right)\sqrt{8r^2+9}}{2\sqrt{8r^2+9}\left(\sqrt{8r^2+9}+3\right)}\left(-\frac{2}{\sqrt{8r^2+9}+3}\right)^n & \text{for } n>0\,. \end{cases} \tag{44}$$

The exponentially decaying factor defines the correlation length:

$$\xi_{\text{spin}}(r)^{-1} \equiv \log\left\{\frac{\sqrt{8r^2+9}+3}{2}\right\}, \tag{45}$$

which monotonically decreases with increase of r (see Fig. 4). In the original VBS limit $r \to 0$, (44) is reduced to yield the well-known results:[5]

$$\langle S_z(i)S_z(i+n)\rangle = \begin{cases} \frac{2}{3} & \text{for } n=0 \\ \frac{4}{3}\left(\frac{-1}{3}\right)^n & \text{for } n>0\,. \end{cases} \tag{46}$$

Similarly, the superconducting correlation function is given by[11]

$$\Delta_{i,i+n} = \langle (a_i b_{i+n} - b_i a_{i+n}) f_i^\dagger f_{i+n}^\dagger \rangle, \tag{47}$$

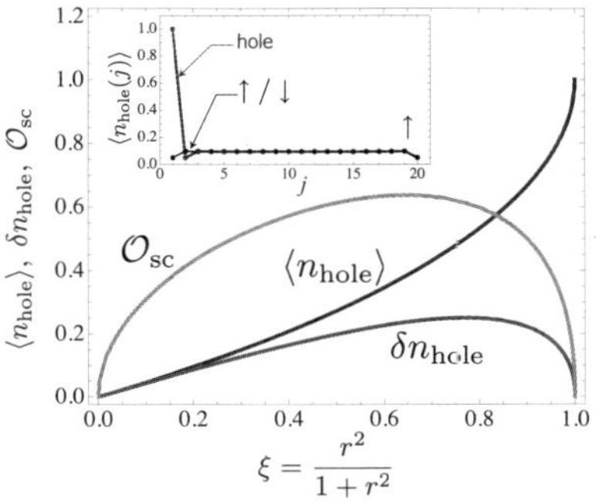

Fig. 5. Behaviors of the superconducting order parameter, the fermion (hole) number and its fluctuation. At $r = 0$ and ∞, SVBS chain is reduced to VB states, *i.e.* antiferromagnetic insulating states. With finite amount of hole-doping ($r \neq 0$), SVBS chain exhibits superconducting property.

and its correlation length is calculated as

$$\xi_{\text{SC}}(r)^{-1} = \log\left\{\frac{\sqrt{8r^2+9}+3}{\sqrt{8r^2+9}-3}\right\}. \tag{48}$$

The two correlation lengths ξspin and ξ_{SC} are plotted in Fig. 4 along with those for other values of M. Intuitively, the superconducting coherence reflects the fermionic property of the system, while the spin-spin coherence reflects the bosonic one. It is interesting to observe the spin correlation length and the superconducting one coincide at the supersymmetric point $r = 1$.

The superconducting order parameter

$$O_{\text{SC}} \equiv \Delta_{i,i+1} = \langle (a_i b_{i+1} - b_i a_{i+1}) f_i^\dagger f_{i+1}^\dagger \rangle \tag{49}$$

and the hole-number fluctuation are plotted in Fig. 5. With finite amount of hole doping, the SVBS state exhibits superconducting property, which indeed demonstrates the favorable property mentioned in subsection 3.2.

6. Hidden String Order

Since the spin correlation function dumps exponentially, the SVBS state is regarded as a disordered spin liquid with no local order parameter. However, if we closely look at the sequence of z-component of spins, we readily see that the spin configuration in fact is not random and suggests the existence of a hidden order (see Fig. 6). One useful way of detecting the hidden order

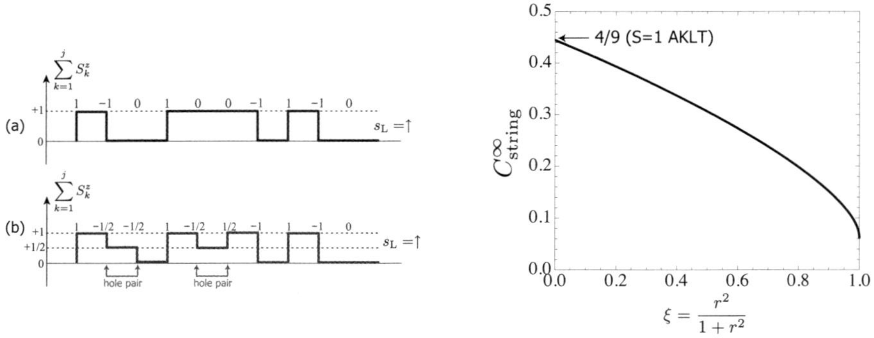

Fig. 6. (left) (a) S_z sequence on VBS chain and (b) that on SVBS. In (a), $S_z = +1$ and -1 alternately appear with neglecting zeros, and sum of S^z fluctuates only between $+1$ and -1. This is the hidden order of VBS state. In (b), $S_z = \frac{1}{2}$ and $-\frac{1}{2}$ can appear, but still the sum of S^z fluctuates only between $+1$ and -1. (right) The behavior of the string order of the SVBS state. The string order decreases with increase of hole doping. At $r = \infty$, the string order takes minimum but non-zero value, representing remaining finite string order of dimer state.

is to use the string correlation function:[17]

$$C_n^{\text{string}} = \langle S_z(i) \exp(i\pi \sum_{k=i}^{i+n-1} S_z(k)) S_z(i+n) \rangle. \tag{50}$$

For $\widetilde{S} = 1$ SVBS chain, the string order parameter is calculated as[12]

$$\mathcal{O}_{\text{string}}(r) \equiv \lim_{n \nearrow \infty} C_n^{\text{string}} = \frac{4(r^4 + 14r^2 + 18 + 2(r^2+3)\sqrt{8r^2+9})}{(8r^2+9)(\sqrt{8r^2+9}+3)^2}. \tag{51}$$

In particular,

$$\mathcal{O}_{\text{string}}(r)\Big|_{r\to 0} = \frac{4}{9}, \qquad \mathcal{O}_{\text{string}}(r)\Big|_{r\to\infty} = \frac{1}{16}, \tag{52}$$

which is consistent with the known result in Refs. 24–26.

7. Lowlying Excitations

The exponentially dumping behavior of spin correlation also suggests the existence of gapped excitations (crackions) on the ground state. We use the single-mode approximation to evaluate the spectra of crackions. On the SVBS states, we have two kinds of crackions: *triplon* and *spinon-holon pair*, which are produced by acting the S_a $(a = x, y, z)$ and K_μ $(\mu = 1, 2)$ operators to the SVBS state, respectively (see also Fig. 7):

$$S_a(k)|\text{SVBS}\rangle, \quad K_\mu(k)|\text{SVBS}\rangle, \tag{53}$$

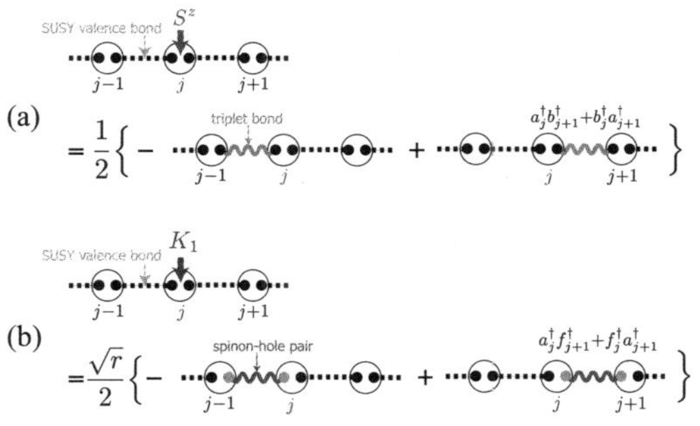

Fig. 7. Making two kinds of crackion: (a) triplon and (b) spinon-hole pair.

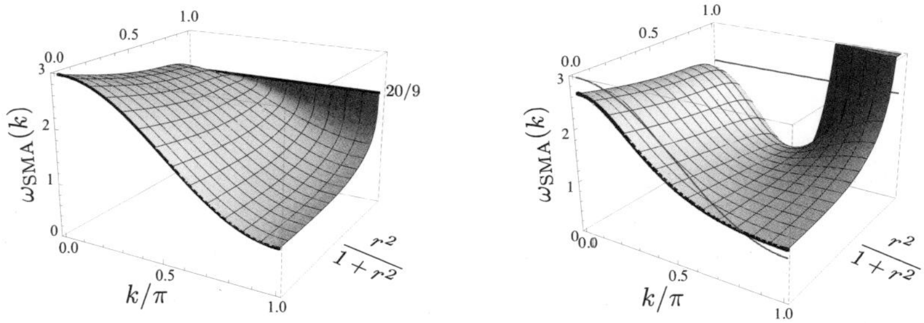

Fig. 8. The spectrum of the triplon. At $r = 0$, the curve exactly reproduces the spectrum of the triplon on the original VBS chain. At $r = \infty$, the dispersion relation becomes trivial, since the ground state becomes the dimer state and the triplon excitation cannot move on the chain.

where $S_a(k)$ and $K_\mu(k)$ are Fourier transform of $S_a(j)$ and $K_\mu(j)$. The approximate dispersion relations of these excitations are shown in Fig. 8.

8. $SU(2|1)$ SVBS State

It is straightforward to extend the previous SVBS states to the ones with larger symmetry. If one more fermion g is added, the following $SU(2|1)$ SVBS state is obtained:

$$|\text{SVBS-II}\rangle = \prod_{<ij>} (a_i^\dagger b_j^\dagger - b_i^\dagger a_j^\dagger - r f_i^\dagger g_j^\dagger - r g_i^\dagger f_j^\dagger)^M |0\rangle. \tag{54}$$

The above SU(2|1) SVBS state** is also antisymmetric under interchange between the sites i and j, and has the SU(2|1) symmetry which is larger than UOSp(1|2). In $SU(2|1)$ SVBS states, we have two species of bosons (a and b) and two species of fermions (f and g) and more symmetric behaviors of bosons and fermions are expected to be observed in SU(2|1) SVBS.

For $M = 1$, the SU(2|1) SVBS chain has the local Hilbert space spanned by the basis elements:

$$\underbrace{\frac{1}{\sqrt{2}}{a^\dagger}^2|0\rangle, \quad a^\dagger b^\dagger|0\rangle, \quad \frac{1}{\sqrt{2}}{b^\dagger}^2|0\rangle,}_{\text{spin-1}}$$
$$\underbrace{a^\dagger f^\dagger|0\rangle, \quad b^\dagger f^\dagger|0\rangle, \quad a^\dagger g^\dagger|0\rangle, \quad b^\dagger g^\dagger|0\rangle,}_{\text{spin-1/2}} \tag{56}$$
$$\underbrace{f^\dagger g^\dagger|0\rangle}_{\text{spin-0}} \ .$$

Expansion of $M = 1$ SU(2|1) SVBS state in terms of the doping parameter r is illustrated in Fig. 9.

Since SVBS states have "many" degrees of freedom, SMPS formalism is crucial for evaluations of physical quantities in SU(2|1) SVBS states. The g matrix is now given by the following 4×4 matrix:

$$g = (\mathcal{R}\hat{\psi})\hat{\psi}^t = \begin{pmatrix} a^\dagger b^\dagger|0\rangle & {b^\dagger}^2|0\rangle & \sqrt{r}b^\dagger f^\dagger|0\rangle & \sqrt{r}b^\dagger g^\dagger|0\rangle \\ -{a^\dagger}^2|0\rangle & -a^\dagger b^\dagger|0\rangle & -\sqrt{r}a^\dagger f^\dagger|0\rangle & -\sqrt{r}a^\dagger g^\dagger|0\rangle \\ -\sqrt{r}a^\dagger g^\dagger|0\rangle & -\sqrt{r}b^\dagger g^\dagger|0\rangle & rf^\dagger g^\dagger|0\rangle & 0 \\ -\sqrt{r}a^\dagger f^\dagger|0\rangle & -\sqrt{r}b^\dagger f^\dagger|0\rangle & 0 & -rf^\dagger g^\dagger|0\rangle \end{pmatrix}, \tag{57}$$

**The SU(2|1) symmetry of (54) is generated by

$$S_a = \begin{pmatrix} a \\ b \end{pmatrix}^\dagger \frac{1}{2}\sigma_a \begin{pmatrix} a \\ b \end{pmatrix}, \quad \Gamma = -f^\dagger f + g^\dagger g,$$
$$K_1 = \frac{1}{2\sqrt{2}x}a^\dagger(f+g) + \frac{x}{2\sqrt{2}}(f^\dagger + g^\dagger)b, \quad K_2 = \frac{1}{2\sqrt{2}x}b^\dagger(f+g) - \frac{x}{2\sqrt{2}}(f^\dagger + g^\dagger)a,$$
$$D_1 = \frac{1}{2\sqrt{2}x}a^\dagger(f-g) - \frac{x}{2\sqrt{2}}(f^\dagger - g^\dagger)b, \quad D_2 = \frac{1}{2\sqrt{2}x}b^\dagger(f-g) + \frac{x}{2\sqrt{2}}(f^\dagger - g^\dagger)a. \tag{55}$$

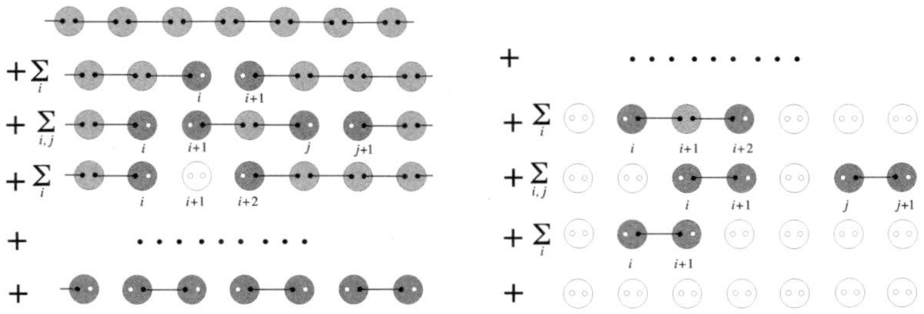

Fig. 9. The expansion of SU(2|1) SVBS states. Unlike the case of UOSp(1|2) SVBS state, SU(2|1) SVBS state can have spinless sites. At $r = 0$, SU(2|1) SVBS state is reduced to the original VBS state, while at $r \to \infty$, SU(2|1) SVBS state reduces to a "hole chain". The dimer state appears in the "middle" of the expansion. Thus, SU(2|1) SVBS state is more symmetric under the interchange between bosons and fermions.

where $\hat{\psi} = (a^\dagger, b^\dagger, \sqrt{r}f^\dagger, \sqrt{r}g^\dagger)^t$ and

$$\mathcal{R} = \begin{pmatrix} 0 & 1 & 0 & 0 \\ -1 & 0 & 0 & 0 \\ 0 & 0 & 0 & -1 \\ 0 & 0 & -1 & 0 \end{pmatrix}. \tag{58}$$

Corresponding to the four edge states ($a^\dagger|0\rangle$, $b^\dagger|0\rangle$, $f^\dagger|0\rangle$, $g^\dagger|0\rangle$) on each edge, the SU(2|1) SVBS state is 16-fold degenerate on a finite open chain.

Using a similar strategy to that used in the previous section, we can evaluate various physical quantities exactly. For instance, the spin- and the superconducting correlation

$$\langle S_i^z S_{i+n}^z \rangle\,, \tag{59a}$$

$$\Delta_n = \langle (a_i b_{i+n} - a_i b_{i+n})(f_i^\dagger g_{i+n}^\dagger + g_i^\dagger f_{i+n}^\dagger) \rangle \tag{59b}$$

decay exponentially with the respective correlation lengths $\xi_{\text{spin}}(r)$ and $\xi_{\text{sc}}(r)$ shown in Fig. 10.

The superconducting order parameter $\mathcal{O}_{\text{sc}} \equiv \Delta_{n=1}$ and the string order parameter $\mathcal{O}_{\text{string}}$ are shown in Fig. 11. Note that $\mathcal{O}_{\text{sc}}$ exhibits more symmetric behavior with respect to $r = 1$ than that in the UOSp(1|2) SVBS case implying higher symmetry between bosons and fermions in SU(2|1) SVBS state.

9. Summary

We proposed the SVBS states with super Lie groups symmetries. In particular, we have investigated SVBS states with UOSp(1|2)-symmetry. The

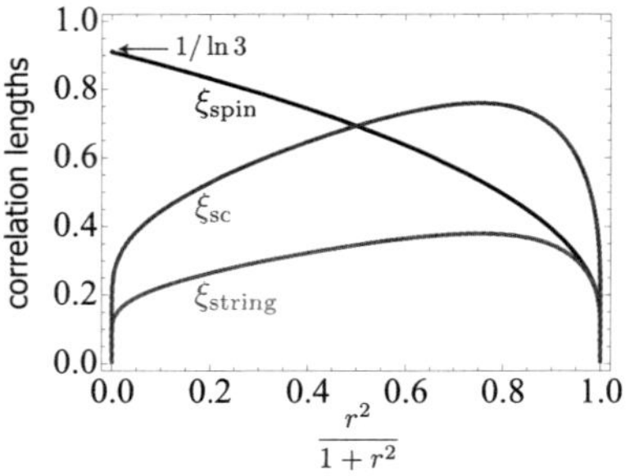

Fig. 10. The spin- and the superconducting correlation lengths ξ_{spin} and ξ_{sc} together with ξ_{string} which characterizes the exponential part of the string correlation eq.(50). The spin and the superconducting correlations exactly coincide at the supersymmetric point $r = 1$ ($r^2/(1+r^2) = 1/2$), again.

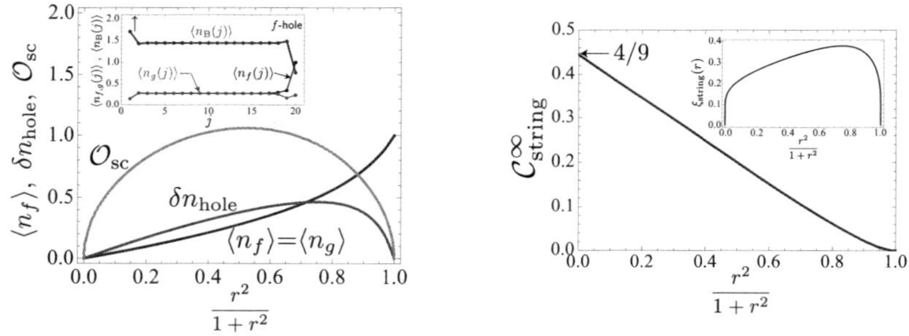

Fig. 11. The behavior of the superconducting order parameter $\mathcal{O}_{\text{sc}}$ (left) and the string order parameter $\mathcal{O}_{\text{string}}$ (right). Note the symmetry of $\mathcal{O}_{\text{sc}}$ with respect to $r = 1$. The string order parameter decreases from 4/9 (VBS value) to 0 (in the $r \to \infty$ limit, spins disappear from the system).

SVBS states contain the arbitrary hole-doping parameter r, but the states respect the supersymmetry regardless of the value of r. To calculate the various physical quantities of the SVBS states, we developed supersymmetric MPS formalism. In the UOSp(1|2) SVBS state, we calculated the spin- and the superconducting correlations to find that at $r = 1$ the correlation lengths of the bosonic- (i.e. spin) and the fermionic (i.e. superconducting) correlation functions coincide with each other. By constructing the supersymmetric parent Hamiltonian, we also evaluated gapful spectra of $S = 1$ triplon and $S = 1/2$ spinon-holon pair. The generalized hidden string order is also proposed for the SVBS states; it monotonically decreases from its VBS value 4/9 to a finite value in the $r \to \infty$ limit, reflecting the dimer-Haldane adiabatic continuity.

We also confirmed that the behaviors of physical quantities on $M = 1$ OSp(1|2) SVBS chain indeed interpolates between those of the original VBS state at $r = 0$ and of the Majumdar-Ghosh dimer state at $r = \infty$. Furthermore, the SVBS chain exhibits the superconducting property at finite r. We also proposed a new type of SVBS state with SU(2|1)-symmetry, which are more symmetric under interchange between bosons and fermions, and applied the SMPS formalism to it. The spin and superconducting lengths coincide at $r = 1$ again, and the superconducting order parameter is more symmetric reflecting the boson-fermion symmetry.

References

1. J. R. Schrieffer, *Theory Of Superconductivity* (Westview Press, 1999).
2. M. Stone (ed.), *Quantum Hall Effect* (World Scientific, 1992).
3. I. Affleck, T. Kennedy, E. Lieb and H. Tasaki, *Phys. Rev. Lett.* **59**, 799 (1987).
4. I. Affleck, T. Kennedy, E. Lieb and H. Tasaki, *Commun. Math. Phys.* **115**, 477 (1988).
5. D. P. Arovas, A. Auerbach and F. Haldane, *Phys. Rev. Lett.* **60**, 531 (1988).
6. F. Verstraete and J. I. Cirac, *Phys. Rev. A* **70**, p. 060302 (2004).
7. M. Fannes, B. Nachtergaele and R. Werner, *Commun. Math. Phys.* **144**, p. 443 (1992).
8. A. Klümper, A. Schadschneider and J. Zittartz, *Z. Phy. B: Condens. Matter* **87**, 281 (1992).
9. K. Totsuka and M. Suzuki, *J. of Phys.: Condens. Matter* **7**, 1639 (1995).
10. D. Perez-Garcia, F. Verstraete, M. M. Wolf and J. I. Cirac, *Quantum Inf. Comput.* **7**, p. 401 (2007).
11. D. P. Arovas, K. Hasebe, X.-L. Qi and S.-C. Zhang, *Physical Review B* **79**, p. 224404 (2009).
12. K. Hasebe and K. Totsuka, *unpublished* (2010).
13. P. Anderson, *Science* **235**, 1196 (1987).
14. S. Zhang, *Science* **275**, p. 1089 (1997).
15. G. Fáth and J. Sólyom, *J. Phys.: condensed matter* **5**, p. 8983 (1993).
16. T. Kennedy, *J. Phys.: Condens. Matter.* **2**, 5737 (1990).
17. M. den Nijs and K. Rommelse, *Phys. Rev.* **B40**, p. 4709 (1989).
18. S. Zhang and D. Arovas, *Phys. Rev. B* **40**, 2708 (1989).
19. K. Penc and H. Shiba, *Physical Review B* **52**, p. R715 (1995).
20. A. Auerbach, *Interacting Electrons and Quantum Magnetism* (Springer-Verlag, Berlin, 1994).
21. C. Majumdar and D. Ghosh, *J. Math. Phys.* **10**, 1399 (1969).
22. C. Majumdar, *J. Phys. C* **3**, 911 (1970).
23. K. Hasebe, *Phys. Rev. Lett.* **94**, p. 206802 (2005).
24. S. Girvin and D. Arovas, *Phys. Scr.* **T 27**, p. 156 (1989).
25. Y. Hatsugai, *Journal of the Physical Society of Japan* **61**, 3856 (1992).
26. K. Hida, *Phys. Rev. B* **45**, p. 2207 (1992).

MATRIX PRODUCT STATES IN QUANTUM INTEGRABLE MODELS

HOSHO KATSURA*

Kavli Institute for Theoretical Physics, University of California, Santa Barbara, CA 93106, USA
E-mail: katsura@kitp.ucsb.edu

ISAO MARUYAMA

Graduate School of Engineering Science, Osaka University, Toyonaka, Osaka 560-8531, Japan
E-mail: maru@mp.es.osaka-u.ac.jp

We review recent developments in the study of matrix product states in quantum integrable models. We first clarify the relation between the matrix product ansatz, proposed by Alcaraz and Lazo, and the algebraic Bethe ansatz. In the spin-$\frac{1}{2}$ Heisenberg chain, this relation turns out to be related to the six-vertex model with domain wall boundary conditions. To obtain explicit finite-dimensional representations for the matrices, factorizing F-matrices are necessary. Secondly, we discuss matrix product states in the Lieb-Liniger model describing a one-dimensional Bose gas with delta-function interaction. Although the model is continuous in space, we can construct continuous but finite-dimensional matrix product states from the integrable version of the Bose-Hubbard model. The factorizing F-matrices are again indispensable in this construction.

Keywords: Matrix product state; Bethe ansatz; Heisenberg chain; Lieb-Liniger model.

1. Introduction

Understanding of entangled many-body states is a central issue both in condensed matter physics and quantum information science. Matrix product states (MPSs) are of considerable current interest as variational states for one-dimensional strongly correlated systems. The archetype of MPS

---

*Present address: Department of Physics, Gakushuin University, Toshima-ku, Tokyo 171-8588, Japan.

can be traced back to Baxter's variational transfer matrix method[1] and the valence-bond-solid (VBS) states in the Affleck-Kennedy-Lieb-Tasaki (AKLT) model.[2–4] The MPSs for a variety of VBS states have been constructed in Refs. 5 and 6. A more mathematically rigorous definition of the MPS can be found in Ref. 7. The MPSs have been used in the density matrix renormalization group (DMRG) method and have turned out to be powerful tools for approximating the ground states of one-dimensional Hamiltonians with local interactions.[8–10] Note that the MPSs are also useful for studying two-dimensional classical lattice models.[11,12] In recent years, the MPSs have been extensively studied from the point of view of quantum information theory, and the criterion for the efficiency of MPS has been investigated. One might think that MPS gives an efficient description of the ground state if the system is non-critical and the von Neumann entropy of a large block is bounded by a constant.[13] In other words, the ground states obey an area law.[14,15] However, a more precise criterion has been found in Ref. 16: if the Rényi entropies S_α with $\alpha < 1$ scales at most logarithmically in the block size, then the ground states can be well approximated by the MPS. A more subtle issue is whether quantum critical ground states can be expressed in terms of MPS. Since the von Neumann entropy ($\alpha = 1$) in a critical system scales logarithmically in the block size,[17] it is unlikely that there exists an efficient way to describe the ground state in the form of MPS. However, this does not deny the existence of the MPSs for the critical ground states if it is permissible to use huge or infinite dimensional matrices. In fact, Alcaraz and Lazo have proposed that the eigenstates of a certain class of quantum integrable models, solved by the Bethe ansatz, can be expressed in the form of MPSs.[18–20] A prototypical example is the spin-$\frac{1}{2}$ Heisenberg chain, which was first solved by Bethe using the coordinate Bethe ansatz and was shown to have gapless excitations.[21] In the formulation initiated by Alcaraz and Lazo, which they call the matrix product ansatz (MPA), one can obtain the spectrum conditions, i.e., Bethe equations, imposing algebraic relations among matrices consisting of the MPS. However, it was unclear how to systematically obtain explicit matrices.

Although their approach looks different from approaches such as the coordinate or functional Bethe ansatz, it is similar to the algebraic Bethe ansatz (ABA). The ABA is an elegant method for solving the eigenvalue problem of quantum integrable models developed in the late 1970s.[22–24] Indeed, we will see that MPS structure is somehow hidden in the ABA formulation and the MPA for the spin-$\frac{1}{2}$ Heisenberg chain used in Refs. 18 and 19 can be rigorously derived from the ABA.[25] A key idea is a change of basis in the tensor product of auxiliary spaces, which was already used in the ABA derivation of the MPS in the asymmetric simple exclusion process (ASEP).[26] In this new basis (F-basis), the matrices appearing in MPS become remarkably simple and one can explicitly obtain finite-dimensional representations for them for a fixed number of down or up spins. Furthermore, in this paper, we also show that this change of basis corresponds to the mapping from the six-vertex model with domain wall boundary conditions[27] to a five-vertex model with them.

Another interesting aspect of MPA is the physical meaning of the matrices appearing in the MPS form of the Bethe ansatz states. Those matrices can be regarded as lattice realizations of the generators of the Zamolodchikov-Faddeev algebra in (1+1)-dimensional field theories.[28] Such a field theory interpretation of matrices may be useful for studying other classes of one-dimensional models. In fact, the recent study by Cirac and Sierra[29] has shown that the ground state of the Haldane-Shastry spin chain,[30,31] another class of quantum integrable model with inverse-square interactions, can be simply expressed by using the chiral vertex operators of conformal field theory instead of using finite-dimensional matrices. Another kind of field theory study has been done by Verstraete and Cirac.[32] They have proposed the use of continuous matrix product states (cMPS) as variational states for one-dimensional continuum field theory such as the Lieb-Liniger model.[33] Shortly after this work, the authors of this article showed that the exact cMPS for the Lieb-Liniger model can be obtained from the continuum limit of the MPS of the Bethe state in the integrable version of the Bose-Hubbard model.[34–36] The original cMPS by Verstraete and Cirac has been further applied to other class of (1+1)-dimensional relativistic field theories.[37]

In this article, we focus on the exact matrix product states in one-dimensional quantum integrable models and review the recent developments in this field. Taking two examples of models, we will illustrate how to obtain the explicit matrices systematically starting from the ABA. One model is the XXZ spin-$\frac{1}{2}$ Heisenberg chain, for which there are previous studies on the MPA. The other is the Lieb-Liniger model, for which there is no previous study on the MPA, but the cMPS is proposed as a variational state for the numerical calculation. We will discuss the difference between previous studies and the exact MPSs, and will emphasize the importance of the factorizing F-matrices. The organization of this paper is as follows. In Section 2, we will provide a brief introduction to the algebraic Bethe ansatz in the XXX and XXZ spin-$\frac{1}{2}$ Heisenberg chains, which are prototypical examples of the quantum integrable models. In Section 3, we first derive formal matrix product expressions for the Bethe ansatz states in the XXZ chain. Although this derivation proves that the Bethe ansatz states have MPS representations, nontrivial change of basis in the tensor product of auxiliary spaces is required to obtain explicit expressions for the matrices as was shown in our previous study.[25] We will summarize the obtained results in this section. In Section 4, we will discuss the continuous matrix product states which describe the eigenstates of the Lieb-Liniger model. They have been obtained from the continuum limit of the integrable Bose-Hubbard model.[38] We will show explicit expressions for the finite-dimensional matrices and compare them with those in the Heisenberg chain. Conclusions and future perspectives are given in the last section.

2. Algebraic Bethe Ansatz for the Spin-$\frac{1}{2}$ XXZ Chain

The XXZ spin-$\frac{1}{2}$ Heisenberg model with the periodic boundary condition is described by the following Hamiltonian:

$$\mathbf{H}_{\text{XXZ}} = \sum_{i=1}^{L} \left\{ \sigma_i^x \sigma_{i+1}^x + \sigma_i^y \sigma_{i+1}^y + \Delta(\sigma_i^z \sigma_{i+1}^z - 1) \right\}, \tag{1}$$

where L denotes the total number of sites and σ_i^α $(\alpha = x, y, z)$ are the Pauli matrices defined on the ith site:

$$\sigma^x = \begin{pmatrix} 0 & 1 \\ 1 & 0 \end{pmatrix}, \quad \sigma^y = \begin{pmatrix} 0 & -\mathbf{i} \\ \mathbf{i} & 0 \end{pmatrix}, \quad \sigma^z = \begin{pmatrix} 1 & 0 \\ 0 & -1 \end{pmatrix}. \tag{2}$$

Here Δ is the anisotropy parameter and, in particular, $\Delta = 1$ and $\Delta = 0$ correspond to the XXX and XY chains, respectively.

The ABA gives an elegant way to diagonalize this Hamiltonian.[22–24,39] We shall briefly review this approach. The R-matrix which ensures the integrability of the model plays a central role in the ABA. The R-matrix of the XXZ model acts on $\mathbb{C}^2 \otimes \mathbb{C}^2$ and is defined by

$$R(\lambda) = \frac{1}{\sinh(\lambda+\eta)} \begin{pmatrix} \sinh(\lambda+\eta) & 0 & 0 & 0 \\ 0 & \sinh\lambda & \sinh\eta & 0 \\ 0 & \sinh\eta & \sinh\lambda & 0 \\ 0 & 0 & 0 & \sinh(\lambda+\eta) \end{pmatrix}, \tag{3}$$

where λ is the spectral parameter, and the relation between η and the anisotropy Δ is given by $\Delta = \cosh\eta$. We next introduce the quantum L-operator represented by a matrix acting on the tensor product of two-dimensional vector spaces $\bar{V}_j \otimes V_i$. Here $V_i = \mathbb{C}^2$ is the Hilbert space at the

ith site (sometimes called quantum space) and is spanned by $|\uparrow\rangle_i$ and $|\downarrow\rangle_i$:

$$|\uparrow\rangle_i = \begin{pmatrix} 1 \\ 0 \end{pmatrix}_{V_i}, \quad |\downarrow\rangle_i = \begin{pmatrix} 0 \\ 1 \end{pmatrix}_{V_i}. \tag{4}$$

The auxiliary space $\bar{V}_j = \mathbb{C}^2$ is spanned by two orthonormal states, which are labeled as $|\leftarrow\rangle_j$ and $|\rightarrow\rangle_j$ for later convenience:

$$|\leftarrow\rangle = \begin{pmatrix} 1 \\ 0 \end{pmatrix}_{\bar{V}_j}, \quad |\rightarrow\rangle = \begin{pmatrix} 0 \\ 1 \end{pmatrix}_{\bar{V}_j}. \tag{5}$$

For the XXZ model, the L-operator $\mathcal{L}_{ji}(\lambda)$ is defined by

$$\mathcal{L}_{ji}(\lambda) = \begin{pmatrix} 1 & 0 & 0 & 0 \\ 0 & b(\lambda) & c(\lambda) & 0 \\ 0 & c(\lambda) & b(\lambda) & 0 \\ 0 & 0 & 0 & 1 \end{pmatrix}, \tag{6}$$

where

$$b(\lambda) = \frac{\sinh\left(\lambda - \frac{\eta}{2}\right)}{\sinh\left(\lambda + \frac{\eta}{2}\right)}, \qquad c(\lambda) = \frac{\sinh\eta}{\sinh\left(\lambda + \frac{\eta}{2}\right)} \tag{7}$$

for the XXZ chain and

$$b(\lambda) = \frac{\lambda - \frac{\mathbf{i}}{2}}{\lambda + \frac{\mathbf{i}}{2}}, \qquad c(\lambda) = \frac{\mathbf{i}}{\lambda + \frac{\mathbf{i}}{2}} \tag{8}$$

for the XXX chain. The latter can be obtained by taking the scaling limit of the former, i.e., $\lambda_{\text{XXZ}} = \epsilon\lambda_{\text{XXX}}$ and $\eta = \mathbf{i}\epsilon$ with the limit of $\epsilon \to 0$. Note that the order of the basis in $\bar{V}_j \otimes V_i$ is fixed as $|\leftarrow\uparrow\rangle, |\leftarrow\downarrow\rangle, |\rightarrow\uparrow\rangle, |\rightarrow\downarrow\rangle$. We then define the monodromy matrix as the ordered product of L-operators:

$$\mathcal{T}_0(\lambda) = \mathcal{L}_{01}(\lambda)\mathcal{L}_{02}(\lambda)...\mathcal{L}_{0L}(\lambda). \tag{9}$$

Note that $\mathcal{L}_{0i}$ acts nontrivially on $\bar{V}_0 \otimes V_i$ and trivially on the other spatial sites. In the basis of $\bar{V}_0$, the monodromy matrix can be expressed as a 2×2 matrix:

$$\mathcal{T}_0(\lambda) = \begin{pmatrix} \mathbf{A}(\lambda) & \mathbf{B}(\lambda) \\ \mathbf{C}(\lambda) & \mathbf{D}(\lambda) \end{pmatrix}, \tag{10}$$

where matrix elements $\mathbf{A}(\lambda)$, $\mathbf{B}(\lambda)$, $\mathbf{C}(\lambda)$, and $\mathbf{D}(\lambda)$ are themselves operators acting on the whole quantum space $\mathscr{H} = \otimes_{i=1}^{L} V_i$. Then, as a direct consequence of the Yang-Baxter equation, we obtain

$$R_{01}(\lambda - \mu)\mathcal{T}_0(\lambda)\mathcal{T}_1(\mu) = \mathcal{T}_1(\mu)\mathcal{T}_0(\lambda)R_{01}(\lambda - \mu). \tag{11}$$

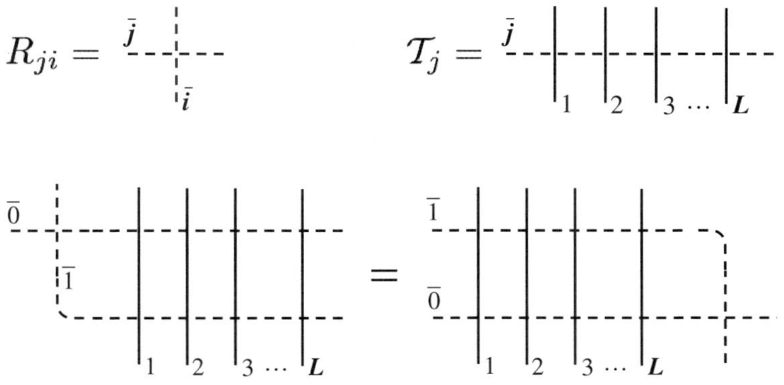

Fig. 1. Graphical representations for the R-matrix, the monodromy matrix, and the relation (11).

Here, R_{01} acts nontrivially on $\bar{V}_0 \otimes \bar{V}_1$ and trivially on the other spaces. The graphical expression for Eq. (11) is shown in Fig. 1. The commutation relations among **A**, **B**, **C**, and **D** can be obtained from this relation.

The Hamiltonian Eq. (1) can then be obtained from $\mathbf{T}(\lambda)$ by the trace identity:

$$\mathbf{H}_{\mathrm{XXZ}} = 2\sinh\eta \frac{\partial}{\partial\lambda} \log \mathbf{T}(\lambda)\Big|_{\lambda=\eta/2} + \text{const.}, \tag{12}$$

where $\mathbf{T}(\lambda) = \mathbf{A}(\lambda) + \mathbf{D}(\lambda)$ is a one-parameter family of transfer matrices acting on $\mathscr{H}$. Since the Hamiltonian commutes with the monodromy matrix, one can construct a simultaneous eigenstate of both $\mathbf{H}_{\mathrm{XXZ}}$ and $\mathbf{T}$. The Bethe eigenstate of these operators is constructed by $\mathbf{B}(\lambda)$'s as

$$|\lambda_1, \lambda_2, ..., \lambda_n\rangle = \mathbf{B}(\lambda_n)...\mathbf{B}(\lambda_1)\,|\Uparrow\rangle\,, \tag{13}$$

where $|\Uparrow\rangle \equiv |\uparrow, \uparrow, ..., \uparrow\rangle$ denotes the reference ferromagnetic state and n denotes the number of down spins. Since one can show that $\mathbf{B}(\lambda_i)$'s commute with each other from the relation (11), this state is invariant under permutations of λ_i's. The spectrum conditions, i.e., the Bethe equations, are then obtained using the commutation relations among **A**, **D**, and **B**.[22–24] For the XXZ model, those equations are given by

$$\left(\frac{\sinh(\lambda_j - \frac{\eta}{2})}{\sinh(\lambda_j + \frac{\eta}{2})}\right)^L \prod_{\substack{k=1 \\ k\neq j}}^{n} \frac{\sinh(\lambda_j - \lambda_k + \eta)}{\sinh(\lambda_j - \lambda_k - \eta)} = 1, \quad j = 1, 2, ..., n. \tag{14}$$

Although physical quantities of interest are expressed in terms of the solutions of the above equations, they are generally hard to solve. But since we are interested in the relation between the MPS and ABA, we shall henceforth assume the existence of solutions and derive the MPS expression for the Bethe state with given λ_i's.

3. Matrix Product State Representation for the Bethe States in the Spin-$\frac{1}{2}$ XXZ Chain

3.1. *Recursion relations*

Let us now derive the MPS representation for the Bethe state from the ABA. Since the creation operator for the Bethe state, $\mathbf{B}(\lambda)$, is a matrix element of the monodromy matrix (see Eq. (10)), we immediately notice

$$\begin{aligned} &\mathbf{B}(\lambda_n)...\mathbf{B}(\lambda_2)\mathbf{B}(\lambda_1) \\ &= \langle\leftarrow|\,\mathcal{T}(\lambda_n)\,|\rightarrow\rangle \ldots \langle\leftarrow|\,\mathcal{T}(\lambda_2)\,|\rightarrow\rangle\,\langle\leftarrow|\,\mathcal{T}(\lambda_1)\,|\rightarrow\rangle \\ &= \mathrm{Tr}_{\bar{\mathscr{H}}}(Q_n\mathcal{T}(\lambda_n)\otimes \ldots \otimes \mathcal{T}(\lambda_2)\otimes\mathcal{T}(\lambda_1)), \end{aligned} \tag{15}$$

where $Q_n = |\rightarrow,\rightarrow \ldots \rightarrow\rangle\,\langle\leftarrow,\leftarrow \ldots \leftarrow| \equiv |\Rightarrow\rangle\,\langle\Leftarrow|$ and $\bar{\mathscr{H}} \equiv \bar{V}^{\otimes n} = \bar{V}_n \otimes \cdots \otimes \bar{V}_1$. Then introducing $\mathcal{L}_i(\lambda_1,...,\lambda_n) \equiv \mathcal{L}_{ni}(\lambda_n)\otimes\cdots\otimes\mathcal{L}_{1i}(\lambda_1)$, which acts on the vector space V_i only in $\mathscr{H}$, we can express the Bethe state as

$$|\lambda_1,\lambda_2,...,\lambda_n\rangle = \mathrm{Tr}_{\bar{\mathscr{H}}}\left(Q_n\left[\prod_{i=1}^{L}\mathcal{L}_i(\lambda_1,...,\lambda_n)\,|\uparrow\rangle_i\right]\right). \tag{16}$$

We now introduce two matrices D_n and C_n via

$$\mathcal{L}_i(\lambda_1,...,\lambda_n)|\uparrow\rangle = D_n(\lambda_1,...,\lambda_n)|\uparrow\rangle + C_n(\lambda_1,...,\lambda_n)|\downarrow\rangle. \tag{17}$$

It should be noted that D_n, C_n, and Q_n are $2^n\times 2^n$ matrices acting on $\bar{V}^{\otimes n}$ but they do not act on the Hilbert spaces V_i. Using D_n, C_n, and Q_n, the Bethe state can be written as

$$|\lambda_1,...,\lambda_n\rangle = \mathrm{Tr}_{\bar{\mathscr{H}}}\left[Q_n\prod_{i=1}^{L}(D_n(\lambda_1,...,\lambda_n)|\uparrow\rangle_i + C_n(\lambda_1,...,\lambda_n)|\downarrow\rangle_i)\right]. \tag{18}$$

Although there is a boundary matrix Q_n in this expression, the form is of the MPS.[40] Note that not only the ground state, but also other excited states can be expressed in this form once we know λ_i $(i = 1,2,...,n)$ for them.

Up to this point, the matrices D_n and C_n have been defined in an abstract way. However, we can derive the recursion relations for them from the fact that $\mathcal{L}_i(\lambda_1, ..., \lambda_n, \lambda_{n+1}) = \mathcal{L}_{n+1,i}(\lambda_{n+1}) \otimes \mathcal{L}_i(\lambda_1, ..., \lambda_n)$:

$$D_{n+1}(\lambda_1, ..., \lambda_n, \lambda_{n+1}) = \begin{pmatrix} D_n(\lambda_1, ..., \lambda_n) & 0 \\ c(\lambda_{n+1})C_n(\lambda_1, ..., \lambda_n) & b(\lambda_{n+1})D_n(\lambda_1, ..., \lambda_n) \end{pmatrix}, \tag{19}$$

$$C_{n+1}(\lambda_1, ..., \lambda_n, \lambda_{n+1}) = \begin{pmatrix} b(\lambda_{n+1})C_n(\lambda_1, ..., \lambda_n) & c(\lambda_{n+1})D_n(\lambda_1 ... \lambda_n) \\ 0 & C_n(\lambda_1, ..., \lambda_n) \end{pmatrix}, \tag{20}$$

where 2×2 matrices acting on $\bar{V}_{n+1}$ have matrix elements that are operators on $\bar{V}^{\otimes n}$. D_{n+1} and C_{n+1} are $2^{n+1} \times 2^{n+1}$ matrices acting on $\bar{V}^{\otimes(n+1)} = \bar{V}_{n+1} \otimes \bar{V}^{\otimes n}$. The initial conditions for Eqs. (19-20) are given by

$$D_1(\lambda) = \begin{pmatrix} 1 & 0 \\ 0 & b(\lambda) \end{pmatrix}, \quad C_1(\lambda) = \begin{pmatrix} 0 & c(\lambda) \\ 0 & 0 \end{pmatrix}. \tag{21}$$

On the other hand, from the definition of the matrix Q_n, we obtain the following recursion with the initial matrix Q_1:

$$Q_{n+1} = \begin{pmatrix} 0 & 0 \\ Q_n & 0 \end{pmatrix}, \qquad Q_1 = \begin{pmatrix} 0 & 0 \\ 1 & 0 \end{pmatrix}. \tag{22}$$

Therefore, one can also express Q_n as

$$Q_n = \bigotimes_{l=1}^{n} \begin{pmatrix} 0 & 0 \\ 1 & 0 \end{pmatrix}. \tag{23}$$

Using the relations (19-21), one can, in principle, obtain the explicit expressions for the matrices D_n and C_n. As expected from general arguments, the dimension of the matrix becomes infinity in the thermodynamic limit with fixed n/L. On the other hand, it becomes finite even in the limit of $L \to \infty$ if the number of down spins is finite.

3.2. *New basis in* $\bar{\mathscr{H}}$

We shall now solve the recursion relations (19-20) and obtain the explicit expressions for the matrices D_n and C_n. To this aim, we introduce the invertible matrix F_n which diagonalizes D_n as $F^{-1}D_nF_n = \mathcal{D}_n$. We call F_n as F-matrix. Although the matrices C_n and Q_n are transformed into $\mathcal{C}_n = F^{-1}C_nF_n$ and $\mathcal{Q}_n = F^{-1}Q_nF_n$, respectively, the Bethe state can still be expressed as the MPS from the cyclic property of the trace:

$$|\lambda_1, \lambda_2, ..., \lambda_n\rangle = \mathrm{Tr}_{\bar{\mathscr{H}}} \left[\mathcal{Q}_n \prod_{i=1}^{L} (\mathcal{D}_n | \uparrow\rangle_i + \mathcal{C}_n | \downarrow\rangle_i) \right]. \tag{24}$$

Here, for simplicity, we have omitted $\lambda_1, ..., \lambda_n$ in the matrices $\mathcal{D}_n$ and $\mathcal{C}_n$. Then, as was shown in Ref. 25, we can construct the matrix F_n which is defined in the following recursive way:

$$F_{n+1} = \begin{pmatrix} F_n & 0 \\ F_n \mathcal{F}_n & F_n \end{pmatrix}, \tag{25}$$

where

$$\mathcal{F}_n = c(\lambda_{n+1}) \mathcal{D}_n^{-1} \sum_{i=1}^{n} \frac{\mathcal{C}_n^{(i)}}{z_i - z_{n+1}}. \tag{26}$$

Here $z_i \equiv b(\lambda_i)$ and $\mathcal{C}_n^{(i)}$ are decompositions of $\mathcal{C}_n$, i.e., $\mathcal{C}_n = \sum_{i=1}^{n} \mathcal{C}_n^{(i)}$, satisfying the following algebraic relations:

$$\mathcal{C}_n^{(i)} \mathcal{D}_n = z_i \mathcal{D}_n \mathcal{C}_n^{(i)}, \tag{27}$$

$$\mathcal{C}_n^{(i)} \mathcal{C}_n^{(j)} = \tilde{S}_{ij}(z_i, z_j) \mathcal{C}_n^{(j)} \mathcal{C}_n^{(i)}, \tag{28}$$

$$\mathcal{C}_n^{(i)} \mathcal{C}_n^{(i)} = 0, \tag{29}$$

where

$$\tilde{S}_{ij}(z_i, z_j) = \frac{b(\lambda_i)}{b(\lambda_j)}\, b\left(\lambda_i - \lambda_j - \frac{\eta}{2}\right) = -\frac{z_i}{z_j} \cdot \frac{z_i z_j + 1 - 2\Delta z_j}{z_i z_j + 1 - 2\Delta z_i}. \tag{30}$$

The above algebraic relations remind us of the Zamolodchikov-Faddeev algebra in (1+1)-dimensional integrable field theories.[28] Since $D_1(\lambda)$ in Eq. (21) is diagonal, we can take F_1 to be the 2×2 identity matrix and hence $\mathcal{D}_1 = D_1$, $\mathcal{C}_1 = C_1$, and $\mathcal{Q}_1 = Q_1$. For $n = 1$, $\mathcal{D}_1$ and $\mathcal{C}_1$ trivially satisfy the relations (27-29).

By an explicit calculation using the matrix $\mathcal{F}_n$ introduced in Eq. (26), one can derive recursions in the new basis in $\mathscr{H}$:

$$\mathcal{D}_{n+1} = \begin{pmatrix} 1 & 0 \\ 0 & z_{n+1} \end{pmatrix} \otimes \mathcal{D}_n, \tag{31}$$

$$\mathcal{C}_{n+1}^{(i)} = \begin{pmatrix} \frac{z_i z_{n+1} + 1 - 2\Delta z_{n+1}}{z_i - z_{n+1}} & 0 \\ 0 & -z_{n+1} \frac{z_i z_{n+1} + 1 - 2\Delta z_i}{z_i - z_{n+1}} \end{pmatrix} \otimes \mathcal{C}_n^{(i)}, \quad i \le n \tag{32}$$

$$\mathcal{C}_{n+1}^{(n+1)} = \begin{pmatrix} 0 & c(\lambda_{n+1}) \\ 0 & 0 \end{pmatrix} \otimes \mathcal{D}_n, \tag{33}$$

$$\mathcal{Q}_{n+1} = \begin{pmatrix} 0 & 0 \\ 1 & 0 \end{pmatrix} \otimes \mathcal{Q}_n. \tag{34}$$

A detailed derivation can be found in Ref. 25. The new recursions are quite simple, and one can easily solve them exactly. As a result, we obtain

$$\mathcal{D}_n = \bigotimes_{l=1}^{n} \begin{pmatrix} 1 & 0 \\ 0 & z_{n-l+1} \end{pmatrix}, \qquad \mathcal{Q}_n = Q_n = \bigotimes_{l=1}^{n} \begin{pmatrix} 0 & 0 \\ 1 & 0 \end{pmatrix} \tag{35}$$

$$\begin{aligned} \mathcal{C}_n^{(i)} &= f(z_i, ..., z_n) c(\lambda_i) \bigotimes_{l=1}^{n-i} \begin{pmatrix} S_{i,n-l+1}(z_i, z_{n-l+1}) & 0 \\ 0 & z_{n-l+1} \end{pmatrix} \\ &\otimes \begin{pmatrix} 0 & 1 \\ 0 & 0 \end{pmatrix} \bigotimes_{l=n-i+2}^{n} \begin{pmatrix} 1 & 0 \\ 0 & z_{n-l+1} \end{pmatrix}, \end{aligned} \tag{36}$$

where

$$S_{ij}(z_i, z_j) = -\frac{z_i z_j + 1 - 2\Delta z_j}{z_i z_j + 1 - 2\Delta z_i} \tag{37}$$

and

$$f(z_i, ..., z_n) = \prod_{l=i+1}^{n} \left(-\frac{z_i z_l + 1 - 2\Delta z_i}{z_i - z_l} \right) \tag{38}$$

for $i = 1, ..., n-1$, and $f(z_n) = 1$. The matrix $\mathcal{D}_n$ is diagonal and $\mathcal{C}_n^{(i)}$ is almost diagonal in $\bar{\mathscr{H}}$, except in the single auxiliary space $\bar{V}_i$. Therefore, we can easily confirm the algebraic relations (27-29) directly using Eqs. (35) and (36). Another remarkable point is that the matrix Q_n is invariant under the similarity transformation by F_n. This is closely related to the six-vertex model with domain wall boundary conditions and will be discussed in the next subsection in more detail.

Finally, let us clarify the relation between our MPS (Eq. (24)) and the MPA found by Alcaraz and Lazo.[18,19] After a little algebra, we find that the Bethe state in the MPA can be written as

$$|\lambda_1, \lambda_2, ..., \lambda_n\rangle = \mathrm{Tr}_{\bar{\mathscr{H}}} \left[\Omega_n \prod_{i=1}^{L} (E_n |\uparrow\rangle_i + A_n |\downarrow\rangle_i) \right], \tag{39}$$

where the matrix A_n is decomposed by n matrices as

$$A_n = \sum_{i=1}^{n} A_{k_i,n} E_n, \tag{40}$$

and matrices Ω_n, E_n, and $A_{k_i,n}$ satisfy

$$A_{k_i,n} E_n = z_i E_n A_{k_i,n}, \tag{41}$$

$$A_{k_i,n} A_{k_j,n} = S_{ij}(z_i, z_j) A_{k_j,n} A_{k_i,n}, \tag{42}$$

$$E_n \Omega_n = e^{-\mathrm{i}P} \Omega_n E_n, \tag{43}$$

with $P = \sum_{i=1}^{n} k_i$. Then, we find that we can reproduce Eqs. (41-43) from Eqs. (35-36) if we take

$$E_n = \alpha \mathcal{D}_n, \quad \Omega_n = \beta \mathcal{Q}_n = \beta Q_n, \quad A_{k_i,n} E_n = \gamma_i \mathcal{C}_n^{(i)}, \tag{44}$$

where α, β, and γ_i $(i = 1, ..., n)$ can be arbitrary numbers. Alcaraz and Lazo have also obtained the explicit expressions for $A_{k_i,n}$ and E_n in Ref. 19. They can be obtained by choosing $\alpha = 1$ and $\gamma_i = z_i[c(\lambda_i) f(z_i, ..., z_n)]^{-1}$:

$$E_n = \bigotimes_{l=1}^{n} \begin{pmatrix} 1 & 0 \\ 0 & z_{n-l+1} \end{pmatrix}, \tag{45}$$

$$A_{k_i,n} = \bigotimes_{l=1}^{n-i} \begin{pmatrix} S_{i,n-l+1}(z_i, z_{n-l+1}) & 0 \\ 0 & 1 \end{pmatrix} \otimes \begin{pmatrix} 0 & 1 \\ 0 & 0 \end{pmatrix} \bigotimes_{l=n-i+2}^{n} \begin{pmatrix} 1 & 0 \\ 0 & 1 \end{pmatrix}. \tag{46}$$

In this way, we have shown that the MPA and ABA are essentially equivalent in the XXX and XXZ spin-$\frac{1}{2}$ Heisenberg chains. Although solving the recursion relations for the matrices is rather involved, the first step towards the MPS (18) is quite straightforward. Therefore, we conjecture that in all the integrable models solved by the algebraic Bethe ansatz, one can obtain the MPS representation for the Bethe states such as Eq. (18).

3.3. *Six-vertex model with domain wall boundary conditions*

We shall now show that the change of basis used in the previous subsection is closely related to the six-vertex model with domain wall boundary conditions (see Fig. 2 (a)). This model was first introduced as a byproduct of the study of the norm of the Bethe state.[27,41]

Let us define the partition function of this two-dimensional statistical mechanics model. For a finite $(n \times n)$ lattice, the partition function apart from an overall factor is defined as a component of the Bethe state:

$$Z_n^{\mathrm{DWBC}} = \langle \Downarrow | \lambda_1, \ldots, \lambda_n \rangle , \tag{47}$$

where $\langle \Downarrow | = \bigotimes_{i=1}^{n} {}_i\langle \downarrow |$. From the elements of the L-operator (6), one can read off the Boltzmann weights as shown in Fig. 2 (c). We have an alternative expression for Z_n^{DWBC}:

$$Z_n^{\mathrm{DWBC}} = \langle \Leftarrow | \left[C_n(\lambda_1, ..., \lambda_n) \right]^n | \Rightarrow \rangle , \tag{48}$$

which can be derived from Eq. (18). As shown in appendices of Ref. 25, one can assign graphical representations for D_n and C_n in Eq. (18), which are vertical objects with specified arrows on the top and bottom. Let us now see what happens when we change the basis in $\bar{\mathscr{H}}$. The important point

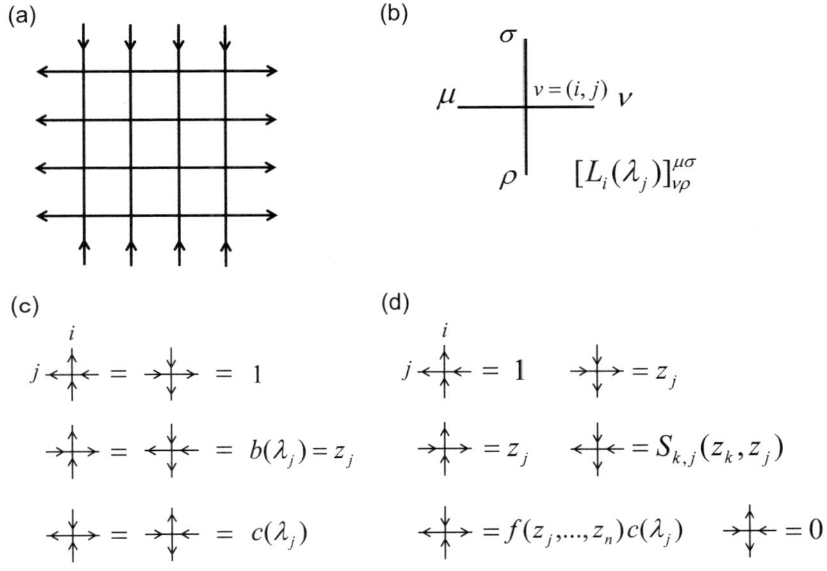

Fig. 2. (a) Domain wall boundary conditions. (b) L-operator associated with the vertex which is the intersection of the ith vertical and jth horizontal lines. (c) Boltzmann weights of six possible configurations of arrows at each vertex. (d) Boltzmann weight in the five-vertex model which can be read off from Eqs. (35-36). In any possible configuration, there should be exactly one fifth-vertex (the bottom-left vertex in (d)) on each vertical line, which determines the index k in the weight of the fourth vertex. The last vertex receives zero weight.

here is that the boundary condition in the horizontal direction remains unchanged, i.e., $\mathcal{Q}_n = Q_n$. Furthermore, as we have seen in the previous subsection, the matrices D_n and C_n become remarkably simple in the new basis. This corresponds to the mapping from the six-vertex model to a five-vertex model since the Boltzmann weight for one out of the six vertices vanishes after the transformation (see Fig. 2 (d)). Although the weights in the five-vertex model are determined in a non-local way, the combinatorial complexity for the calculation of Z_n^{DWBC} is greatly reduced by the similarity transformation. This kind of simplification reminds us of the Drinfel'd twist[42,43] which have been used to calculate the correlation functions in the XXZ chain.[44–47] In fact, our F-matrix defined in Eqs. (25) and (26) can be regarded as a factorizing matrix in the auxiliary space ($\bar{\mathscr{H}}$) instead of $\mathscr{H}$. Note that not only the component of the state $\langle \Downarrow |$ but also the other components of the Bethe state can be interpreted as partition functions of the six-vertex model with different boundary conditions.[25]

4. Continuous Matrix Product State Representation for the Bethe States in the Lieb-Liniger Model

In this section, we shall construct a continuous matrix product state (cMPS) representation of the Bethe states for the Lieb-Liniger model, which describes the one-dimensional Bose gas with delta-function interactions.[33] This model has recently attracted considerable interest owing to the recent experimental realization of the trapped one-dimensional Bose gas.[48–50] We will see that the F-matrix is indispensable for obtaining an exact cMPS.

The Hamiltonian of the Lieb-Liniger model on a periodic ring of length ℓ is given by

$$H = \int_0^\ell [\partial_x \psi^\dagger(x)\partial_x \psi(x) + \kappa \psi^\dagger(x)\psi^\dagger(x)\psi(x)\psi(x)]\mathrm{d}x, \tag{49}$$

where $\kappa > 0$ (repulsive) and units are chosen such that $\hbar = 1$ and the mass of each particle is $1/2$. The bosonic creation and annihilation operators satisfy the canonical commutation relation $\left[\psi(x), \psi^\dagger(y)\right] = \delta(x-y)$ (and zero otherwise). One can derive the Schrödinger equation

$$\left[-\sum_{j=1}^{n} \frac{\partial^2}{\partial x_j^2} + 2\kappa \sum_{1\le j<j'\le n} \delta(x_j - x_{j'})\right] \Psi_0(x_1, ..., x_n) = E\Psi_0(x_1, ..., x_n) \tag{50}$$

for a state with a fixed number of particles,

$$|\Psi_0\rangle = \int_0^\ell \cdots \int_0^\ell \mathrm{d}x_1 \cdots \mathrm{d}x_n \Psi_0(x_1, \ldots, x_n)\psi^\dagger(x_1)\cdots\psi^\dagger(x_n)\,|\mathrm{vac}\rangle\,, \tag{51}$$

where $|\mathrm{vac}\rangle$ is the vacuum state, i.e., $\psi(x)|\mathrm{vac}\rangle = 0$.

4.1. *Matrix product states for the integrable lattice model*

To derive the exact cMPS, we start from the integrable lattice discretization of the Lieb-Liniger model,[35,36] whose integrability is guaranteed by the Yang-Baxter relation. In this model, the Hilbert space at the ith site, V_i, is spanned by $|m\rangle_i = \frac{1}{\sqrt{m!}}(\psi_i^\dagger)^m\,|0\rangle_i$ with $m \ge 0$, where $|0\rangle_i$ is the local vacuum. Here, we have introduced the lattice regularizations of $\psi(x)$ and $\psi^\dagger(x)$ as ψ_i and $\psi_i^\dagger$, which satisfy $[\psi_i, \psi_l^\dagger] = \delta_{il}$ with all other commutators vanishing. Note that ψ_i is dimensionless and $\psi_i \to \psi(x)/\sqrt{a}$ with $x = ia$ in the continuum limit. Although V_i is infinite dimensional due to the bosonic nature, the auxiliary space $\bar{V}_j$ is two-dimensional and spanned by $|{\leftarrow}\rangle_j$ and $|{\rightarrow}\rangle_j$ in the same way as the XXZ model discussed in Sec. 2. We now define the L-operator acting on $\bar{V}_j \otimes V_i$ as

$$\mathcal{L}_{ji}(\lambda) = \begin{pmatrix} 1 - \frac{\mathbf{i}\lambda a}{2} + \frac{\kappa a}{2}\psi_i^\dagger\psi_i & -\mathbf{i}\sqrt{\kappa a}\psi_i^\dagger\rho_i \\ \mathbf{i}\sqrt{\kappa a}\rho_i\psi_i & 1 + \frac{\mathbf{i}\lambda a}{2} + \frac{\kappa a}{2}\psi_i^\dagger\psi_i \end{pmatrix}, \tag{52}$$

where $\rho_i = \left(1 + \frac{\kappa a}{4}\psi_i^\dagger\psi_i\right)^{1/2}$. Note that the matrix elements of $\mathcal{L}_{ji}$ in Eq. (52) and ρ_i are operators on V_i. Then, the monodromy matrix $\mathcal{T}(\lambda_j)$ is obtained through Eq. (9) along the same lines as the XXZ case:

$$\mathcal{T}(\lambda) = \mathcal{L}_{j1}(\lambda)\mathcal{L}_{j2}(\lambda)...\mathcal{L}_{jL}(\lambda) = \begin{pmatrix} \mathbf{A}(\lambda) & \mathbf{B}(\lambda) \\ \mathbf{C}(\lambda) & \mathbf{D}(\lambda) \end{pmatrix}, \tag{53}$$

where L is the total number of sites being related to $\ell = La$. The lattice Hamiltonian defined by $\frac{1}{\mathbf{i}}\frac{\mathrm{d}}{\mathrm{d}\lambda}\log[\mathbf{A}(\lambda) + \mathbf{D}(\lambda)]$ is known to have non-local interactions which disappear in the continuum limit.[36] Similar to Sec. 2, for a given set of rapidities, we can express the Bethe state as

$$|\Psi_a(\lambda_1, \lambda_2, ..., \lambda_n)\rangle = \mathbf{B}(\lambda_n)...\mathbf{B}(\lambda_2)\mathbf{B}(\lambda_1)\,|0\rangle\,, \tag{54}$$

where n denotes the total number of particles and $|0\rangle \equiv \bigotimes_{i=1}^{L} |0\rangle_i$. Note that all the rapidities are real numbers if $\kappa > 0$.[23] Following the strategy in Sec. 3.1, we introduce the tensor product of the auxiliary spaces as $\bar{\mathscr{H}} \equiv \bar{V}_n \otimes \cdots \otimes \bar{V}_1$. Similar to Eq. (18), we obtain

$$|\Psi_a(\lambda_1, ..., \lambda_n)\rangle = \mathrm{Tr}_{\bar{\mathscr{H}}}\left[Q_n \prod_{i=1}^{L}(\sum_{m=0}^{n} C_{n,m}(\lambda_1, ..., \lambda_n)|m\rangle_i)\right]. \tag{55}$$

where $Q_n = |\Rightarrow\rangle\langle\Leftarrow|$ and

$$C_{n,m}(\lambda_1, ..., \lambda_n) = \langle m|\,\mathcal{L}_i(\lambda_1, ..., \lambda_n)\,|0\rangle_i \tag{56}$$

with $\mathcal{L}_i(\lambda_1, ..., \lambda_n) = \mathcal{L}_{ni}(\lambda_n) \otimes \cdots \otimes \mathcal{L}_{1i}(\lambda_1)$. Compared to the XXZ case, a complication arises from the fact that V_i is not two-dimensional. Fortunately, for a fixed number of particles (n), m is restricted to be $0 \leq m \leq n$ and the recursion relations for $C_{n,m}$ can be obtained in a closed form:

$$C_{n+1,m} = \begin{pmatrix} b_{n+1,m}C_{n,m} & c_{m-1}C_{n,m-1} \\ c_m^* C_{n,m+1} & b_{n+1,m}^* C_{n,m} \end{pmatrix}, \tag{57}$$

where

$$b_{j,m} = 1 - \frac{\mathbf{i}\lambda_j a}{2} + \frac{\kappa a}{2}m, \qquad c_m = -\mathbf{i}\sqrt{(m+1)\kappa a\left(1 + \frac{m\kappa a}{4}\right)}. \tag{58}$$

From a physical point of view, $C_{n,m}$ corresponds to the creation operator for the m-particle state and is presumably related to the single particle creation $C_{n,1}$. Indeed, as we will see below, $C_{n,m}$ can be expressed in terms of $C_{n,0}$ and $C_{n,1}$.

We now solve the recursion Eq. (57) by introducing F-matrix along the same lines as in Sec. 3.2. The recursive definition of the F-matrix is given by

$$F_{n+1} = \begin{pmatrix} F_n & 0 \\ F_n \mathcal{F}_n & F_n \end{pmatrix} \quad \text{with} \quad \mathcal{F}_n = -\mathbf{i}\kappa \sum_{i=1}^{n} \frac{\mathcal{F}_n^{(i)}}{\lambda_i - \lambda_{n+1}}, \tag{59}$$

where $\mathcal{F}_n^{(i)}$ $(i = 1, 2, ..., n)$ are $2^n \times 2^n$ matrices defined by

$$\mathcal{F}_n^{(i)} = g(\lambda_i, ..., \lambda_n) \bigotimes_{l=1}^{n-i} \begin{pmatrix} S_{i,n-l+1}(\lambda_i, \lambda_{n-l+1}) & 0 \\ 0 & 1 \end{pmatrix} \otimes \begin{pmatrix} 0 & 1 \\ 0 & 0 \end{pmatrix} \bigotimes_{l=n-i+2}^{n} \begin{pmatrix} 1 & 0 \\ 0 & 1 \end{pmatrix} \tag{60}$$

with

$$g(\lambda_i, ..., \lambda_n) = \prod_{l=i+1}^{n} \frac{\lambda_i - \lambda_l + \mathbf{i}\kappa}{\lambda_i - \lambda_l} \quad \text{and} \quad S_{il}(\lambda_i, \lambda_l) = \frac{\lambda_i - \lambda_l - \mathbf{i}\kappa}{\lambda_i - \lambda_l + \mathbf{i}\kappa}. \tag{61}$$

Since $C_{0,0} = 1$, the initial condition for the F-matrix is $F_0 = 1$. Then, from the recursive definition of the F-matrix, one can easily confirm that the domain wall boundary condition is not altered, i.e., $\mathcal{Q}_n = F_n^{-1} Q_n F_n = Q_n$. As discussed in Ref. 38, one can confirm that the above F-matrix diagonalizes $C_{n,0}$ as

$$\mathcal{D}_n = F_n^{-1} C_{n,0} F_n = \bigotimes_{l=1}^{n} \begin{pmatrix} b_{n-l+1,0} & 0 \\ 0 & b^*_{n-l+1,0} \end{pmatrix}, \tag{62}$$

and transforms $C_{n,m>0}$ into upper triangular matrices. The following algebraic relations among $\mathcal{D}_n$ and $\mathcal{F}_n^{(i)}$ are useful for finding the explicit expressions for $\mathcal{C}_{n,m} = F_n^{-1} C_{n,m} F_n$:

$$\mathcal{F}_n^{(i)} \mathcal{D}_n = z_i \mathcal{D}_n \mathcal{F}_n^{(i)}, \tag{63}$$

$$\mathcal{F}_n^{(i)} \mathcal{F}_n^{(j)} = S_{ij}(\lambda_i, \lambda_j) \mathcal{F}_n^{(j)} \mathcal{F}_n^{(i)}, \tag{64}$$

$$\mathcal{F}_n^{(i)} \mathcal{F}_n^{(i)} = 0, \tag{65}$$

where $z_j = b^*_{j,0}/b_{j,0}$. One can directly confirm the above relations using Eqs. (60) and (62). To obtain $\mathcal{C}_{n,m}$, we decompose $\mathcal{C}_{n,1}$ in a parallel way as the XXZ case, i.e., $\mathcal{C}_{n,1} = \sum_{i=1}^{n} \mathcal{C}_n^{(i)}$. Then, by induction on n, we find that the following relations are true:

$$\mathcal{C}_n^{(i)} = \frac{c_0}{b_{i,0}} \mathcal{D}_n \mathcal{F}_n^{(i)} \tag{66}$$

and find the explicit finite dimensional representation for $\mathcal{C}_n^{(i)}$ as follows:

$$\mathcal{C}_n^{(i)} = g(\lambda_i, ..., \lambda_n)\frac{c_0}{b_{i,0}}\left(\prod_{l=1}^{n} b_{l,0}\right)\bigotimes_{l=1}^{n-i}\begin{pmatrix} S_{i,n-l+1}(\lambda_i, \lambda_{n-l+1}) & 0 \\ 0 & z_{n-l+1} \end{pmatrix}$$
$$\otimes \begin{pmatrix} 0 & 1 \\ 0 & 0 \end{pmatrix} \bigotimes_{l=n-i+2}^{n} \begin{pmatrix} 1 & 0 \\ 0 & z_{n-l+1} \end{pmatrix} \tag{67}$$

The obtained expression is essentially the same as Eq. (36), apart from the overall factors and the definition of z_l. Therefore, the operators satisfy the algebraic relations similar to Eqs. (27-29) by replacing $\tilde{S}_{ij}(z_i, z_j)$ with

$$\tilde{S}_{ij}(\lambda_i, \lambda_j) = \frac{z_i}{z_j} \cdot \frac{\lambda_i - \lambda_j - \mathbf{i}\kappa}{\lambda_i - \lambda_j + \mathbf{i}\kappa}. \tag{68}$$

Moreover, one can show

$$\mathcal{C}_{n,m} = \frac{1}{m!}\left(\prod_{j=0}^{m-1}\frac{c_j}{c_0}\right)\mathcal{D}_n(\mathcal{D}_n^{-1}\mathcal{C}_{n,1})^m, \tag{69}$$

for $m > 0$, and $\mathcal{C}_{n,0} = \mathcal{D}_n$. This implies the continuity of the many-body wavefunction, i.e., $\Psi_0(x_1, ..., x, x, ..., x_n) = \lim_{y\to x}\Psi_0(x_1, ..., x, y, ..., x_n)$, in the continuum limit.

4.2. *Derivation of continuous matrix product states*

We are now in a position to derive the cMPS. In the new basis, the MPS for the Bethe states is written as

$$|\Psi_a\rangle = \mathrm{Tr}_{\mathscr{H}}\left[\mathcal{Q}_n \prod_{i=1}^{L}(\sum_{m=0}^{n}\mathcal{C}_{n,m}(\lambda_1, ..., \lambda_n)|m\rangle_i)\right]. \tag{70}$$

where $\mathcal{C}_{n,0} \equiv \mathcal{D}_n$. Let us first consider the continuum limit of the diagonal part. In Eq. (70), $\mathcal{D}^{i_2-i_1}$ is picked up when particles are absent in the interval $i_1 \leq i < i_2$. Therefore, in the continuum limit, we obtain

$$\lim_{a\to 0}\mathcal{D}^{i_2-i_1} = \exp[\mathbf{i}(x_2 - x_1)\mathcal{K}_n], \tag{71}$$

where $x_2 - x_1 = (i_2 - i_1)a$ and the matrix $\mathcal{K}_n$ is defined by

$$\mathcal{K}_n = \sum_{j=1}^{n}\bigotimes_{l=1}^{n-j}\begin{pmatrix} 1 & 0 \\ 0 & 1 \end{pmatrix} \otimes \begin{pmatrix} -\frac{\lambda_j}{2} & 0 \\ 0 & \frac{\lambda_j}{2} \end{pmatrix}\bigotimes_{l=n-j+2}^{n}\begin{pmatrix} 1 & 0 \\ 0 & 1 \end{pmatrix}. \tag{72}$$

Next we examine the off-diagonal part which creates the m-particle states. In the limit of $a \to 0$ with a finite n, one can show $\sqrt{m!}\mathcal{C}_{n,m} \simeq (\mathcal{C}_{n,1})^m$. Therefore, we find

$$\sum_{m=1}^{n} \frac{1}{\sqrt{m!}} \mathcal{C}_{n,m} (\psi_i^\dagger)^m \simeq \mathcal{D}_n \exp(\mathcal{D}_n^{-1} \mathcal{C}_{n,1} \psi_i^\dagger)$$
$$\simeq \mathcal{D}_n \exp[\mathcal{R}_n^\dagger \psi^\dagger(x) a], \tag{73}$$

where

$$\mathcal{R}_n^\dagger = -\mathbf{i}\sqrt{\kappa} \sum_{i=1}^{n} \mathcal{F}_n^{(i)}, \tag{74}$$

which is the analogue of the reflection operator.[51] Finally, introducing the path-ordered exponential ($P\exp$), we arrive at the integral expression for Eq. (70):

$$|\Psi_0\rangle = \lim_{a\to 0} |\Psi_a\rangle = \mathrm{Tr}_{\bar{\mathscr{H}}} \left[\mathcal{Q}_n P \exp[\int_0^\ell \mathrm{d}x (\mathbf{i}\mathcal{K}_n + \mathcal{R}_n^\dagger \psi^\dagger(x))] \right] |\mathrm{vac}\rangle \, . \tag{75}$$

At first glance, the above expression appears to be a coherent state with fluctuating particle numbers. However, it is not the case because the matrix $\mathcal{Q}_n$ projects onto the subspace in which the number of particles is fixed. This appearance of $\mathcal{Q}_n$ is a significant difference from the variational cMPS proposed by Verstraete and Cirac.[32]

5. Conclusion

In summary, we have studied exact MPSs in one-dimensional quantum integrable models solvable by means of the Bethe Ansatz. In contrast to other classes of models such as the AKLT model, the exact MPSs with explicit matrix elements can be obtained not only for the ground states but also for the excited states. For a fixed number of degrees of freedom (say n), the dimension of the matrices is 2^n, and it can be potentially infinite in the thermodynamic limit. In the XXX and XXZ spin-$\frac{1}{2}$ Heisenberg chains, we have shown that the MPS representation for the Bethe states, which is equivalent to the MPA introduced by Alcaraz and Lazo,[18,19] can be derived from the ABA. With the aid of the similarity transformation, we have obtained finite-dimensional representations for the matrices. This transformation has a natural interpretation in terms of the mapping from the six-vertex model to the five-vertex model with non-local Boltzmann weights. Similar to the case of the Drinfel'd twist in the quantum space ($\mathscr{H}$), the matrices in $\bar{\mathscr{H}}$

have a quite simple structure (see Eqs. (35) and (36)). This observation has led us to study the continuous MPS in the Lieb-Liniger model which is continuous in space. From the lattice version of the Bose-Hubbard model which is integrable by means of the Bethe ansatz, we have derived the exact MPS for the Bethe states in this model. The exact cMPS can then be obtained as a continuum limit of the above lattice MPS. The similarity transformation by the F-matrix is indispensable for obtaining the cMPS in the form of the path-ordered exponential as shown in Eq. (75).

An interesting direction for future research would be the application of the obtained matrices to the calculation of correlation functions. However, a technical difficulty arises from the boundary matrix $Q_n(=\mathcal{Q}_n)$, which corresponds to the domain wall boundary condition in the six-vertex model. For instance, in the case of the XXZ model, the expectation value of the local operator O_i at the ith site yields

$$\langle\Psi_n| O_i |\Psi_n\rangle = \frac{\langle L| \mathcal{E}_n^{i-1}\mathcal{O}_n\mathcal{E}_n^{L-i} |R\rangle}{\langle L| \mathcal{E}_n^{L} |R\rangle}, \tag{76}$$

where $|\Psi_n\rangle = |\lambda_1, ..., \lambda_n\rangle$ in Eq. (24) and $|L\rangle$ and $|R\rangle$ are $|{\Leftarrow}\rangle \otimes |{\Leftarrow}\rangle$ and $|{\Rightarrow}\rangle \otimes |{\Rightarrow}\rangle$, respectively. Here, the operators in $\bar{\mathscr{H}} \otimes \bar{\mathscr{H}}$ are defined by

$$\mathcal{E}_n = \sum_{\sigma=\uparrow,\downarrow} \mathcal{M}_{n,\sigma}^* \otimes \mathcal{M}_{n,\sigma}, \quad \mathcal{O}_n = \sum_{\sigma,\sigma'=\uparrow,\downarrow} \langle\sigma| O_i |\sigma'\rangle_i \mathcal{M}_{n,\sigma}^* \otimes \mathcal{M}_{n,\sigma'}, \tag{77}$$

where the asterisk denotes the complex conjugate, and $\mathcal{M}_{n,\uparrow}$ and $\mathcal{M}_{n,\downarrow}$ are $\mathcal{D}_n$ and $\mathcal{C}_n$, respectively. Compared to the standard MPS, which is translationally invariant, it is difficult to utilize the transfer matrix $\mathcal{E}_n$ since this matrix is a 2^{2n}-dimensional matrix, and the numerator and denominator in Eq. (76) cannot be written in the form of trace.

Another interesting direction is fermionic description instead of matrices. At the special anisotropy point $\Delta = 0$, the matrices in Eq. (46) turn out to be Jordan-Wigner fermions in the auxiliary space because $S_{i,j} = -1$ for any i and j. The same simplification also occurs at the infinitely strong repulsion ($\kappa \to \infty$) in the Lieb-Liniger model. In this limit, called the Tonks-Girardeau limit,[52] the matrices $\mathcal{F}_n^{(i)}$ in Eq. (60) again become the Jordan-Wigner fermions in $\bar{\mathscr{H}}$. This seems to be related to the kink-like configuration of arrows on a vertical line in the graphical representations of $\mathcal{C}_n^{(i)}$ (see Fig. (A. 10) in Ref. 25). Therefore, it would be interesting to explore a fermionic description of matrices away from the free-fermion point, which is perhaps related to the hidden Grassmann structure discussed in another context.[53–55]

Acknowledgments

The authors are grateful to Z. C. Gu, V. E. Korepin, A. W. W. Ludwig, C. Matsui, T. Nishino, M. Suzuki, Hong-Hao Tu, and F. Verstraete for their valuable comments and discussions. This work was supported in part by Grant-in-Aids (No. 20740214) from the Ministry of Education, Culture, Sports, Science and Technology of Japan. HK is supported by the JSPS Postdoctoral Fellowships for Research Abroad and and NSF Grant No. PHY05-51164.

References

1. R. J. Baxter, *J. Stat. Phys.* **19**, p. 461 (1978).
2. I. Affleck, T. Kennedy, E. H. Lieb and H. Tasaki, *Phys. Rev. Lett.* **59**, p. 799 (1987).
3. I. Affleck, T. Kennedy, E. H. Lieb and H. Tasaki, *Commun. Math. Phys.* **115**, p. 477 (1988).
4. A. Klümper, A. Schadschneider and J. Zittarz, *Z. Phys. B* **87**, p. 281 (1992).
5. K. Totsuka and M. Suzuki, *J. Phys.: Condens. Matter* **7**, p. 1639 (1995).
6. V. Karimipour and L. Memarzadeh, *Phys. Rev. B* **77**, p. 094416 (2008).
7. M. Fannes, B. Nachtergaele, and R. F. Werner, *Commun. Math. Phys.* **144**, p. 443 (1992).
8. S. R. White, *Phys. Rev. Lett.* **69**, p. 2863 (1992).
9. S. R. White, *Phys. Rev. B* **48**, p. 345 (1993).
10. U. Schollwöck, *Rev. Mod. Phys.* **77**, p. 259 (2005).
11. T. Nishino, *J. Phys. Soc. Jpn.* **64**, p. 3598 (1995).
12. T. Nishino and K. Okunishi, *J. Phys. Soc. Jpn.* **64**, p. 4084 (1995).
13. G. Vidal, J. I. Lattore, E. Rico and A. Kitaev, *Phys. Rev. Lett.* **90**, p. 227902 (2003).
14. M. B. Hastings, *J. Stat. Mech* , p. P08024 (2007).
15. J. Eisert, M. Cramer and M. B. Plenio, *Rev. Mod. Phys.* **82**, p. 277 (2010).
16. N. Schuch, M. M. Wolf, F. Verstraete and J. I. Cirac, *Phys. Rev. Lett.* **100**, p. 030504 (2008).
17. P. Calabrese and J. Cardy, *J. Stat. Mech.* , p. P06002 (2004).
18. F. C. Alcaraz and M. J. Lazo, *J. Phys. A: Math. Gen.* **37**, p. L1 (2004).
19. F. C. Alcaraz and M. J. Lazo, *J. Phys. A: Math. Gen.* **39**, p. 11335 (2006).
20. F. C. Alcaraz and M. J. Lazo, *J. Phys. A: Math. Gen.* **37**, p. 4149 (2004).
21. H. A. Bethe, *Z. Phys.* **71**, p. 205 (1931).
22. L. D. Faddeev, *Les Houches 1982, Recent Advances in Field Theory and Statistical Mechanics* (Elsevier Science, Amsterdam, 1984), p. 561.
23. V. E. Korepin, N. M. Bogoliubov and A. G. Izergin, *Quantum Inverse Scattering Method and Correlation Functions* (Cambridge University Press, 1993).
24. R. I. Nepomechie, *Int. J. Mod. Phys. B* **13**, 2973 (arXiv:hep-th/9810032) (1999).
25. H. Katsura and I. Maruyama, *J. Phys. A: Math. Theor.* **43**, p. 175003 (2010).
26. O. Golinelli and K. Mallick, *J. Phys. A: Math. Gen.* **39**, p. 10647 (2006).

27. V. E. Korepin, *Commun. Math. Phys.* **86**, p. 391 (1982).
28. A. B. Zamolodchikov and A. B. Zamolodchikov, *Annals of Phys., NY* **120**, p. 253 (1979).
29. J. I. Cirac and G. Sierra, *Phys. Rev. B* **81**, p. 104431 (2010).
30. F. D. M. Haldane, *Phys. Rev. Lett.* **60**, p. 635 (1988).
31. B. S. Shastry, *Phys. Rev. Lett.* **60**, p. 639 (1988).
32. F. Verstraete and J. I. Cirac, *Phys. Rev. Lett.* **104**, p. 190405 (2010).
33. E. H. Lieb and W. Liniger, *Phys. Rev.* **130**, p. 1605 (1963).
34. L. Amico and V. E. Korepin, *Ann. Phys.* **314**, p. 496 (2004).
35. A. G. Izergin and V. E. Korepin, *Nucl. Phys. B* **205**, p. 401 (1982).
36. V. O. Tarasov, L. A. Takhtadzhyan and L. D. Faddeev, *Theor. Math. Phys.* **57**, p. 1059 (1983).
37. J. Haegeman, J. I. Cirac, T. J. Osborne, H. Verschelde and F. Verstraete, *Phys. Rev. Lett.* **105** p. 251601 (2010).
38. I. Maruyama and H. Katsura, *J. Phys. Soc. Jpn.* **79**, p. 073002 (2010).
39. T. Deguchi, *Classical and Quantum Nonlinear Integrable Systems: Theory and Applications* (Institute of Physics Publishing, 2003), p. 113.
40. S. Östlund and S. Rommer, *Phys. Rev. Lett.* **75**, p. 3537 (1995).
41. A. G. Izergin, *Sov. Phys. Doklady* **32**, p. 878 (1987).
42. J. M. Maillet and J. S. de Santos, *q-alg/9612012* (1996).
43. N. Kitanine, J. M. Maillet and V. Terras, *Nucl. Phys. B* **554**, p. 647 (1999).
44. N. Kitanine, J. M. Maillet and V. Terras, *Nucl. Phys. B* **567**, p. 554 (2000).
45. N. Kitanine, J. M. Maillet, N. A. Slavnov and V. Terras, *Nucl. Phys. B* **641**, p. 487 (2002).
46. N. Kitanine, J. M. Maillet, N. A. Slavnov and V. Terras, *Nucl. Phys. B* **712**, p. 600 (2005).
47. N. Kitanine, J. M. Maillet, N. A. Slavnov and V. Terras, *Nucl. Phys. B* **729**, p. 558 (2005).
48. B. Paredes, A. Widera, V. Murg, O. Mandel, S. Folling, I. Cirac, G. V. Shlyapnikov, T. W. Hansch and I. Bloch, *Nature* **429**, p. 277 (2004).
49. T. W. T. Kinoshita and D. S. Weiss, *Science* **305**, p. 1125 (2004).
50. P. Calabrese and J.-S. Caux, *Phys. Rev. Lett.* **98**, p. 150403 (2007).
51. H. Itoyama and H. B. Thacker, *Phys. Rev. Lett.* **58**, p. 1395 (1987).
52. M. Girardeau, *J. Math. Phys.* **1**, p. 516 (1960).
53. H. Boos, M. Jimbo, T. Miwa, F. Smirnov and Y. Takeyama, *Commun. Math. Phys.* **272**, p. 263 (2007).
54. H. Boos, M. Jimbo, T. Miwa, F. Smirnov and Y. Takeyama, *Commun. Math. Phys.* **286**, p. 875 (2009).
55. M. Jimbo, T. Miwa and F. Smirnov, *J. Phys. A: Math. Theor.* **42**, p. 304018 (2009).

A SYSTEMATIC WAY TO FIND AND CONSTRUCT EXACT FINITE DIMENSIONAL MATRIX-PRODUCT STATIONARY STATES

Y. HIEIDA

Computer and Network Center, Saga University,
Saga 840-8502, Japan
E-mail: hieida@cc.saga-u.ac.jp

T. SASAMOTO

Department of Mathematics and Informatics,
Faculty of Science, Chiba University,
1-33 Yayoi-cho, Inage, Chiba 263-8522, Japan

We explain how to construct matrix-product stationary states which are composed of finite-dimensional matrices. Our construction explained in this article was first presented in a part of [Hieida and Sasamoto: J. Phys. A: Math. Gen. **37** p. 9873 (2004)] for general models. In this article, we give more details on the treatment than in the above-mentioned reference, for one-dimensional asymmetric simple exclusion process(ASEP).

Keywords: Stationary state; matrix-product state; asymmetric simple exclusion process(ASEP); PASEP.

1. Introduction

Evolution of the probabilities of configurations in a class of one-dimensional stochastic models is described by the master equation like this:

$$\frac{d}{dt}\vec{P}_L(t) = -H\vec{P}_L(t), \tag{1}$$

where H is a transition rate matrix and $\vec{P}_L(t)$ is a column vector whose component $P(C;t)$ is the probability of finding the L-site system in a configuration C at time t. If $\vec{P}_L(t)$ is independent of t, we obtain from Eq. (1)

$$0 = H\vec{P}_L. \tag{2}$$

We call the solution $\vec{P}_L$ of this equation the **stationary state**. Usually, as $t \to \infty$, $\vec{P}_L(t)$ reaches a stationary state $\vec{P}_L$. So, once we have the stationary state, we can calculate, for example, the density profile, correlations, etc. in the long-time limit. From this viewpoint, obtaining stationary states is important in nonequilibrium statistical mechanics.

For small L, one can solve Eq. (2) numerically and study its properties exactly. But, in general, obtaining stationary state for **any** system size L as a function of L is a difficult task.

It is known that, for some one-dimensional stochastic models, their stationary states can be written in a special form, namely, in the form of matrix products. We call this the matrix-product stationary state(**MPSS**).[2] We consider the situation in which this matrix is independent of system size; in this case, we can construct its stationary state for **any** system size.

Thus we want to find matrices which compose a MPSS. Although, the dimensions of those matrices are generally infinite for a model whose interaction range is finite,[3,4] the dimensions can be finite for some models including the ASEP under some conditions for model parameters.[5,6] We call a MPSS for the former(resp. latter) case **an infinite(resp. a finite)-dimensional MPSS**. To our knowledge, there is no systematic way of constructing an infinite-dimensional MPSS. But, for a finite-dimensional MPSS, we found a systematic way of finding them,[1] which we explain in this article.

Our construction explained in this article was first presented in a part of Ref. 1, where the above-mentioned construction and a systematic way of checking the validity of the obtained MPSS for an arbitrary system size were explained for general stochastic models on the one-dimensional lattice. Main purpose of this article is to give more details on the treatment in Ref. 1. In order to explain our construction concretely, we use in this article only the one-dimensional asymmetric simple exclusion process(**ASEP**).[7]

It is interesting that numerical matrix-product states in stochastic models (for example, Ref. 8) also play an important role in the method of density matrix renormalization group (DMRG).[9–11] This article, however, does not treat such a numerical method.

2. Model

The ASEP in this article [a] is defined on the one-dimensional lattice whose size is L(Fig. 1). Each site can take two states, namely, a site is either empty or occupied by one particle. The time of the model is continuous. A particle in the bulk hops to the left(resp. right) neighbor site with a rate q(resp. 1)

[a]The ASEP in this article is also called Partially ASEP(PASEP).

if the left(resp. right) neighbor site is empty. At the leftmost site, a particle is injected with a rate α if the site is empty. A particle at the rightmost site is removed with a rate β if the site is occupied by a particle.

Fig. 1. The ASEP on the one-dimensional lattice whose size is $L = 5$. A particle in the bulk hops to the left(resp. right) neighbor site with a rate q(resp. 1) if the left(resp. right) neighbor site is empty. At the leftmost site, a particle is injected with a rate α if the site is empty. A particle at the rightmost site is removed with a rate β if the site is occupied by a particle.

As for the master equation, for example, $\vec{P}_{L=4}(t)$ in Eq. (1) has the following components:

$$\vec{P}_{L=4}(t) = \begin{pmatrix} P(0,0,0,0;t) \\ P(0,0,0,1;t) \\ P(0,0,1,0;t) \\ P(0,0,1,1;t) \\ P(0,1,0,0;t) \\ P(0,1,0,1;t) \\ P(0,1,1,0;t) \\ P(0,1,1,1;t) \\ P(1,0,0,0;t) \\ P(1,0,0,1;t) \\ P(1,0,1,0;t) \\ P(1,0,1,1;t) \\ P(1,1,0,0;t) \\ P(1,1,0,1;t) \\ P(1,1,1,0;t) \\ P(1,1,1,1;t) \end{pmatrix}. \tag{3}$$

where τ_k $(1 \leq k \leq 4)$ in the component $P(\tau_1, \tau_2, \tau_3, \tau_4; t)$ of Eq. (3) represents the number of a particle at the k-th site. The transition rate matrix in Eq. (1) for the ASEP is defined as follows:

$$H := h^{(\mathrm{L})} + \sum_{k=1}^{L-1} h_k + h^{(\mathrm{R})}. \tag{4}$$

In this equation, $h^{(\mathrm{L})}$(resp. $h^{(\mathrm{R})}$) which expresses injections of particles at the leftmost site(resp. removals of particles at the rightmost site), is defined by

$$h^{(\mathrm{L})} := h^{\mathrm{left}} \otimes I^{\otimes(L-1)} \qquad \text{and} \qquad h^{(\mathrm{R})} := I^{\otimes(L-1)} \otimes h^{\mathrm{right}} \quad , \tag{5}$$

where

$$h^{\mathrm{left}} := \begin{pmatrix} \alpha & 0 \\ -\alpha & 0 \end{pmatrix} \tag{6}$$

$$h^{\mathrm{right}} := \begin{pmatrix} 0 & -\beta \\ 0 & \beta \end{pmatrix} \quad . \tag{7}$$

Eq. (6)(resp. Eq. (7)) is expressed in a basis of states whose order is $\tau_1 = 0, 1$(resp. $\tau_L = 0, 1$). In Eq. (5), we introduce a shorthand notation of direct products of the two-dimensional identity matrix I:

$$I^{\otimes n} := \overbrace{I \otimes I \otimes \cdots \otimes I}^{n}. \tag{8}$$

In Eq. (4), h_k, which describes hopping process between k-th and $(k+1)$-th sites is defined

$$h_k := I^{\otimes(k-1)} \otimes h_{\mathrm{int}} \otimes I^{\otimes(L-k-1)} \quad , \tag{9}$$

where

$$h_{\mathrm{int}} := \begin{pmatrix} 0 & 0 & 0 & 0 \\ 0 & q & -1 & 0 \\ 0 & -q & 1 & 0 \\ 0 & 0 & 0 & 0 \end{pmatrix}. \tag{10}$$

This is expressed in a basis of states whose order is $(\tau_k, \tau_{k+1}) = (0,0), (0,1), (1,0), (1,1)$.

3. Matrix-product Stationary States (MPSSs)

It is possible that some models including the ASEP have a special form of a stationary state, namely, a matrix-product stationary state(**MPSS**). For example, "the vector Eq. (3) can be represented by a MPSS" means that there exist $\langle W \mid$, E, D and $\mid V\rangle$ such that Eq. (3) is equal to

$$\frac{1}{Z_4}\langle W| \begin{pmatrix} E \\ D \end{pmatrix}^{\otimes 4} |V\rangle. \tag{11}$$

Generally, the MPSS for the ASEP is the stationary state which is equal to the following form

$$\frac{1}{Z_L}\langle W| \begin{pmatrix} E \\ D \end{pmatrix}^{\otimes L} |V\rangle. \tag{12}$$

In Eq. (11) and Eq. (12),

- $\langle W|$ and $|V\rangle$ are an M-dimensional row vector and an M-dimensional column vector, respectively.
- E and D are M-dimensional square matrices.
- Z_L is the normalization constant whose definition is

$$Z_L := \langle W| (E+D)^L |V\rangle. \tag{13}$$

Hereafter, we call Eq. (12) **an M-dimensional MPSS**. It should be noted that M is not always finite(in general, M is ∞) as explained in Sec. 1.

In the following section 4, we derive necessary conditions for the existence of an $M(=1,2)$-dimensional MPSS for the ASEP by using its stationary states for finite system sizes. In the subsequent section(Sec. 5), we explain how to find $\langle W|$, E, D and $|V\rangle$ in a two-dimensional MPSS assuming the existence of the MPSS.

4. Necessary Conditions for the Existence of an $M(=1,2)$-dimensional MPSS

In this section, we derive necessary conditions for the existence of an $M(=1,2)$-dimensional MPSS for the ASEP [b].

As a tool of our derivation, we introduce a matrix form of a stationary state $\vec{P}_L$. This form has the same components as $\vec{P}_L$. The difference between the matrix form and $\vec{P}_L$ exists in ways of arranging the components. For example, the matrix form $P^{2,2}$ is defined by rearranging components of

[b] Ref. 1 describes also derivation for general M-dimensional MPSSs which some models with N states per site have. The ASEP in this article corresponds to $N = 2$.

$\vec{P}_{L=4}$(see the components of Eq. (3) with t ignored) as

$$P^{2,2} := \begin{pmatrix} P(0,0,0,0) & P(0,0,0,1) & P(0,0,1,0) & P(0,0,1,1) \\ P(0,1,0,0) & P(0,1,0,1) & P(0,1,1,0) & P(0,1,1,1) \\ P(1,0,0,0) & P(1,0,0,1) & P(1,0,1,0) & P(1,0,1,1) \\ P(1,1,0,0) & P(1,1,0,1) & P(1,1,1,0) & P(1,1,1,1) \end{pmatrix}. \tag{14}$$

According to this definition, the matrix form $P^{2,2}$ converted from the MPSS(Eq. (12) with $L = 4$) is

$$P^{2,2} = \frac{1}{Z_4} \begin{pmatrix} \langle W|EEEE|V\rangle & \langle W|EEED|V\rangle & \langle W|EEDE|V\rangle & \langle W|EEDD|V\rangle \\ \langle W|EDEE|V\rangle & \langle W|EDED|V\rangle & \langle W|EDDE|V\rangle & \langle W|EDDD|V\rangle \\ \langle W|DEEE|V\rangle & \langle W|DEED|V\rangle & \langle W|DEDE|V\rangle & \langle W|DEDD|V\rangle \\ \langle W|DDEE|V\rangle & \langle W|DDED|V\rangle & \langle W|DDDE|V\rangle & \langle W|DDDD|V\rangle \end{pmatrix}. \tag{15}$$

Generally, the matrix form $P^{m,n}$ converted from the MPSS(Eq. (12)) can be written down as

$$P^{m,n} := \frac{1}{Z_{m+n}} \langle W| \begin{pmatrix} E \\ D \end{pmatrix}^{\otimes m} (E\ D)^{\otimes n} |V\rangle, \tag{16}$$

with $m + n = L$.

The rank of a matrix form is important to our derivation. Suppose, by elementary transformations, the $2^m \times 2^n$ matrix $P^{m,n}$ can be transformed into a matrix whose form is

$$\left(\begin{array}{c|c} I_r & B \\ \hline \multicolumn{2}{c}{O} \end{array}\right), \tag{17}$$

where I_r, O and B represents the $r \times r$ identity matrix, an $(2^m - r) \times 2^n$ zero matrix and an $r \times (2^n - r)$ matrix, respectively. Then, we can tell that the rank of $P^{m,n}$ is r because elementary transformations do not change ranks of matrices and the rank of Eq. (17) is r. Therefore, the rank of $P^{2,2}$ in Eq. (14) is generally greater than 2. However, we can prove the fact that

> "if the stationary state can be written in a two(resp. one)-dimensional MPSS, the rank of $P^{2,2}$ is at most 2(resp. 1)." (18)

We give a proof of this fact only for the case of a two-dimensional MPSS. [c]

[c]This proof is slightly different from one in Ref. 1.

From Eq. (16) with $m = n = 2$, we obtain

$$Z_4 P^{2,2} := \langle W| \begin{pmatrix} E \\ D \end{pmatrix}^{\otimes 2} \left(E\ D\right)^{\otimes 2} |V\rangle. \tag{19}$$

The right hand side of this equation can be written as a matrix product

$$AB \tag{20}$$

where

$$A := \langle W| \begin{pmatrix} E \\ D \end{pmatrix}^{\otimes 2}, \quad B := \left(E\ D\right)^{\otimes 2} |V\rangle. \tag{21}$$

A is a 4×2 matrix and B is a 2×4 matrix. Please note that both of the rank of A and the rank of B is at most 2. And linear algebra tells us that the rank of AB is less than or equal to the smaller number of the rank of A and the rank of B. This concludes the rank of $P^{2,2}$ is not greater than 2.

Using the fact(18), let us calculate concretely the necessary condition of existence of an $M(=1,2)$-dimensional MPSS for

$$\alpha > 0\,,\ \beta > 0\,. \tag{22}$$

First, we obtain $\vec{P}_{L=4}$ by solving Eq. (2) with $L = 4$. And then we convert the vector form $\vec{P}_{L=4}$ into the matrix form $P^{2,2}$. In the following, we denote the element in the i-th row and the j-th column of a matrix A by $(A)_{i,j}$. According to Maple, which is one of computer algebra systems, we get

$$\left(Z_4 P^{2,2}\right)_{1,1} = \left(\frac{\beta}{\alpha}\right)^4, \tag{23}$$

Because this is nonzero element(see Eq. (22)), we can multiply the first row of $P^{2,2}$ by the inverse of Eq. (23). Then, we subtract the first row of the resultant matrix multiplied by $\left(P^{2,2}\right)_{i,1}$ $(i = 2,3,4)$ from the i-th row. Thus we obtain a matrix $A^{(2)}$ like this:

$$\begin{pmatrix} 1 & * & * & * \\ 0 & * & * & * \\ 0 & * & * & * \\ 0 & * & * & * \end{pmatrix}, \tag{24}$$

where the elements denoted by "$*$" mean some expressions. The second diagonal element of $A^{(2)}$ is this:

$$\left(A^{(2)}\right)_{2,2} = -\frac{\beta^2 (\alpha + q)(\alpha - 1 + \beta + q)}{\alpha^2} \tag{25}$$

$$\propto (\alpha - 1 + \beta + q)\,. \tag{26}$$

case 1 First, we treat the case

$$\alpha - 1 + \beta + q \neq 0. \tag{27}$$

Multiplying the second row of $A^{(2)}$ by the inverse of $\left(A^{(2)}\right)_{2,2}$ in Eq. (25), we obtain the matrix $\widetilde{A}^{(2)}$. Then, we subtract the second row (of $\widetilde{A}^{(2)}$) multiplied by $\left(A^{(2)}\right)_{i,2}$ $(i = 1, 3, 4)$ from the i-th row of $\widetilde{A}^{(2)}$. Thus we obtain the matrix, which we call $A^{(3)}$, like this:

$$\begin{pmatrix} 1 & 0 & * & * \\ 0 & 1 & * & * \\ 0 & 0 & 0 & 0 \\ 0 & 0 & * & * \end{pmatrix}. \tag{28}$$

It should be noted that any of the elements of the third row of this matrix is zero. $\left(A^{(3)}\right)_{4,3}$ is

$$\left(A^{(3)}\right)_{4,3} = \frac{(q+1)\left(\alpha\beta + \alpha q - q + q^2 + q\beta\right)(\alpha - 1 + \beta + q)\,\beta}{\alpha + q} \tag{29}$$

$$\propto \left(\alpha\beta + \alpha q - q + q^2 + q\beta\right)(\alpha - 1 + \beta + q). \tag{30}$$

This is nonzero if we assume Eq. (27) and

$$\alpha\beta + \alpha q - q + q^2 + q\beta \neq 0. \tag{31}$$

We interchange the 3rd row and the 4th row of $A^{(3)}$ and call the resultant matrix $A'^{(3)}$.

Because $\left(A'^{(3)}\right)_{3,3} = \left(A^{(3)}\right)_{4,3} \neq 0$, we can multiply the 3rd row of $A'^{(3)}$ by the inverse of $\left(A'^{(3)}\right)_{3,3}$. We call the obtained matrix $\widetilde{A'}^{(3)}$. Then, we subtract the 3rd row (of $\widetilde{A'}^{(3)}$) multiplied by $\left(A'^{(3)}\right)_{i,3}$ from the $i(=1, 2, 4)$-th row. Thus we obtain the matrix

$$A^{(4)} := \begin{bmatrix} 1 & 0 & 0 & 0 \\ 0 & 1 & 0 & -\frac{\beta+q}{\beta} \\ 0 & 0 & 1 & \beta^{-1} \\ 0 & 0 & 0 & 0 \end{bmatrix}. \tag{32}$$

Comparing this with Eq. (17), we know the rank of $A^{(4)}$ is 3.

case 2 We go back to Eq. (30) and treat the case where Eq. (31) does not hold. In this case,

$$\alpha\beta + \alpha q - q + q^2 + q\beta = 0. \tag{33}$$

We solve this with respect to α and obtain

$$\alpha = -\frac{q\,(-1+\beta+q)}{\beta+q}. \tag{34}$$

Substituting this into $A^{(3)}$, we obtain

$$\begin{bmatrix} 1 & 0 & -\frac{q(-1+\beta+q)^2}{\beta\,(\beta+q)} & -\frac{q(-1+\beta+q)^2}{\beta^2(\beta+q)} \\ 0 & 1 & -\frac{q\beta-\beta-2\,q+q^2}{\beta+q} & -\frac{2\,q^2+3\,q\beta-2\,q-\beta+\beta^2}{\beta\,(\beta+q)} \\ 0 & 0 & 0 & 0 \\ 0 & 0 & 0 & 0 \end{bmatrix}. \tag{35}$$

Comparing this with Eq. (17), we know the rank of Eq. (35) is 2.

case 3 We go back to Eq. (26) and treat the case where Eq. (27) does not hold. In this case,

$$\alpha - 1 + \beta + q = 0. \tag{36}$$

We solve this with respect to α and obtain

$$\alpha = 1 - \beta - q. \tag{37}$$

Substituting this into $A^{(2)}$, we obtain

$$\begin{bmatrix} 1 & -\frac{-1+\beta+q}{\beta} & -\frac{-1+\beta+q}{\beta} & \frac{(-1+\beta+q)^2}{\beta^2} \\ 0 & 0 & 0 & 0 \\ 0 & 0 & 0 & 0 \\ 0 & 0 & 0 & 0 \end{bmatrix}. \tag{38}$$

Comparing this with Eq. (17), we know the rank of Eq. (38) is 1.

In summary,

- the rank of $P^{2,2}$ is 3 if $\alpha - 1 + \beta + q \neq 0$ and $\alpha\beta + \alpha q - q + q^2 + q\beta \neq 0$ are satisfied. [d]

[d]The reason why the rank is not 4 in this case is that the ASEP has infinite-dimensional MPSSs with

$$(DE - qED) \propto (E + D)\,, \quad D \mid V\rangle \propto \mid V\rangle \tag{39}$$

and thus $EE \mid V\rangle$,$ED \mid V\rangle$,$DE \mid V\rangle$ and $DD \mid V\rangle$ can be written as linear combinations of (at most) three independent vectors.

- the rank of $P^{2,2}$ is 2 if $\alpha - 1 + \beta + q \neq 0$ and
$$\alpha\beta + \alpha q - q + q^2 + q\beta = 0 \tag{40}$$
are satisfied.
- the rank of $P^{2,2}$ is 1 if
$$\alpha - 1 + \beta + q = 0 \tag{41}$$
is satisfied.

So we can conclude that according to the fact(18), Eq. (40)(resp. Eq. (41)) is the necessary condition of existence of a two(resp. one)-dimensional MPSS. These conditions agree with the known result.[5,6,12,13]

5. How to Find Two-dimensional Matrices and Vectors in MPSSs for the ASEP

For the ASEP, the MPSS is constructed according to Eq. (12) with matrices(E and D) and vectors($\langle W|$ and $|V\rangle$). Hereafter we call the matrices and vectors in the MPSS "**the set of matrices**".

In this section, we explain our way of finding the set of matrices in the two-dimensional MPSS for the ASEP when $\vec{P}_L$ can be expressed as a MPSS. It should be noted that our way of finding is not restricted to two-dimensional MPSS for the ASEP. Our way can be applied to finite $M(\geq 2)$-dimensional MPSSs for models which have the finite numbers $N(\leq M)$ of states per site.[1]

A similarity transformation plays a crucial role in our way of finding. So we begin with an explanation of the transformation in the MPSS.

Let us consider a similarity transformation for the set of matrices
$$\widetilde{E} := S^{-1}ES\ , \quad \widetilde{D} := S^{-1}DS\ , \quad \langle\widetilde{W}| := \langle W|S\ , \quad |\widetilde{V}\rangle := S^{-1}|V\rangle, \tag{42}$$
where S is a 2×2 matrix. Please note that, in this section, E and D are 2×2 matrices and $\langle W|$ and $|V\rangle$ are two-dimensional vectors. The transformation Eq. (42) does not change a matrix form Eq. (16) including, for example, $P^{2,2}$ in Eq. (15).

We make a choice S in Eq. (42) as [e]
$$S := \begin{pmatrix} \kappa & \mu \\ \lambda & \nu \end{pmatrix}, \tag{44}$$

[e]The following set of Eqs. (44) and (45) is what the equation
$$S := \left(E|V\rangle\ D|V\rangle\right), \tag{43}$$
which is Eq. (2.28) in Ref. 1, means.

where

$$\begin{pmatrix}\kappa\\ \lambda\end{pmatrix} := E \mid V\rangle\,, \quad \begin{pmatrix}\mu\\ \nu\end{pmatrix} := D \mid V\rangle. \tag{45}$$

Our choice, namely, the set of Eqs. (44) and (45), plays another crucial role in our way of finding. Because of $S^{-1}S = \begin{pmatrix}1\,0\\ 0\,1\end{pmatrix}$,

$$S^{-1}\begin{pmatrix}\kappa\\ \lambda\end{pmatrix} = \begin{pmatrix}1\\ 0\end{pmatrix}\,, \quad S^{-1}\begin{pmatrix}\mu\\ \nu\end{pmatrix} = \begin{pmatrix}0\\ 1\end{pmatrix}. \tag{46}$$

Using Eqs. (42) and (45) and (46), we can derive [f]

$$\widetilde{E}|\widetilde{V}\rangle = S^{-1}E|V\rangle = S^{-1}\begin{pmatrix}\kappa\\ \lambda\end{pmatrix} = \begin{pmatrix}1\\ 0\end{pmatrix}. \tag{48}$$

$$\therefore\, \widetilde{E}|\widetilde{V}\rangle = \begin{pmatrix}1\\ 0\end{pmatrix}. \tag{49}$$

Similarly, we can derive also the equation

$$\widetilde{D}|\widetilde{V}\rangle = \begin{pmatrix}0\\ 1\end{pmatrix}. \tag{50}$$

Using Eqs. (49) and (50), we can derive the equations for $\widetilde{E}$ and $\widetilde{D}$ [g]

$$Z_3\; P^{1,2}[1:2,1:2] = \left(Z_2P^{1,1}\right)\widetilde{E} \tag{52}$$

and

$$Z_3\; P^{1,2}[1:2,3:4] = \left(Z_2P^{1,1}\right)\widetilde{D}. \tag{53}$$

In Eqs. (52) and (53), we introduce the notation $A[b:c,d:e]$ for a submatrix of a matrix A, which is constructed by selecting the row range from the b-th row to the c-th row and the column range from the d-th column and e-th column.

[f]The following set of Eqs. (49) and (50) is what the equation

$$\begin{pmatrix}1\,0\\ 0\,1\end{pmatrix} = \left(\widetilde{E}|\widetilde{V}\rangle\; \widetilde{D}|\widetilde{V}\rangle\right), \tag{47}$$

which is Eq. (2.29) in Ref. 1, means.
[g]The following set of Eqs. (52) and (53) is what the equation

$$Z_3P^{1,2} = Z_2P^{1,1}\left(\widetilde{E}\;\widetilde{D}\right), \tag{51}$$

which is Eq. (2.35) in Ref. 1, means.

The derivation of Eqs. (52) and (53) is postponed to the end of this section. [h]

Obtaining $\widetilde{E}$ and $\widetilde{D}$ from Eqs. (52) and (53) under Eq. (40) [i] is, of course, easy: By solving Eq. (2) for $L = 2$ under Eq. (40), we obtain $\vec{P}_{L=2}$. Then we convert the matrix form $Z_2\ P^{1,1}$ from $Z_2\ \vec{P}_{L=2}$ as we converted Eq. (14) from $\vec{P}_{L=4}$. Similarly, we obtain the matrix form $Z_3\ P^{1,2}$. By multiplying the inverse of the 2×2 matrix $Z_2 P^{1,1}$ from the left of both sides of Eq. (52)(resp. Eq. (53)), we can obtain the solution $\widetilde{E}$(resp. $\widetilde{D}$).

Now that we know $\widetilde{E}$ and $\widetilde{D}$, we can obtain $|\widetilde{V}\rangle$(resp. $\langle\widetilde{W}|$) from Eq. (49) or Eq. (50)(resp. the set of Eqs. (59) and (56) or the set of Eqs. (59) and (60). We can show[1] that the thus obtained $\widetilde{E}$, $\widetilde{D}$, $|\widetilde{V}\rangle$ and $\langle\widetilde{W}|$ are equivalent to the known results.[5,6,12,13]

In the remaining part of this section, we derive only Eq. (52)(we can derive Eq. (53) in the similar way).

By the definition Eq. (16) and the similarity transformation Eq. (42), we have

$$Z_2 P^{1,1} = \langle\widetilde{W}| \begin{pmatrix} \widetilde{E} \\ \widetilde{D} \end{pmatrix} \begin{pmatrix} \widetilde{E} \ \widetilde{D} \end{pmatrix} |\widetilde{V}\rangle. \tag{54}$$

Therefore

$$Z_2 P^{1,1} = \begin{pmatrix} \langle\widetilde{W}|\widetilde{E}^2|\widetilde{V}\rangle & \langle\widetilde{W}|\widetilde{E}\widetilde{D}|\widetilde{V}\rangle \\ \langle\widetilde{W}|\widetilde{D}\widetilde{E}|\widetilde{V}\rangle & \langle\widetilde{W}|\widetilde{D}^2|\widetilde{V}\rangle \end{pmatrix}. \tag{55}$$

Using

$$\langle\widetilde{W}|\widetilde{E} =: (x_1, y_1)\,, \tag{56}$$

$\langle\widetilde{W}|\widetilde{E}^2|\widetilde{V}\rangle$ in Eq. (55) is equal to x_1 in Eq. (56) because

$$\langle\widetilde{W}|\widetilde{E}^2|\widetilde{V}\rangle = \langle\widetilde{W}|\widetilde{E} \begin{pmatrix} 1 \\ 0 \end{pmatrix} = x_1, \tag{57}$$

where the first (resp. second) equal sign in Eq. (57) holds because of Eq. (49)

[h]The derivation starts at the paragraph containing Eq. (54).

[i]In this article, we assume that the necessary condition Eq. (40) for existence of a two-dimensional MPSS is also sufficient condition.

(resp. Eq. (56)). In the similar manner, using also Eq. (50), we can show [j]

$$Z_2 P^{1,1} = \begin{pmatrix} x_1 \; y_1 \\ x_2 \; y_2 \end{pmatrix}, \tag{59}$$

where we introduced

$$\langle \widetilde{W} | \widetilde{D} =: (x_2, y_2) \,. \tag{60}$$

By the same way as the derivation of Eq. (54), we have

$$Z_3 P^{1,2} = \langle \widetilde{W} | \begin{pmatrix} \widetilde{E} \\ \widetilde{D} \end{pmatrix} \left(\widetilde{E} \; \widetilde{D} \right)^{\otimes 2} | \widetilde{V} \rangle. \tag{61}$$

From this equation, we obtain the left part of $Z_3 P^{1,2}$

$$Z_3 P^{1,2}[1:2, 1:2] = \begin{pmatrix} \langle \widetilde{W} | \widetilde{E}^3 | \widetilde{V} \rangle & \langle \widetilde{W} | \widetilde{E}^2 \widetilde{D} | \widetilde{V} \rangle \\ \langle \widetilde{W} | \widetilde{D} \widetilde{E}^2 | \widetilde{V} \rangle & \langle \widetilde{W} | \widetilde{D} \widetilde{E} \widetilde{D} | \widetilde{V} \rangle \end{pmatrix}. \tag{62}$$

Using Eq. (49), the left upper element in the right hand side of this equation can be transformed as

$$\langle \widetilde{W} | \widetilde{E}^3 | \widetilde{V} \rangle = \left(\langle \widetilde{W} | \widetilde{E}^2 \right) \begin{pmatrix} 1 \\ 0 \end{pmatrix}. \tag{63}$$

Using Eq. (50), the right upper element in the right hand side of Eq. (62) can be transformed as

$$\langle \widetilde{W} | \widetilde{E}^2 \widetilde{D} | \widetilde{V} \rangle = \left(\langle \widetilde{W} | \widetilde{E}^2 \right) \begin{pmatrix} 0 \\ 1 \end{pmatrix}. \tag{64}$$

Eqs. (63) and (64) mean that components of the first row of $Z_3 P^{1,2}[1:2, 1:2]$ is equal to components of $\langle \widetilde{W} | \widetilde{E}^2 \left(= (x_1, y_1) \, \widetilde{E} \right)$. Similarly, we can show that components of the second row of $Z_3 P^{1,2}[1:2, 1:2]$ is equal to components of $(x_2, y_2) \, \widetilde{E}$. Therefore we obtain

$$Z_3 P^{1,2}[1:2, 1:2] = \begin{pmatrix} (x_1, y_1) \, \widetilde{E} \\ (x_2, y_2) \, \widetilde{E} \end{pmatrix}, \tag{65}$$

[j]The set of Eqs. (59), (56) and(60) is what the equation

$$Z_2 P^{1,1} = \begin{pmatrix} \langle \widetilde{W} | \widetilde{E} \\ \langle \widetilde{W} | \widetilde{D} \end{pmatrix} \tag{58}$$

which is Eq. (2.30) in Ref. 1, means.

that is,

$$Z_3 P^{1,2}[1:2,1:2] = \begin{pmatrix} x_1 & y_1 \\ x_2 & y_2 \end{pmatrix} \widetilde{E}. \tag{66}$$

This equation and Eq. (59) means Eq. (52). In the similar way, we can show Eq. (53).

6. Summary

In this article, after a brief introduction, we described the asymmetric simple exclusion process (ASEP) in Sec. 2 and a matrix-product stationary state(MPSS) for the ASEP in Sec. 3.

Furthermore, we have explained:

(i) a systematic way to find necessary conditions for the existence of an $M(=1,2)$-dimensional MPSS(Eq. (12)) in Sec. 4;
(ii) a systematic way by which the two-dimensional matrices(E and D) and vectors($\langle W|$ and $|V\rangle$) in the MPSS(Eq. (12)) can be constructed from the stationary states for the two-site system and the three-site system if the condition in (i) is also a sufficient condition.

The method (i)(resp. (ii)) is applicable to $M(\geq 2)$-dimensional MPSSs(Eq. (12)) for not only the ASEP but also models which have $N(\geq 2)$(resp. $N(\leq M)$) states per site.[1]

For other examples and details, please see Ref. 1. This reference explains also a systematic way to check the validity of the obtained MPSS for arbitrary system sizes in the restricted models.

Acknowledgments

The authors would like to thank Tomotoshi Nishino for informing us of the workshop, where partial contents written in this article was presented.

References

1. Y. Hieida and T. Sasamoto, *Journal of Physics A: Mathematical and General* **37**, p. 9873 (2004).
2. R. A. Blythe and M. R. Evans, *Journal of Physics A: Mathematical and Theoretical* **40**, p. R333 (2007).
3. K. Krebs and S. Sandow, *Journal of Physics A: Mathematical and General* **30**, p. 3165 (1997).
4. K. Klauck and A. Schadschneider, *Physica A: Statistical Mechanics and its Applications* **271**, 102 (1999).

5. F. H. L. Essler and V. Rittenberg, *Journal of Physics A: Mathematical and General* **29**, p. 3375 (1996).
6. K. Mallick and S. Sandow, *Journal of Physics A: Mathematical and General* **30**, p. 4513 (1997).
7. G. M. Schütz, *Exactly solvable models for many-body systems far from equilibrium in Phase Transitions and Critical Phenomena vol 19 ed C Domb and J L Lebowitz* (Academic, London, 2001).
8. Y. Hieida, *Journal of the Physical Society of Japan* **67**, 369 (1998).
9. S. R. White, *Phys. Rev. Lett.* **69**, 2863(Nov 1992).
10. S. R. White, *Phys. Rev. B* **48**, 10345(Oct 1993).
11. Peschel I et al (ed), *Lecture Note in Physics: Density-Matrix Renormalization* (Springer, Berlin, 1999).
12. T. Sasamoto, *Journal of Physics A: Mathematical and General* **32**, p. 7109 (1999).
13. T. Sasamoto, *Journal of the Physical Society of Japan* **69**, 1055 (2000).